Democratizing Brazil

DEMOCRATIZING BRAZIL

Problems of Transition and Consolidation

Edited by Alfred Stepan

New York Oxford
OXFORD UNIVERSITY PRESS
1989

Oxford University Press

Oxford New York Toronto
Delhi Bombay Calcutta Madras Karachi
Petaling Jaya Singapore Hong Kong Tokyo
Nairobi Dar es Salaam Cape Town
Melbourne Auckland

and associated companies in
Berlin Ibadan

Published by Oxford University Press, Inc.,
200 Madison Avenue, New York, New York 10016

Library of Congress Cataloging-in-Publication Data
Democratizing Brazil: problems of transition and consolidation / edited by Alfred Stepan.
p. cm. Includes bibliographies and index.
Contents: Brazil's slow road to democratization / Thomas E. Skidmore—
"Authoritarian Brazil" revisited / Bolivar Lamounier
A tale of two presidents / Albert Fishlow—Brazil's debt / Edmar L. Bacha and Pedro S. Malan
The "people's church," the Vatican, and Abertura / Ralph Della Cava
Grassroots popular movements and the struggle for democracy / Scott Mainwaring
Politicizing gender and engendering democracy / Sonia E. Alvarez
The new unionism in the Brazilian transition / Margaret E. Keck
Associated-dependent development and democratic theory / Fernando Henrique Cardoso
Why democracy? / Francisco Weffort—The Brazilian "New Republic" / Maria do Carmo Campello de Souza.
ISBN 0-19-505151-3. ISBN 0-19-505152-1 (pbk)
1. Representative government and representation—Brazil. 2. Brazil—Economic policy.
3. Democracy. I. Stepan, Alfred C.
JL2411.D46 1989 321.8′0981—dc19 88-22738 CIP

9 8 7 6 5 4 3 2 1

Printed in the United States of America
on acid-free paper

Contents

PART IV
DEMOCRATIC DISCOURSE AND PRAXIS: EVOLUTION AND FUTURE

Introduction

ALFRED STEPAN

This book about democracy in Brazil does not celebrate a successfully achieved political condition.* That Brazil is instead "democratizing" suggests tasks, some done, many not yet done. Writing three years after the military yielded formal control of the presidency, it is readily apparent to all observers that the Brazilian transition to political democracy is still incomplete. One crucial indicator of this is that direct elections for the presidency have not yet been held. Most observers are also aware that the multiple tasks of consolidating democracy will be increasingly complex, and probably much more problematic, than they had once thought.

Democratizing Brazil, the successor to *Authoritarian Brazil*,[1] was designed to include two crucial dimensions seldom combined in modern social science—sustained analytic attention to a structural-historical process as it evolved, and self-conscious reflection by a community of scholars about the strengths and weaknesses of their earlier conceptual approaches to this structural-historical process.

Authoritarian Brazil was written in 1971–72, when the military regime that had ruled Brazil since 1964 was at the zenith of its authoritarian power and "economic miracle." *Democratizing Brazil* is very different—it analyzes the slow decomposition of the authoritarian development project and the gradual composition of a still incomplete democratizing alternative. The story of this composition has four distinctive interactive elements and it is the analysis of these interactive elements that gives the volume its internal organization.

*This volume grew out of a series of conferences and exchanges held between 1983 and 1987 at Yale University, Columbia University, and CEBRAP in São Paulo. The support given by the Tinker Foundation for these activities is acknowledged gratefully. In addition to the authors, I would like to thank those colleagues who participated in one or more of these exchanges and made a significant contribution to our intellectual enterprise, including Marcelo Cavarozzi, Carlos Díaz-Alejandro, Manuel Antonio Garretón, Albert Hirschman, Juan J. Linz, Guillermo O'Donnell, and Paulo Sérgio Pinheiro. Margaret E. Keck provided invaluable intellectual and editorial support. At different moments along the way I was greatly assisted by my graduate research assistants at Columbia University, Edward Gibson, Scott Martin, Biorn Maybury-Lewis, and William Nylen.

In Part One the central endeavor is to identify where and why liberalization of the authoritarian regime began. Our argument is that liberalization began within the State apparatus and was given immediate stimulus by some of the institutions of the authoritarian political situation itself. This argument, if accepted by others, raises new issues for most contemporary theories of the state, which devote little attention to predicting, much less explaining, such a process of State-led liberalization. Important questions we explore are: Why did leading figures of the regime initiate for their own reasons a process of controlled liberalization from above? What divisions within the State coercive apparatus, and what distinctive features of the regime's political institutions, made it difficult for the regime to control and restrict the liberalization process once it had begun? And what powerful institutional instruments and routines of the authoritarian regime (and the authoritarian society) still persist and represent a challenge to democracy?

In Part Two our focus is on the national and international political economy of the authoritarian regime and how its contradictions contributed to the decomposition of the regime. We are interested in exploring what, despite the often widely heralded advantages of authoritarian planning, emerged as persistent weaknesses of the planning model. Why did Brazil plunge from economic miracle to the most indebted country in the developing world? What are the analytical, structural, and policy implications of this process and its legacy for the political economy of democratizing Brazil in an era of continuing debt crises?

In Part Three our analytic attention is devoted to the unexpectedly creative but nonetheless fragile civil society that emerged once liberalization began. For theorists and practitioners interested in the fundamental question of how societies generate the capacity to resist authoritarianism while—and this is seldom the case—*simultaneously* creating democratic patterns of action in civil society, the Brazilian transition opens difficult but promising terrain. How in fact did the advances toward empowering new areas of civil society occur? What was the dialectic between regime concession and societal conquest? And, finally, the critical question, can the energy created in the anti-authoritarian resistance be sustained, and contribute to the tasks of completing the still incomplete democratic transition, consolidating political democracy, and eventually expanding this new democracy's economic and social content?

In Part Four we re-examine democratic theory. From the perspective of world history and comparative politics, it is indispensable to insist upon both the generic and the specific historical dimensions of democracy. Democracy has an indispensable generic dimension that entails a minimalistic core of procedural guarantees. Without these guarantees, no matter what else may exist, political democracy does not. However, the reasons why different social classes come to accept or espouse democracy, the emotional support they attach to it, and the specific institutions that conflicting coalitions craft over time to implement democracy, vary immensely. Democracy in England, the United States, India, and Japan took root in extremely distinctive ways. This is not an atheoretical statement but a theory-based injunction to examine the historic evolution of democratic politics as theory and praxis. In the case of Brazil, what is of great interest for democratic theory is why, in an authoritarian regime in a condition of associated-dependent development, opponents of the regime began an unprecedented debate about democracy. What did political actors and thinkers concerned with democratic praxis believe were the essential elements of democracy in a country under conditions found in Brazil? What did they tend to neglect? Let us look at each of the parts that make up this volume in greater detail.

Part One is devoted to contradictions and ambivalences (only partially noted in *Authoritarian Brazil*) within the authoritarian political model itself, and how these internal vulnerabilities helped to generate the regime's efforts at controlled change from above. This point is crucial. The Brazilian transition began as "regime-initiated liberalization."[2] Of all the recent transitions in southern Europe and South America, those of Spain and Brazil contain the strongest elements of elite initiation and elite transaction. In 1973 when the process of Brazilian liberalization began, there was no significant political opposition, no economic crisis, and no collapse of the coercive apparatus due to defeat in war. "Liberalization" is distinct from "democratization." In an authoritarian setting, "liberalization" may involve a mix of policy and social changes, such as less censorship of the media, somewhat greater latitude for the organization of autonomous working-class activities, the reintroduction of some legal safeguards for individuals such as *habeas corpus*, the release of most political prisoners, the return of political exiles, possibly measures for improving the distribution of income, and, most important, the toleration of political opposition. "Democratization" entails liberalization but is a wider and more specifically political concept. Democratization requires open contestation for the right to win control of the government, and this in turn requires free elections, the results of which determine who governs. Liberalization refers fundamentally to the relationship between the State and civil society. Democratization refers fundamentally to the relationship between the State and political society. From this discussion of the distinction between "liberalization" and "democratization" it is clear that liberalization does not necessarily entail democratization. Yet the controlled liberalization of 1973 ended in the controlled and incomplete democratization of 1985.

The historian Thomas Skidmore gives us an indispensable overview of these twelve years, an overview that enables the other authors to develop critical sub-themes, and the reader to integrate these sub-themes into the overall argument. Skidmore concluded his article in *Authoritarian Brazil* with an analysis of how much Brazil's institutional arrangements of the Vargas Estado Novo corporatist regime of 1937–45 were left undismantled by the governments of 1946–64. These arrangements both constrained popular forces and were in place to be utilized by the new authoritarian regime that began in 1964. In this volume Skidmore evaluates how much of this legacy still remained intact at the end of military rule in 1985, thus posing a confining condition on those seeking democratic consolidation. He concludes with a comparative evaluation of the transitions of 1946 and 1985.

A strategy of controlled liberalization from above, described by Skidmore and elsewhere by myself, would not have set so many forces in motion if there had not already been contradictions and ambiguities in the authoritarian political model itself.[3] These contradictions and ambiguities were understudied and undervalued in *Authoritarian Brazil.* Bolivar Lamounier's article in *Democratizing Brazil* is a telling critique of the neglect of purely political factors in *Authoritarian Brazil.* Unlike the recent authoritarian regimes found in Greece, Spain, Portugal, Argentina, or Uruguay, Brazil kept legislative elections, presidential terms, and some role for a legislative-based electoral college in the presidential selection (or ratification) process. Despite profound ambivalence within the regime about liberal democratic principles, it is also true that regime ideologues in Brazil never articulated—as did their counterparts in Portugal, Spain, Chile, or Argentina—a systematic attack on democracy as a system of government. The initial claim of the regime's founders was that they would try to create conditions so that democracy could work. Bolivar Lamounier argues that the only author in *Authoritarian Brazil* to appreciate fully the latent weight of the electoral dynamics and the ideological ambivalence of the authoritarian regime was Juan Linz. In fact, not only is it true that, from the mid-1960s to the late 1970s, Brazil and all the Southern Cone political systems fell into the grip of new types of more pervasive military regimes, but it is also true that the social science community in these countries so focused on the economic and coercive logic of these regimes that they attached relatively little weight to political institutions as a significant sphere of action or analysis. Writing against this trend, Bolivar Lamounier pioneered electoral analysis and political surveys in the 1970s, and his article assesses why, once the military as government embarked on a modest degree of liberalization, the pre-existing electoral mechanisms rapidly began to acquire a greater degree of prominence in Brazil than in any other contemporary case of democratic transition.

Part Two is devoted to political economy. The articles by Fishlow and by Bacha and Malan are exemplary in their effort to identify critical turning points in the world economy, the range of policy alternatives available to developing countries in the face of these turning points, and the social and economic consequences of policies taken. In the process, enduring myths are dismissed and new arguments are elaborated. Though Brazil was the developing world's leading oil importer, this fact does not go very far to explain why it became the most indebted country by the 1980s. In comparison, as Fishlow demonstrates, the East Asian developing countries suffered a much more severe deterioration of their terms of trade than did Brazil, and yet managed to make a stable adjustment. In a context of low and at times negative real interest rates, Brazil followed a debt-led growth model. As Brazil's debt-to-export ratio grew, its asymmetric integration into the world economy increased its vulnerability. Bacha and Malan show how this vulnerability brutally materialized in the 1980–85 period after interest rates—which had changed from predominantly fixed to predominantly floating in the 1970s—soared to unprecedented real heights. Fishlow argues that the military-technocratic policy to cope with the debt crisis, which called for a sharp cut in capital goods imports while trying to increase exports no matter what the true economic and social costs, had inherent developmental and even trade limitations. Brazil's cutback on capital goods imports (a policy the Asian developing countries did not elect), while immediately contributing to improved trade balances, raised questions about Brazil's long-term ability to sustain the industrial dynamism necessary to increase its share of world trade.

A persistent stream of the development literature calls attention to the technocratic advantages of apolitical management. Another body of observers, though critical of authoritarian technocracy, are skeptical about newly democratic governments' capabilities to initiate, much less sustain, the standard austerity programs thought necessary to cope with the debt crisis. Both streams of analysis contribute to the anxiety and pessimism about the political economy of redemocratization.

While not underestimating the problems created by the debt crisis, several articles in this volume cast issues of political economy in a different light. The mounting disarray of the military-technocratic Brazilian development model had some favorable consequences for eroding military rule. It both weakened military resolve to stay in power and undermined the myth of the inherent superiority of authoritarian decision making. Fishlow argues further that the political weaknesses of the State meant that the large State sector neither "commanded nor effectively cooperated with the private sector. It had no secure control over real resources." State intervention in fact increasingly reflected State weakness rather than State strength. Unfortunately, as the first three years of civilian rule have made painfully clear, the Brazilian State apparatus was in such a stage of decomposition that the attempt to utilize it, without making serious changes in its structure, values, and responsiveness, served only to deepen developmental crises.

Part Three explores how and why democratizing pressures from below in civil society added new weight and greater depth to the initial liberalization efforts. While I think it is absolutely correct to insist upon regime concession as an important ingredient of the Brazilian transition, it is even more important to understand that what gave the Brazilian transition its special character was the dialectic between regime concession and societal conquest. In historical terms, Brazil has long stood out as the major Latin American country where State power has most structured and controlled civil society, especially the popular sectors. Brazil has never had the strong working-class-based unions or parties such as those found in Chile or Argentina, or a peasant-based upheaval to compare with Mexico's. Brazil is also the country in Latin America in which State-structured corporatist mechanisms of labor control have been most entrenched. As the 1980s come to a close, Brazil still is a polity marked by an extremely weak sense of the rights of citizenship and by a degree of income inequality not found in any contemporary democracy in the world.

In the mid-1970s, however, Brazil witnessed an historic awakening of new forms of social creativity and resistance in virtually all components of civil society, such as entrepreneurs, the press, lawyer associations, church organizations, labor unions, and women's groups. The social construction of new forms of solidarity and resistance is central to our democratizing theme. If the Brazilian liberalization had been restricted to regime concessions, instead of the dialectic of regime concession and societal conquest, what had been given could have been taken away, or what had been given could have stopped at liberalization. Just as important, any eventual effort at democratic consolidation would have had to have been addressed without the new ideological, organizational, and human resources generated by the democratizing movements of civil society. Because the Argentinian military regime collapsed so rapidly after its defeat in Malvinas, nothing comparable to the Brazilian emergence of civil society developed in that country. In Uruguay, the two traditional parties constituted the core of a basically passive, but always available, alternative to the authoritarian regime, but virtually no new social movements formed, and pre-existing groups were quiescent and mutually isolated.

The mid-1970s in Brazil, however, saw not only the formation of numerous new social movements but intricate and creative *horizontal relations of civil society with itself*, relations that helped interweave the weft and warp of civil society and give it a more variegated, more resistant fabric. From this vast panorama, four elements in civil society have been selected for attention—the Church, grassroots popular movements, new women's movements, and the new trade unionism.

Ralph Della Cava's sweeping and incisive article on the Church places the Brazilian Church in its correct international and national context. In the last fifteen years, the Brazilian Church and the Polish Catholic Church have ranked among the most politically influential in the world. The power of the Polish Church, often the embodiment of the nation in those historical periods when there was no Polish State, derived, to a great extent, from its ideological resources vis-à-vis a Leninist, foreign-sponsored State and its institutional capacity to mobilize and protect the Polish people.

For the future of world-wide Catholicism, however, the Brazilian Church posed far deeper and more novel alternatives. By the mid-1970s the Brazilian Church had become the most theologically progressive and institutionally innovative Catholic Church in the world. With the initial support of the Vatican, it also became in Brazil the most legitimate, most nation-wide, and most useful organizational resource for the opposition forces of civil society. The principle of national collegiality of bishops embodied in the National Conference of Brazilian Bishops (CNBB) was extended into uncharted doctrinal and social domains in Brazil. The CNBB created numerous new types of action-oriented vehicles to assist the poor, such as the Pastoral of the Land. Church-sponsored base-community movements helped stimulate new degrees of consciousness. The support the Brazilian Church gave for the "preferential option for the poor," for the spiritual and material critique of alienated State power, for new modes of participation, and for new definitions of authority and mission within the Church not only weakened the leadership claims and developmental priorities of the Brazilian authoritarian regime but eventually were seen by the Polish-born Pope, John Paul II, as posing a threatening alternative vision for the third millennium of Roman Catholicism.

Della Cava, in his path-breaking analysis, argues that the mid-1970s' unity, exceptional in Latin America, between Vatican, national hierarchy, priests, nuns, grassroots movements, and the democratizing opposition owed its vitality to special conjunctural conditions that were rapidly eroding in the last years of the authoritarian regime. Indeed, if Della Cava's conjunctural analysis is correct, the Church will in the future have fewer national and international allies, compared with the earlier period, and will play a less politically central role in the post-military period that began in Brazil in 1985. The full implications of the new conjuncture are still not clear, but the ingredients for a serious analysis of this new political, theological, and global context are provided for the first time by Della Cava.

Scott Mainwaring's study of a grassroots neighborhood association complements Della Cava's article. Mainwaring illustrates concretely the themes of self-empowerment by the poor, the horizontal relations of civil society with itself, the role of the Church in facilitating new forms of organization, and the spillover of new behavior patterns learned in one part of civil society, such as a base community, to another, such as a neighborhood association. In addition, a growing concern for theorists and activists alike is how social movements which are important to the period of protest against authoritarian regimes can adapt themselves to new circumstances. Mainwaring's article is an excellent analysis of the paradoxical threats and new challenges social movements face in the transition from authoritarian to democratic politics.

This latter problem is also a major theme of the article by Sonia E. Alvarez. Building upon important new approaches in feminist theory and in post-structural Marxist theory, and on her own previous research in Cuba, Nicaragua, and Argentina, she argues that both the State and civil society are indispensable arenas for the struggle to "politicize gender and engender democracy." Her substantive focus is a rich comparative analysis of day-care and family-planning issues under authoritarian and "micro-democratic" governments in São Paulo. Her article highlights major theoretical and political dilemmas that a democratizing social movement faces as it attempts to achieve institutionalization while avoiding cooptation.

Since the late 1930s, Brazilian unions, whether under authoritarian or democratic regimes, have been encapsulated and confined within the complex corporatist system instituted under Vargas. The persistence of these structures was due partly to ambivalence on the part of elements of the left who believed they could use the State-structured system to extract rewards for and from labor. Some wags have even said that the longest love affair in Brazil was that of the left for the State. However, one of the most fundamental of the democratizing endeavors in Brazil was that conducted by the new trade unionists, who struggled to formulate a counter-State ideology and to forge direct collective bargaining arrangements with capital, thus eroding the Ministry of Labor's long-standing wage-setting prerogatives. The painful and intricate maneuvers of the new unionists' long march toward empowerment in the 1980s went largely unobserved because, like the Church and the neighborhood associations, unions began to lose political prominence as campaigns for elections became more important in the period 1982–85. However, Margaret Keck's detailed analysis demonstrates that it was precisely in the 1980s, more than in the well-publicized 1970s, that trade unions established themselves as more internally organized and increasingly autonomous parts of civil society. Indeed, in Brazil's fragile democracy the component of Brazilian civil society that has demonstrated the greatest organizational capacity to continue to militate against those still repressive authoritarian features of the Brazilian State and Brazilian social life has been the trade unions, and the article by Margaret Keck provides valuable insight as to why this is so.

In Part Four, the general question of democracy is addressed. More than in any other country which has had a recent democratization movement, Brazil witnessed a profound debate about democracy itself. In Uruguay, the country was virtually silenced until just before the transition, and most of the debates revolved around the details of democratic restoration. In Argentina, the transition occurred so rapidly after the military defeat in Malvinas that the core debate concerned how to handle the legacy of military human-rights abuses. The Portuguese revolution caught its citizens by surprise, and the theoretical issue most at stake was the possibility of revolution from above by the military. Spain and Greece witnessed important transitions but the main effort was to try to approximate slowly, however difficult it might be, the pre-existing models in other European Common Market and NATO democracies. In Brazil, however, the combination of the sheer length of the opening from 1973 to 1985, the relative freedom of the press, and the unprecedented flowering of civil society produced a debate of the greatest breadth and depth. In retrospect, it is also clear, as the concluding chapter demonstrates, that the debate neglected some very important issues.

We are fortunate that two of the major political actors in the Brazilian democratizing opposition are also leading social scientists who have written articles for this volume. In *Authoritarian Brazil*, Fernando Henrique Cardoso first advanced his now well-known concept of "associated-dependent development." In his article for *Democratizing Brazil*, more explicitly than ever before, Cardoso attempts to confront the following challenge: As a theorist of dependency and as an active democratic politician, can he coherently reconcile democratic theory with dependency theory? Consistent with the methodology of his approach to dependency, Cardoso's response builds upon structural-historical considerations. In the case of Brazil, he explores four questions. What types of previously unseen social formations and economic structures have actually been produced by the world's most industrialized instance of associated-dependent development? Given these, what new collective actors must be incorporated into the democratizing effort if it is to take hold? In this new setting, what types of collective goods and social policies are necessary to give democracy deeper socio-economic as well as political meaning? And, finally, given that democracy is not only generic but also specific to concrete structural-historical processes, what types of democratic political parties are not really possible in Brazil, and what types are most feasible and promising?

Francisco Weffort addresses a different theme. In Europe in the 1970s, one of the most important ideological developments within the left was the critique of Leninism from the perspective of Marxist democratic theory. During the last decade in the more advanced countries of South America, with the partial exception of Chile, the most interesting ideological development within the left was the critique of violence whether from the left or the right. The self-criticism by the left explicitly called into question its previous attitude toward its own participation in democratic politics. This previous attitude included virtually unqualified support of strong State prerogatives, preoccupation with strategies for rapidly capturing the State by non-constitutional, often violent means to exercise those prerogatives, and relative neglect of strategies for the long-term democratic development of civil and political society so that the correlation of forces could be peacefully but powerfully altered. Most important, the debate on the left acknowledged the previous ambivalence about democracy itself, and advanced arguments that constituted a major revalorization of democracy not just as a temporary tactic but as a permanent value. The article by Weffort is, I believe, the most explicit and extensive examination of this South American debate about violence and democracy to be published in English.

In the concluding essay of the volume, one of Brazil's best political analysts, Maria do Carmo Campello de Souza, conducts a sobering review of the first three years of civilian rule. She argues that, in the process of this century's longest transition, the democratic opposition made a number of "meta-assumptions" about the post-military rule period—none of which have been borne out. One "meta-assumption" was that the new movements of civil society would sustain their push for deepening democratic values, practices, and institutions. Unfortunately, in the absence of the *galvanizing* effect of direct military rule many of the groups of civil society—with the partial exception of rural and urban trade unionists—have experienced internal divisions and, in the case of the church and human rights groups, outright *attacks* from the ranks of their former allies. In the political realm the mistaken "meta-assumption" was that the forces of the democratic opposition would assume control of the executive and of the legislature. Due to the tragic death by illness of the opposition leader Tancredo Neves, the compromise vice-president, José Sarney, who had previously been president of the military-controlled party, ironically became the first president of what was supposed to be Brazil's democratic "New Republic." The leading party of the democratic opposition, the PMDB, did become the largest party in the Brazilian Congress after the November 1986 election, but never in any real sense did it become the ruling party of the New Republic or assume co-responsibility for governing. The last "meta-assumption" to be proved wrong was that military prerogatives would inevitably contract in the democratic regime. In fact, when one compares the first three years in Brazil with comparable periods in Argentina, Uruguay, Spain, and Greece, what is clear is that Brazil is the only one of these countries where *no* substantial military prerogatives were challenged. The Brazilian military have reconstituted themselves so as to be able to be powerful actors in the new context of more open politics. However, to date the new civilian rulers have not developed a political strategy with a view toward attempting to reduce military power and presence in the State, and indeed in the polity.

When one considers the implications of these three false "meta-assumptions," together with the continuing debt crisis and the continuing decomposition of the State apparatus as a source of policy direction, one understands why there has been more "system blame" against the new regime, and more worries about "governability" in Brazil than are found in comparable periods in Spain, Uruguay, or Argentina. As we go to print the drama concerning the fate of democratization in Brazil is intensifying.

A final editorial note. The time frame of each article in the volume varies according to its analytic purpose and empirical focus. The Bacha-Malan piece is fundamentally a new interpretation of the emergence of the crisis in Brazil's international borrowing markets, so it essentially closes with the outbreak of that crisis in 1983.

Weffort wrote his article in 1983 when he was the Secretary General of the important trade-union-based party, the Partido dos Trabalhadores (PT). Cardoso also wrote his article in 1983, when he was the senator for São Paulo and the São Paulo president of Brazil's leading opposition party to the authoritarian regime, the Partido do Movimento Democrático Brasileiro (PMDB). When these two articles were written most close observers of Brazilian politics thought the authoritarian regime would be able to control the 1985 presidential succession and thus rule until 1991. The two articles are therefore important documents by democratic theorists and activists written in the context of a still powerful authoritarian regime. In order to retain the full documentary quality of the two articles, I have left them as written.

Three authors, Skidmore, Fishlow, and Lamounier, address political and economic processes that occurred between the "miracle" days of the authoritarian regime under Médici and the regime's demise under Figueiredo, so their time frame is 1971–85.

The articles by Della Cava, Mainwaring, Alvarez, and Keck analyze the democratic pressures that emerged during the most repressive days of the dictatorship, pressure that intensified during the abertura. However, to varying degrees, all four are also concerned about the *sustainability* of popular pressures in the absence of the catalytic effect of direct military rule. Thus, all end their essays with some reflections about the Brazilian experience with the critical problem of how to maintain democratic pressures from below so as to consolidate and deepen democracy.

The book concludes with the chapter by Campello de Souza, whose explicit theme is the problem of consolidation and whose time frame is thus the first three years of civilian rule, from 1985–88.

Columbia University
March 1988

NOTES

1. Alfred Stepan, ed., *Authoritarian Brazil: Origins, Policies and Future* (New Haven and London: Yale Univ. Press, 1973).

2. I discuss this category in greater detail in my "Paths Toward Redemocratization: Theoretical and Comparative Considerations," in Guillermo O'Donnell, Philippe C. Schmitter, and Laurence Whitehead (eds.), *Transitions from Authoritarian Rule: Comparative Perspectives* (Baltimore and London: Johns Hopkins Univ. Press, 1986), 64–84.

3. Alfred Stepan, *Rethinking Military Politics: Brazil and the Southern Cone* (Princeton: Princeton Univ. Press, 1988).

Democratizing Brazil

PART I

The Opening of Authoritarianism: Origins and Dynamic

1

Brazil's Slow Road to Democratization: 1974-1985

THOMAS E. SKIDMORE

Background: The Return of Brazilian Authoritarianism: 1964-1974

In March 1964 a coalition of military and civilian conspirators deposed President João Goulart and opened a new chapter in the history of Brazilian authoritarianism.* Their clearest precedent was November 1937, when President Getúlio Vargas suspended the election that was to choose his successor. Vargas had created a semi-corporatist authoritarian regime (the *Estado Nôvo*) based largely on the military. In 1945 Vargas himself was deposed by the military, and Brazil adopted a competitive multi-party system. The reversion to authoritarianism in 1964 was again the work of the military, who decided to continue the presidential system, but under military tutelage. The President would be a general designated by the Army officer corps for election by the Congress (later enlarged into an electoral college) for a set term.

Brazil's new authoritarian experiment has attracted much scholarly analysis, including the volume from the 1971 conference at Yale on *Authoritarian Brazil.*[1] One of the most influential papers was by Juan Linz, who suggested that "the Brazilian case represents an authoritarian *situation* rather than an authoritarian *regime*."[2] Linz thought it unlikely that the

*For financial support I am indebted to the Fulbright program for a Faculty Research Abroad Fellowship, to the University of Wisconsin Graduate School Research Committee, and to the Nave Fund at the University of Wisconsin. For helpful comments and criticism I am grateful to Bolivar Lamounier, Peter Evans, Pedro Malan, and Fernando Henrique Cardoso.

For a much fuller account of these years, see my *The Politics of Military Rule in Brazil, 1964-1985* (New York: Oxford University Press, 1988).

Brazilian military government would institutionalize a new system comparable to that of Franco's Spain. Rather, he saw an unstable regime as most probable. Subsequent history has confirmed Linz's analysis. This essay offers a schematic account of how the military government, starting about 1974, moved back from arbitrary rule and slowly toward a competitive multi-party system under a restored rule of law.[3]

The first of the general-presidents after the 1964 coup was General Humberto de Alencar Castelo Branco.[4] From 1964 to 1967 Castelo Branco faced the thankless task of carrying out an economic stabilization program. His economic team sharply reduced the rate of inflation and successfully renegotiated the foreign debt, but failed to resume economic growth. At the same time, the Castelo Branco government faced great pressure from the "hardline" military who wanted to suspend elections and constitutional guarantees to facilitate purges and arbitrary procedures. Castelo Branco held off the hardliners with concessions (prolonging his term for a year, assuming exceptional powers by means of "Institutional Acts"). Purges were carried out in the bureaucracy, the military, the universities, and trade unions. Meanwhile, the President and his key advisers worked on a new constitution so that the next President could assume power in a "normal" constitutional regime.

Constitutionalism seemed to reign when General Artur da Costa e Silva assumed office in 1967. The new President's pledge to "humanize" the military's 1964 "Revolution" seemed to bode well. In the economic sphere, the news could not have been better. Brazil began a burst of economic growth that averaged an extraordinary 11 percent per year from 1968 through 1974. In politics, however, the scene was less happy. The fragile structure of the Constitution of 1967 could not withstand the growing political radicalization. In 1968 there were huge student demonstrations, and two important industrial strikes.[5] The government fought back with police measures that became highly repressive, especially since hardliners often commanded the security forces. In December 1968 the military was infuriated by a Congressional vote that protected a deputy whom the military wished to prosecute. The President decreed the Fifth Institutional Act (AI-5), which authorized the suspension of normal civil rights, such as *habeas corpus*, justifying the measure by the need to protect national security.[6] Unlike its predecessors, this new act bore no expiration date.

After President Costa e Silva died of a stroke in 1969,[7] his successor, designated after intense debate among the officer corps, was General Emílio Garrastazú Médici, who served until 1974.[8] Médici's presidency was the most authoritarian since 1964. Although elections were held[9] and Congress continued to function (with a suspension in 1969–71, broken only to ratify Médici's succession in early 1970), Brazil was in the grip of the security forces, which were locked in battle with several small guerrilla movements. Even after the guerrillas were ruthlessly liquidated, the arbitrary procedures and dictatorial practices continued.[10] Yet the Médici presidency had won at least a partial *de facto* legitimacy from members of the middle and upper classes, because of the record economic growth rates and the reign of "law and order."

The Médici years ended with the government apparently stronger than at any time since 1964. The armed threat from the left was gone—the guerrillas were dead, in prison, or in exile. The economy was booming, with 1973's growth rate of 14 percent the highest for any year since 1928. Brazil seemed to be the model for many seeking the secret to economic growth in the Third World. Notwithstanding this apparent success, important changes were in store.

One group of military officers had for some time been arguing the need to return to the rule of law. They were closely identified with former President Castelo Branco (killed in an air crash only a few months after leaving office in 1967) and the original declared intention to return as soon as possible to constitutional government. Prominent among this group was General Golbery Couto e Silva, chief of Castelo's civilian staff and a key member of the military "Sorbonne," which wanted to modernize Brazil rapidly. Golbery had fought against the designation of Costa e Silva as successor to Castelo Branco, and since his departure from government in 1967 had been planning a return to power. Equally important, Golbery had an ally in General Ernest Geisel, who was the army's consensus candidate to succeed Médici.

1974: A New Presidency—Liberalization from Within?

The presidential succession of 1973–74 came off more smoothly than any since 1964. During 1973 officer opinion had coalesced around the name of Ernesto Geisel, the President of Petrobrás, who had the added virtue of being the brother of the Army Minister. The Electoral College duly followed instructions in January 1974 and elected Geisel President by 400 out of 503 votes. The MDB's overwhelming loss in a highly manipulated body was in small part compensated by the hope that the new President wanted to (and actually could) ease the repression. In late February, Geisel announced his cabinet, in which key posts went to technocrats. Delfim Neto was replaced at the Treasury by Mário Simonsen, another economics professor with close ties to São Paulo. Delfim was sent off as Ambassador to France, in part to get his formidable personality as far as possible from the new policymakers. The Planning Secretariat remained in the hands of João Paulo Reis Velloso, the economist from the Northeast now proving to be the most durable technocrat.[11]

The air was full of talk that the new government planned to make up for past social inequities. Better income distribution was a prime concern: between 1960 and 1970 the richest 10 percent of the population had increased their income share from 40 percent to 47 percent, while the share of the poorest fifty percent had dropped from 17 percent to 15 percent. A new Ministry of Social Welfare was to consolidate the ill-coordinated welfare programs created by past governments.

A more equitable distribution of the benefits required continued high growth if no one was to lose out in absolute terms. The OPEC oil price shock in 1973, however, meant that Brazil, which imported 80 percent of its oil, had few options: cut back on non-oil imports, run down its foreign exchange reserves, or borrow more capital abroad. To cut back on imports would have slowed growth, and the world recession due to the oil price shock precluded a rise in export earnings. The only way to absorb the price rise and maintain growth was to use reserves or borrow abroad. Brazil did both, doubling its net external debt in 1974 alone (from $6.2 billion to $11.9 billion).[12]

Bolstered by foreign loans, the Geisel government decided to maintain the huge state investment projects it had inherited (the Itaipú dam, the Açominas steel project, etc.) while committing itself to a set of costly new investment projects to increase domestic capacity in capital goods and basic raw materials. In short, the new government undertook a challenge even greater than its predecessor had attempted—and in a far less favorable international climate.

General Golbery Couto e Silva had now returned to power, regaining the position he had held under Castelo Branco—chief of the President's civilian staff (*casa civil*). Golbery and Geisel had worked closely together in the Castelo Branco government, and since 1967 had been maneuvering to regain power for what they saw as the "authentic revolutionaries."[13]

Public expectations about the new government, especially among the elite, centered on the hope that the Geisel government would bring the repressive apparatus under control, especially by ending torture. In late February, Geisel nourished those hopes by conferring with Cardinal Arns of São Paulo, an outspoken critic of the government's violation of human rights. Emissaries of the National Conference of Bishops of Brazil (CNBB) were encouraged by their talks with General Golbery, who seemed genuinely responsive to the call for a return to the rule of law. Optimism grew when Geisel expressed his hope for a gradual redemocratization, beginning with a *distensão* (decompression), although he also pointedly warned that national "security" was indispensable to ensure development.[14]

It was clear from the outset that any moves towards redemocratization and a return to the rule of law would depend on the President's ability to mobilize support within the officer corps of the three services, especially the army. The hardliners could be expected to oppose and perhaps even sabotage any liberalization. There was speculation as to how much the President in fact controlled the security apparatus. An incident in the Northeast illustrated the problem. Soon after Geisel's inauguration, the Fourth Army command in Recife arrested Carlos Garcia, a respected journalist who headed the Recife office of *O Estado de São Paulo*. After interrogation and torture, Garcia was released. This kind of brutal treatment, typical of the Médici era, was what the opposition was hoping Geisell could end. The owners of the newspaper, who had been courageous in attacking military rule, vigorously protested the mistreatment of their reporter, against whom no public charges were ever brought.

In late May the federal censors launched a crackdown on the media,[15] reacting in part to more aggressive criticism from the opposition. The Brazilian Bar Association (OAB) expressed concern over the government's failure to account for missing persons believed to have been apprehended by the security forces.[16] In July the MDB formally requested that Justice Minister Armando Falcão comment on the whereabouts of persons believed held by the government. In August the Bar Association stepped up its campaign, with its national convention in Rio devoted entirely to "The Lawyer and the Rights of Man."[17]

In October, two further incidents showed that liberalization was hardly imminent. The arrest and torture by the Fourth Army in Recife of Fred Morris, a United States citizen and former Methodist missionary, created a crisis in U.S.-Brazilian relations, and reinforced U.S. interest in Brazilian liberalization.[18] Hardliners also flexed their muscle in the case of Francisco Pinto, a prominent radical MDB deputy from Bahia, who was deprived of his Congressional seat and stripped of his political rights. The cause was a radio speech in March denouncing Chile's President Pinochet, when the latter came to participate in Geisel's inauguration.

November 1974: The Shock of Electoral Defeat

All these struggles between government and critics turned out to be a minor prelude to the Congressional elections of November 1974, Brazil's most important electoral test at the federal level in the decade since the "Revolution." The Geisel government had received a contradictory electoral legacy from its predecessor. While governors were elected indirectly by state legislatures which were easily manipulated by Brasília, the Congressional elections were to be direct. The Médici regime had solved the problem of such challenges in 1970 through massive intimidation of the electorate and harassment of the opposition. But what if the elections were relatively free?

Geisel's first test proved no problem. ARENA swept the indirect election of governors in October 1974, hardly surprising since ARENA controlled all the legislatures that did the electing. Yet the victory may have seriously misled political strategists in the presidential palace, who had consistently underestimated the breadth and depth of electoral opposition. ARENA leaders confidently predicted victory in the Congressional elections, and as late as early October few informed political observers would have bet against them. But the political climate changed rapidly at the end of October. To everyone's surprise, the government decided to allow all candidates relatively free access to television. Suddenly the electorate began to wonder if their votes *could* make a difference. Perhaps the MDB *did* represent a real alternative. Perhaps the President *was* prepared to cooperate with the opposition. The MDB had for some months been arguing that it was more in tune with the President's liberalization intentions than was the government party. The last fortnight before election day brought a rush of enthusiasm for the opposition.

The election results stunned everyone,[19] including the most optimistic MDB strategists. The MDB had almost doubled their representation in the lower house, (the total number of seats had been increased from 310 to 364), jumping from 87 to 165. ARENA dropped from 223 to 199. The results in the Senate were equally dramatic. The MDB delegation went from seven to twenty, as ARENA dropped from fifty-nine to forty-six. While ARENA had won by a small margin in the total vote for federal deputies, the MDB won in the total vote for senators, which was the best indicator of national opinion. In addition, the MDB won control of the state legislatures in key states where the urban electorate was crucial: São Paulo, Rio Grande do Sul, Rio de Janeiro (including the city of Rio), Paraná, Acre, and Amazonas. Previously they had controlled only the state legislature in greater Rio de Janeiro (then still the state of Guanabara).

What did this opposition victory mean? The MDB had focused its campaign on three issues: social justice (denouncing the trend toward a more unequal income distribution), civil liberties (the human rights violations which so worried elite critics), and denationalization (denouncing foreign penetration in the Brazilian economy).[20] The MDB leaders argued that their victory showed the electorate had endorsed their position. At a minimum, everyone agreed the elections showed a quite unexpected lack of support for the "Revolution." The economic slowdown of 1974 had certainly played a role. But there was no doubt the voters had warned the government: they wanted a change. The Planalto could no longer harbor any illusions about ARENA's ability to win relatively free elections.

Whither "Decompression?"

The idea of a government-led liberalization was now greatly complicated. The Geisel government had hoped to carry out a gradual and carefully controlled opening. Now its very legitimacy had been challenged. The clash of government and opposition—which it had hoped to mute—was accentuated. There was a danger of MDB victories in the direct gubernatorial elections scheduled for 1978. Even if indirect elections were maintained, opposition control of the state legislatures in key states made the danger of government defeat a quasi-certainty.

By late December, Geisel's advisers were faced with a political panorama far less under control than the one they had inherited. The dilemma for the moderates in the military government was underlined anew: how to find civilian politicians who were both attractive to the electorate and acceptable to the military hardliners?

President Geisel moved quickly to give assurance that election results would be respected. He was strengthened by the fact that the opposition had not gained control of any executive power. Furthermore, the Congress had long since been stripped of the power to take any initiative. One new threat from the MDB, however, was its ability—because it now held more than one-third of the lower house seats—to block constitutional amendments.

In 1975 the debate over human rights continued full pitch. In January censorship of *O Estado de São Paulo* was lifted in time for the paper's centennial. This concession marked the end of a long battle between the paper's owners, the Mesquita family, and the military government.[21] None of the other publications subject to prior censorship was removed from the list. The government had made a very limited gesture to improve the political atmosphere.

By February 1975 the opposition was again pressing for action on political prisoners. The MDB and the Roman Catholic Church activists renewed their demands for an accounting on the "disappeared." But in March a crackdown on the Communist Party, considered by the hardliners to have been a major element in the MDB's electoral success, brought a large number of arrests and new charges of torture of detainees. The critics from the Church and the MDB continued to protest against torture and arbitrary police action. The Commander of the Third Army, General Oscar Luís da Silva, responded to the opposition by citing the Portuguese example, where protest, he said, had led directly to the threat of Communist control.

At the beginning of August 1975, Geisel made an important speech defining his government's attitudes toward liberalization. He argued that any such change had to be slow and sure. Significantly, he gave notice that the government had no intention of giving up its exceptional powers, as codified in Institutional Act No. 5. The more optimistic opposition spokesmen had hoped that some way could be found to phase out that act. Geisel warned that opposition pressures would generate "counter pressures," which he could not afford to ignore. His meaning was clear: any new government policy had to be the product of a compromise between vying political viewpoints within the military. Only if the military felt confident about national security, however they defined it, could the opposition hope to achieve its goal of a return to the rule of law. At the same time, Geisel attacked Communist infiltration of the labor unions and warned that tighter censorship would be necessary in view of the subversive campaign against his government. Church officials were among the very few opposition figures who could effectively contest the government's continuing denials of torture and arbitrary behavior. By mid-September, Church spokespersons were again on the attack, citing government mistreatment of the Indians and torture of political prisoners.[22]

Death in São Paulo

Such complaints were subsequently overshadowed by the sensational case of Vladimir Herzog, a prominent journalist who was the director of a widely respected noncommercial television channel, operated under the sponsorship of the state of São Paulo. Herzog was a Yugoslav-born Jew who had immigrated to Brazil as a young man. In October 1975 he learned from friends that the Second Army security forces were looking for him. In an effort to be cooperative, he voluntarily reported to the army barracks. He apparently had no notion that the Second Army intelligence considered him a direct link to Communist plotters.

The next day the army command announced that he had committed suicide in his prison cell. São Paulo was stunned. No one believed the suicide explanation. A prominent member of the media elite was suddenly dead, surely from torture. Students went on strike at the University of São Paulo, and the Bar Association issued a statement charging the government with torture. The fact that Herzog was Jewish undoubtedly added to the fearful reaction of the Paulista elite, since there had been hints of anti-Semitism in past hardliner behavior. Forty-two São Paulo bishops signed a statement denouncing the government violence.

Cardinal Arns, in spite of the fear of Herzog's family and much of São Paulo's cultural elite, organized and presided over a dramatic joint Catholic-Jewish ceremony for Herzog in the São Paulo cathedral. The police resorted to petty harassment to frighten away participants, running "safety checks" on all cars coming into the central city. Yet the service was held—a minor triumph in a community gripped by fear.

The shock of Herzog's death was all the greater since many had hoped that President Geisel was bringing the military security apparatus under control. Geisel ordered an immediate investigation of the journalist's death. Cynical observers felt confirmed in their opinion when the military investigators announced in late December that the death had indeed been a suicide.[23]

The sense of shock lingered in São Paulo. The Second Army commander, General Ednardo d'Avila Melo, was a hardliner who gave the local security apparatus a virtual free hand. As 1975 ended, many thought decompression was doomed. Yet another incident in São Paulo further electrified the atmosphere. In early January 1976 the Second Army was interrogating Manoel Fiel Filho, a worker active in the metalworkers' union. Suddenly the news leaked out that he was dead. Could anyone doubt that he had been killed under torture? President Geisel was livid. He now looked ridiculous for his defense of General d'Avila in the Herzog case. After satisfying himself of the facts, he fired d'Avila and replaced him with General Dilermando Gomes Monteiro, a known moderate and a close associate of Geisel.

Most significant of all, Geisel acted without convening the Army Higher Command, normally consulted on such an important change in command. His ability to do so showed the President's great power within the army officer corps. The most important effect of these two deaths in Second Army headquarters was to put the hardliners on the defensive. The security specialists could no longer be certain that higher authority would cover for them in case of public uproar over torture.

The increase in Geisel's power within the military—as indicated by his ability to remove General Ednardo d'Avila—had its price. To maintain support he had to demonstrate that he would brook no backsliding toward the civilian corruption and subversion that hardliners felt had necessitated the "Revolution of 1964." Thus, ironically, the President's ability to carry out political liberalization would depend in the short run on his ability to demonstrate his authority.

The Constitutional Realities

From the beginning of the Geisel government, the opposition pressure for a return of the rule of law clashed with the President's insistence on retaining the arbitrary powers granted by AI-5. Still, in 1974–75 it was used in relatively minor cases, leading to a hope that the Act might simply fade away from lack of use.

The opposition was disabused of that hope in early 1976, soon after the sacking of General Ednardo d'Avila. Throughout 1975, the federal Justice Minister had engaged in a hunt for the Communists he thought had played a key role in the November 1974 Congressional elections. In March 1975, there were trials in which defendants were convicted of attempting to rebuild the Communist Party; others were convicted of having belonged to guerrilla Carlos Marighela's once active ALN. In October, military spokesmen announced the arrest of seventy-six Communists, of whom sixty-three were said to be military police. In January 1976, Geisel used AI-5 to strip two Paulista state legislators of their mandates. They were alleged to have welcomed Communist support in previous elections. At the end of March the same step was taken against two federal deputies who had virulently attacked the government and the military.

In response, a fiery MDB Congressman decided to go out fighting. Lysaneas Maciel, an ordained Methodist minister and a leading critic of the military regime, made an eloquently bitter speech in Congress, defending the two deputies about to lose their mandates. He charged that only force had allowed the government to stay in power. Word of his attack reached the Presidential office in time to add his name in the order revoking the two deputies' mandates. The order hit the floor of Congress with Lysaneas still at the podium. Wild insults were traded in an ugly scene that had been building for some time. The MDB leadership, deeply frustrated, issued a statement attacking the government for its resort to "violence."

Few politicians had been surprised when the ax fell on these young deputies, who were notorious for skating near the edge of military tolerance. Yet the President's resort to AI-5 against minor federal Congressmen suggested that the hoped-for "dialogue" would be clouded by the President's power to silence political voices he disliked.

Other signs of government nervousness surfaced. In early April 1976, the federal censor banned from television a broadcast of the Bolshoi Ballet.

Since censorship had recently eased, the government looked silly as jokes about possible "contamination" from televised Communist dancers circulated freely. In late June the government again showed its fear of the power of the media when it pushed through Congress a bill (known as the "Falcão Law" after the Justice Minister) restricting the use of radio or television for the 1976 municipal elections to the broadcast of still photos of candidates with innocuous voice-over résumés of each. This was a tardy reaction to the 1974 campaign, when the MDB successfully used television to generate great political momentum in the closing weeks. President Geisel now referred to the MDB as "enemies." The MDB leaders were divided on how to respond.

The import of the opposition party's electoral victory in 1974 had taken time to sink in. When it did, the Geisel government found itself, less than a year into its term, in a political cul-de-sac. The promise of "decompression" had been overshadowed by the government's fear of electoral defeats that might prematurely undermine military control of the executive. The MDB leadership had wisely decided to look moderate and wait to collect the fruits of victory in the upcoming 1978 elections.

The final months of 1978 saw the eruption of a force that many Brazilians had long feared—rightist terrorism. In September a bomb exploded at the Rio de Janeiro headquarters of the Brazilian Press Association, and in early October there were firebombs and telephone threats aimed at churchmen known for their criticism of the government. A bishop was kidnapped, beaten, and his car blown up. A group calling itself the "Brazilian Anti-Communist Alliance," sounding ominously like the Argentine terrorist group, claimed credit for the bishop's persecution. But the incidents did not develop into a larger campaign at this time, and those responsible (undoubtedly linked to the police and the military) were either restrained by their superiors or decided to lie low.

The most important political developments of this period came from the municipal elections of November 15, 1976. ARENA was expected to do well, at least in the less-developed regions, where few dared to run against the government party, and this proved to be the case. But in larger urban areas the MDB ran very strong, winning control of municipal councils in Rio de Janeiro, São Paulo, Belo Horizonte, Pôrto Alegre, Salvador, Campinas, and Santos. The opposition victory in the most important cities boded ill for ARENA's future. The election also had its lighter side: in Rio, some 150,000 voters cast ballots for "black beans," the working class staple whose disappearance from stores a month earlier had occasioned riots. It was another danger sign for the regime.

Another Twist of the Constitution: The "April Package"

In early 1977 the Geisel government took a decisive move to deal with the political problem created by the 1974 election results. Having lost its two-thirds majority in Congress, it still had AI-5; but the power to rule by decree could only be used if Congress were not in session. Geisel therefore closed Congress (on April 1, 1977), using as his pretext the MDB's refusal to support a government bill to reform the judiciary. Under the arbitrary powers of AI-5, Geisel then announced a series of major constitutional changes known as the "April Package" (*pacôte de abril*), all aimed at strengthening ARENA's hand in future elections. Constitutional amendments would henceforth need only majority approval in Congress; all state governors and a third of the federal senators would be elected indirectly by 1978 by electoral colleges (which would include municipal councilors, where ARENA predominated); federal deputies would be allocated on the basis of population rather than registered voters (as in 1970 and 1974); and finally, the 1976 Lei Falcão was extended to the Congressional elections.

The MDB complained bitterly at this new manipulation of the rules of the political game. The press was rife with commentary on Geisel's apparent betrayal of his commitment to liberalization. The President responded by calling for the Congress to reconvene on April 15.

The rest of 1977 saw the Geisel government severely tested. Those who doubted its commitment to liberalization found no shortage of evidence. In June, Geisel purged the MDB leader of the Chamber, Alencar Furtado, and stripped him of his political rights for ten years. This was in response to Furtado's denunciation of the President in a television broadcast (under the law the opposition got one hour a year) earlier in June.[24] It appeared that Geisel, who was known for his autocratic and closed temperament, had taken personal offense at Furtado's speech and responded by using arbitrary power.

Earlier in 1977 there were signs of opposition from a more familiar source—students. Protests in March over academic issues expanded into anti-government demonstrations in May at several university campuses. Justice Minister Falcão banned all further demonstrations; an attempted strike at the University of Brasília led to repression, and 850 students were arrested at a "national" student meeting in Belo Horizonte calling for the restoration of democracy. Although repression occurred in most cases, the police often showed hesitation. For the first time since 1968, activist students felt that they could afford to defy the security apparatus.

Another source of tension was protest from intellectuals over the issue of censorship. Although prior censorship had been lifted from the major newspapers in 1975, it remained in effect for other publications, and was especially vigilant on radio and television, the media with greatest popular impact. In May, Falcão announced that censorship would be extended to all imported printed matter. Some 2,750 journalists issued a nationwide protest, following an earlier anti-censorship manifesto by a thousand intellectuals in January.

Human Rights, Nuclear Technology, and Uncle Sam

The relationship between the Geisel government and its opposition took on an international dimension when the newly inaugurated Carter administration decided to make human rights a central concern in U.S. foreign policy.[25] When in 1977 the State Department issued its first human rights report on countries receiving U.S. military assistance—required by Congress under the Harkin Amendment to the 1976 foreign aid bill—the report was strongly critical of the human rights situation in Brazil. The Geisel government was furious over what it considered intolerable interference in Brazil's internal affairs. It immediately cancelled a military aid agreement which dated back to 1952, and by the end of September had cancelled all remaining formal military cooperation agreements. There would thus be no occasion for further State Department reports.

Hoping to repair the damage, Carter scheduled a visit to Brazil for late November, but negative signals from Brasília forced its cancellation. The trip was rescheduled for March. First Lady Rosalyn Carter did make an official visit to Brazil in July 1977 and was immediately swept into the human rights controversy. Brazilian student representatives gave her a letter denouncing human rights violations. During a stop in Recife, she met with two U.S. missionaries who gave her a hair-raising account of their mistreatment by police only a few days earlier. She promised to raise the matter with her husband. Mrs. Carter's trip dramatized anew the issue of the U.S. attempt to influence Brazilian government policies. The hardline military used foreign criticism to exploit nationalist sentiments among their more moderate colleagues.

The Carter government chose to stress another issue that also aroused recriminations between Washington and Brasília. This was the question of nuclear technology. Since 1945 the U.S. had been campaigning to prevent the proliferation of nuclear weapons technology. The non-proliferation policy was particularly aimed at smaller (at least in technological capacity) countries wanting to develop atomic power as an energy source. The U.S. encouraged these countries to import nuclear technology, but wanted to ensure that the recipient countries did not gain the capacity to enrich uranium—a necessary step for the production of atomic weapons as well as for maintaining a supply of reactor fuel. Thus the U.S., as Brazil's supplier of nuclear technology, had guaranteed Brazil a supply of enriched uranium and had vetoed Westinghouse Electric's request to build enrichment facilities in Brazil, on the grounds that Brazil had refused to sign the Nuclear Non-Proliferation Treaty.

In 1974, however, the U.S. Atomic Energy Commission announced that due to limited capacity it might not in the future be able to meet Brazil's full needs for enriched uranium. This withdrawal of the fuel guarantee came just in time to abort a huge Brazilian contract with Westinghouse, which would have provided for up to twelve reactors and $10 billion in sales.

Shock at the ease with which the U.S. had revoked its commitment reinforced the position of nationalist critics of Brazil's dependence on a foreign source in such a critical area of technology. It also opened the door for Brazil to look elsewhere to secure its projected atomic fuel needs.

In June 1975, after intensive negotiations between the Brazilian government and a consortium of West German suppliers, an agreement was signed which provided for Brazil's purchase of two to eight giant reactors, costing from $2 billion to $8 billion. More important, the technology to be supplied included the capacity to produce nuclear weapons. If fully carried out, this deal would have been the largest transfer ever of nuclear technology to a developing country. The Brazilian government was triumphant about its imminent move into the exclusive club of nations that possessed the complete atomic fuel cycle.

Earlier the Ford administration had been concerned over the deal, but chose not to do more than register protests in Bonn and Brasília. Newly elected President Carter, however, decided to make a major issue of nuclear proliferation, especially the Brazilian case. In early 1977, the Carter administration mounted an offensive for a repeal of the agreement both in West Germany and Brazil. But U.S. pressure only stiffened the resolve of Bonn and Brasília, and by late 1977 the U.S. had given up. The triumph over the nuclear issue disarmed domestic critics of Brazil's nuclear policy and gave the Geisel government some welcome popularity.[26]

Geisel Defeats the Hard Line

The year 1977 had even greater surprises in store. Within the Geisel government the struggle had intensified between the hardliners and the moderates (although the positions were more varied than that simple dichotomy suggests). The Army Minister, General Sylvio Frota, had emerged as the hardline leader. He believed Brazil was in imminent danger of Communist subversion and considered liberalization a ruse to facilitate subversives, many of whom masqueraded under supposedly respectable party labels. Furthermore, Frota considered himself a presidential candidate, and he was prepared to challenge Geisel's choice, Figueiredo, directly. The fight came to a head in early October.

On October 12 there was a sudden announcement that Frota had been dismissed. The next day Frota issued a fiery manifesto accusing the government of "criminal complacency" over "Communist infiltration," even at high levels of government. He attacked Brazil's 1974 recognition of the People's Republic of China and the 1975 recognition of the Marxist government in Angola. The purged Army Minister was appealing to the militant opponents of *abertura*, but his most important audience, the military, gave every evidence of rallying around the winner.

Frota's dismissal had great significance. First, it demonstrated that Geisel had accumulated more personal power than any previous President. He had taken an unprecedented step—firing the Army Minister before consulting the assembled High Command—thus showing that unlike previous military Presidents, his power within the army had not declined.[27]

The struggle between Geisel and Frota was really over the 1978 presidential succession and the continuation of the Geisel-Golbery political strategy. From the day he entered office, Geisel had made clear to his intimates that his candidate was General João Batista Figueiredo, director of the SNI. In early 1978, Geisel made his choice public (which was tantamount to nomination by the party). For the vice-presidential candidate Geisel selected Aureliano Chaves, an electrical engineer, nuclear energy specialist, and former governor of Minas Gerais. In April the National Convention of ARENA obediently endorsed the nominees.

Further military fallout from the designation of Figueiredo came when General Hugo Abreu, chief of the presidential military household and the paratroop commander who had liquidated the guerrilla threat in the Amazon, resigned in protest over the choice.[28] When the hardliners again did not visibly rally, Geisel showed his mastery over the military, at least on the question of the succession.

The presidential election turned out to be more interesting than it at first promised. The opposition had a plausible candidate, General Euler Bentes Monteiro, the former director of SUDENE, the development agency for the Northeast. Euler was endorsed by the MDB convention, along with MDB Senator Paulo Brossard (Rio Grande do Sul) as his vice-presidential running mate. There was a campaign, but it remained shadow boxing, since the election was in the hands of an ARENA-dominated electoral college. Figueiredo promised to carry forward a gradual democratization, while Euler called for a Constituent Assembly to rewrite the Constitution. The MDB also called for a change in economic policy to correct the glaring economic inequities. Figueiredo was elected by the electoral college on October 14.

A month later came the Congressional elections. The electoral law changes in Geisel's 1977 "April Package" succeeded in preventing the MDB from winning a Congressional majority. A new provision made a third of the Senate elected indirectly (irreverent Brazilians christened these Senators the "bionics") and a revision of the formula on representation in the Chamber of Deputies gave ARENA continued control of both houses. But the trend in direct voting was obvious—in the direct Senate elections the MDB won 52 percent and Arena 34 percent, with 14 percent spoiled or blank ballots. Sooner or later the *abertura* seemed bound to escape the government's control.

In late 1978, Geisel followed through on his promise to phase out key elements of the authoritarian structure. In September the Congress approved a set of extensive reforms, with the MDB boycotting the final vote, on the grounds that the government proposals did not go far enough. Most important among these measures was the abolition of AI-5, thereby depriving the President of the authority to declare a Congressional recess, remove Congressmen, or strip citizens of their rights. In addition, *habeas corpus* was reinstituted for political detainees, and prior censorship was lifted from radio and television. At the same time, new "safeguard" powers were given to the executive, including authority to declare a limited state of emergency without Congressional approval.

The government also proposed a revised version of the National Security Law, which many observers considered as important a source of arbitrary power as AI-5. Although the number of possible crimes against State security was reduced, and the penalties softened, the law still allowed for political prisoners to be held incommunicado for eight days (instead of ten). Since torture was most likely to occur in the days immediately following arrest, and since definitions of violations remained broad enough to include virtually any oppositional activity, human rights advocates rejected the proposed revisions as a sham. In fact, Congress never voted on the bill; it was promulgated in December under the *decurso de prazo* clause which provided for automatic approval of any bill not acted upon within forty days.[29]

At the very end of 1978, Geisel took another step to promote political reconcilation. He revoked the banishment orders on more than 120 political exiles, most of whom had left Brazil in 1969–70 in exchange for foreign diplomats who had been kidnapped by the guerrillas. Eight exiles were specifically excluded, however, including Leonel Brizola and Luís Carlos Prestes, the long-time Secretary General of the Brazilian Communist Party.

What kind of legacy was Figueiredo about to receive from his strong-minded mentor?[30] Geisel and Golbery had pursued liberalization farther than almost all observers would have thought possible in 1974. But important arbitrary powers remained, especially in the National Security Law. And since 1964, the federal executive had sharply increased its economic and legal powers. The federal Congress, for example, was deprived of what in democratic countries had long been considered a primary legislative power—control over appropriations. The Brazilian Congress could neither initiate a spending bill nor increase an appropriation requested by the President. Furthermore, the huge security apparatus remained untouched. Opposition politicians thus had good reason to go on considering politics a dangerous occupation.[31]

The Figueiredo Years Begin

The new President was an unknown, notwithstanding his national campaign in 1978. At his inauguration in March 1979 he showed much more informality than the stiff Geisel. Figueiredo went out of his way to compliment a palace guard, giving off the kind of *bonhomie* his political advisers thought a big asset.

The new cabinet showed more continuity than change. The key minister was Mário Simonsen, Geisel's Finance Minister, now Minister of Planning in a new "superministry" for economic policy. Delfim Neto returned from Paris to become the Agriculture Minister. Few observers thought Delfim would be satisfied in that post, despite the new President's claim that agriculture was to become a much more important ministry. The only politically interesting name was Petrônio Portella, the new Justice Minister, an ARENA Senator from the Northeastern state of Piauí who had gained wide respect for his leadership and conciliatory skills.[32] The most important figure of all, at least from the standpoint of influencing the President, was General Golbery, who retained his position as head of the President's civilian household. His strong hand in the Planalto seemed to guarantee that liberalization would continue, presumably along gradual and tightly controlled lines.

On the economic scene there were storm signals. Brazil had managed to maintain high growth since the oil shock of 1973, but only at the cost of rapidly rising foreign indebtedness and an overheated economy. Mário Simonsen, designated as the key economic policy maker, thought Brazil had no choice but to slow down its economy because of the increasing balance of payments pressure. One clear symptom of the problem was the inflation rate, now picking up steam from its 1978 rate of 40 percent.

Simonsen's Five-Year Plan, which forced him to explain the need for a slowdown, made him the target of wrath from every quarter. Political strategists within the new government refused to accept that the "miracle" was over. The authoritarian government's political legitimacy had partly depended on high growth since 1967. The business community was equally upset, but for more immediate reasons. Any significant recession was bound to wipe out the many Brazilian firms that operated with perilously little working capital.

Many thought Delfim could work another miracle, and Delfim himself openly encouraged that idea. Simonsen's inhibited personality made him ineffective at both bureaucratic intrigue (at which Delfim was a master) and in public appearances. This further undermined his effort to deliver the bad news that a dose of slow growth (perhaps close to zero in per capita terms) was inevitable if Brazil were not to face rocketing inflation and even more crippling balance of payments deficits.

Delfim now closed in on Simonsen, whose post he obviously coveted. Under pressure from every side and showing signs of indecision, Simonsen resigned in August, only five months into the Figueiredo presidency. Delfim was immediately named the new Minister of Planning. The São Paulo business community was jubilant, certain that Delfim could find a way to continue high growth.

Delfim claimed that he could satisfy the expectations of the government and business community, and from August to December he continued a rapid growth policy.[33] His timing was less than ideal, however, since OPEC had just administered the second "oil shock" to major oil importers such as Brazil. Earlier in the year the government had launched an ambitious program to develop alcohol as a substitute fuel. But it would require time and resources—such as land and distilling plants—which would have to be taken from other sectors, such as domestic food production.

The new government also saw important developments on the labor front.[34] In 1978 autoworkers in the ABC industrial belt of São Paulo, led by São Bernardo Metalworkers' Union president Luís Inácio "Lula" da Silva,[35] struck in the first significant labor action since the repression of the 1968 Osasco and Contagem strikes. A year later they struck again, setting off one of the largest strike waves in the history of Brazil, affecting some 3,000,000 workers. In São Paulo a massive support campaign collected money and food for the strikers.

The Church, led by Cardinal Arns, gave important moral and material help, offering the churches as meeting halls after the government had taken over the Metalworkers' Union headquarters. After waiting a fortnight, the police cracked down on the striking metalworkers, arresting at least 200 people. In response to Church mediation, Lula agreed to suspend the strike, which gave time for a negotiated settlement which included a 63 percent pay raise.

A series of precedents had been set. First, a new cadre of union leadership had emerged, contemptuous of the government-endorsed figures who since 1964 had negotiated worker compliance with the repressive labor regulations and with work agreements seriously prejudicing workers' interests. The new leadership also emphasized increased organization of and contact with the rank-and-file union members. A second precedent involved the discovery that some employers might be prepared to negotiate directly with the workers. The parent firms of Volkswagen and General Motors in their home countries were accustomed to open collective bargaining, yet the legal structure of Brazilian labor relations made direct negotiations virtually impossible. The third precedent was the degree of solidarity shown the workers by other elements in civil society, such as the Church and middle-class professionals who had the courage to come forward and help the strikers. With over a hundred strikes in fifteen states, by the end of 1979 labor protest had become a truly national question.

Amnesty for Whom?

One of the issues on which the opposition had mobilized wide support was amnesty. Geisel's December 1978 reversal of most of the earlier banishment orders was now followed by Justice Minister Petrônio Portella's amnesty bill, approved by Congress in August 1979. Amnestied were all those imprisoned or exiled for political crimes since September 2, 1961 (the date of the last amnesty—they had been quite regular in the history of the Republic). Excluded were those guilty of "blood crimes" during armed resistance to the government. The law also restored political rights to politicians who had lost them under the Institutional Acts.

The new law brought back a flood of exiles, including Leonel Brizola and Luís Carlos Prestes. Back in Brazil now also were such *bête noires* of the military as Miguel Arraes, Márcio Moreira Alves, and Francisco Julião, along with key figures in the PCB and the PC do B. The amnesty was a powerful tonic in the political atmosphere, giving an immediate boost to the President's popularity. It showed Figueiredo's confidence that he could withstand hardline objections to having so many "subversives" back in politics. It also helped to reduce to normal size some myths from the past. With the old-line Communists and Trotskyists back in Brazil, and with the press virtually uncensored—though still subject to pressures, threats, and even occasional violence—Brazil was looking more like an open political system than at any time since 1968.

The amnesty movement was not content with the new law alone, however. It demanded also an accounting for the 197 Brazilians believed to have died at the hands of the security forces since 1964. For many there were detailed dossiers, including eyewitness accounts by other prisoners. Here the opposition was pressing on a very sensitive nerve—the military fear that a judicial investigation might some day attempt to fix responsibility for the torture and murder of prisoners. A good example of this reaction came in March 1979, when the military took steps to close *Veja* magazine because it had published an exposé on alleged torture camps, complete with photographs. Police also seized copies of *Em Tempo*, a leftist biweekly that reprinted a list of accused torturers.

In fact, this issue had been settled by inclusion in the amnesty law of a definition that included perpetrators of both "political crimes" and "connected crimes." The latter euphemism was generally understood to cover the torturers. It was a political trade off: the opposition leaders knew that they could move toward an open regime only with the cooperation of the military.

Golbery and the Multiparty Solution

Figueiredo's first year in office was also marked by a reorganization of the political party system. Golbery had long since drawn the obvious conclusion from the 1974 election results: in practice, the compulsory two-party system tended to consolidate the opposition and make it difficult to defeat in even a halfway open election. The solution seemed simple—keep the government party (under a new name) and facilitate the creation of multiple parties among the opposition. The government might thereby maintain its hold, either by splitting the opposition vote, or by forming a coalition with the more conservative elements of the opposition.[36]

A bill to accomplish this purpose was sent to Congress and passed in November. By the end of 1979 the new parties had been formed. ARENA had regrouped as the Partido Democrático Social (PDS), while much of the former MDB coalesced in the PMDB (Partido do Movimento Democrático Brasileiro), a verbal trick that both met the new rules (forbidding reuse of any previous party label and having "party" in the name) and irritated the government, because the opposition had preserved its name recognition.

Golbery's strategy was vindicated, at least in the short run, when other opposition parties emerged. Most publicized was the struggle for the party name of the PTB, which Leonel Brizola coveted, and with good credentials. But the electoral authorities, probably on government cue, awarded it to Ivete Vargas, a minor political figure who was a grand-niece of Getúlio Vargas. Brizola then founded his own Partido Democrático Trabalhista (PDT). To the left of both of those parties was the Partido dos Trabalhadores (PT) led by Lula.[37] Rounding out the field was the Partido Popular (PP), a conservative opposition party led by bankers, which merged with the PMDB in November, 1981.

No sooner was the new party alignment clear than maneuvering began for the 1982 elections, although few thought that the Planalto had finished tinkering with the electoral machinery. Government and opposition tried to guess each other's next move, especially in terms of possible campaign coalitions. The ARENA leadership suffered a heavy loss, when Justice Minister Petrônio Portella, a key figure in building a network of party strength around the country, died in early January 1980. He had no adequate successor, either in the Ministry or in the ARENA leadership.[38]

Fearing government losses if elections came too soon, in spite of the party reorganization, in May the government cancelled the nationwide municipal elections scheduled for the end of the year. The vast majority of mayors and municipal councilors were ARENA members, and the government wanted to keep them that way. Ironically, the MDB leaders also favored postponement because they feared they could not organize quickly enough under the 1979 party law. In September the Congress passed a law postponing these elections until 1982, when direct elections were to be held for state governors (for the first time since 1965), a third of the Senate, the entire Chamber of Deputies, and all state legislatures. By piling up so many elections in one day the government was gambling that its party, not the opposition, would benefit from the linkage.

Another government action apparently contrary to liberalization was the tough new law regulating the entry and residence of foreigners in Brazil. It gave the government greater authority to bar or expel foreigners, which included the large number of foreign clergy in Brazil, amounting to about 40 percent of Brazil's total religious.[39] Some foreign clergy had angered local and State authorities by leading popular protests, such as the resistance to land takeovers on the agricultural frontier in the West and Northwest and the Amazon.[40] The CNBB and human rights activists fought to revise the bill, arguing that it was a radical change in the country's long-standing relatively open door to foreigners.

The government remained firm, however, as the PMDB and even some PDS Congressmen fought the bill in Congress. The opponents were able to delay action past the forty-day limit, after which the government, in early August, declared the bill to be law by *decurso de prazo*. Here was a clear example of how limited were the legislators' powers. The government's aggressiveness on this front was shown again in October, when it deported an Italian priest for actions allegedly contrary to the national interest.

Delfim Finds Rough Sailing

The end of 1979 brought an important shift in economic policy. Delfim Neto had encountered problems—primarily balance of payments deficits and inflation—far more vexing than those he faced in 1967–74. Brazil ended 1979 with 77 percent inflation, the highest of any year since 1964. No longer able to pursue the high growth policy he announced when taking over from Simonsen in August, Delfim decided to gamble. In December 1979 he devalued the cruzeiro by 30 percent, and then in January preset the rate of devaluation and of indexation (on key financial obligations) for all of 1980. If inflationary expectations fell, that could reverse the inflationary momentum. But if inflation exceeded the preset schedule, the overvalued cruzeiro would encourage imports, discourage exports, and stimulate investors to avoid financial instruments paying negative real rates. In adopting this policy Delfim was in part following the example of Argentine Finance Minister Martínez de Hoz, whose similar policy had attracted a strong capital inflow. The odds were not favorable for Brazil, however, since the forces behind inflation and the balance of payments deficit lay deep in the structure of the Brazilian economy and its relationship with the world economy.[41]

On another front, Labor Minister Murillo Macêdo convinced the President that major changes had to be made in the government's formula for calculating annual wage adjustments.[42] All analysts agreed that workers had suffered a significant loss in the real value of the minimum wage since 1964: at least 25 percent, and DIEESE, an independent labor union research institute, placed the loss much higher.[43] Macedo pushed through two significant changes. First, wage adjustments would be semiannual, of crucial importance with inflation headed toward 100 percent. Second, wage adjustments would be scaled in such a way as to redistribute wage income towards the bottom of the wage pyramid. In addition to income redistribution, Macêdo's intention was to undercut the growing mobilization of workers shown in the strikes of 1978 and 1979.

The militant sectors of labor were not mollified by Macêdo's new formula. The flashpoint of protest was in the São Paulo industrial heartland, where the metalworkers of ABC still held the initiative. In April 1980 a strike began involving 300,000 workers. The Church and a sympathetic public once again supported the strikers, who were asking a 15 percent real (above inflation) pay increase, the right to have union representation on the shop floor, the forty-hour work week, and job security. The last was an urgent priority, because employers promoted high labor turnover in order to keep down the number of better paid workers with seniority.

This time the government was more repressive. Pickets were attacked by the police and hundreds of strikers were arrested, including Lula, who was jailed for a month. The government intervened in the union, replacing elected representatives with appointees and banning the former officers from running for union election in the future. In early May, after forty-one days out, the strikers returned to work with their demands unmet. One reason was the lack of a strike fund sufficient to finance a long strike. Another was the threat that the employers might start hiring replacements for striking workers.

The results of this strike were very negative for the workers. First, government repression had returned in force, despite the talk of *abertura* and the Labor Minister's new pay formula. Second, the limits of the "new unionism" became clear. Without larger financial reserves, it was impossible to sustain a long strike. In addition, the goal of direct negotiation with employers remained distant as long as government intervention always took the employers off the hook.

On the political front, liberalization was still on course. In November 1980, the Congress passed a government-sponsored Constitutional amendment to reintroduce direct elections for state governors and all Senators (although the term of the "bionics" would not end until 1986). This was a partial undoing of Geisel's "April Package," which had allowed ARENA to survive the 1978 elections with majorities in both houses of the federal Congress. Given these majorities, the government now felt confident in pushing on with *abertura.*

The Enemies of *Abertura* Fight Back

Not everyone was in favor of liberalization, and its clandestine opponents were mounting a campaign of violence. During 1980 and early 1981, Brazil was wracked by a series of violent incidents. Newsstand vendors, for example, were sent threatening notes ordering them to stop selling leftist publications. Some who refused had their stands firebombed in the night. Not all the violence was bloodless. A letter bomb sent to the Brazilian Bar Association killed the secretary who opened it. Few doubted that the attack had come from the right.

On April 30, 1981, came an even bigger explosion. The incident began when an army lieutenant and sergeant (in plainclothes) from the DOI-CODI political police drove into the parking lot of a Rio theater (Rio-Centro) where a concert to benefit leftist causes was under way. Suddenly a bomb exploded in the car, killed the sergeant and gravely wounded the lieutenant. Although army authorities later released stories claiming otherwise, all evidence indicated that the duo were bringing the bomb to disrupt the concert, perhaps even to create mass panic. This suspicion was reinforced by the fact that another bomb had exploded inside the building near a power generator. This was apparently a desperate move by rightist military who felt the political process had gotten out of hand. The army immediately assumed jurisdiction over the investigation and carried out a clumsy cover-up, made more difficult by the fact that the civil authorities had already issued a medical report on the dead sergeant. All parties in Congress condemned the terrorism and many worried that the cover-up was evidence that the hardliners might still be able to sabotage the liberalization.

No one had a greater political stake in *abertura* than Golbery. He immediately pressed behind the scenes for a no-holds-barred inquiry into the RioCentro incident. As the whitewash became obvious, and as he found himself increasingly isolated in the presidential palace, he resigned in August 1981.[44] His departure generated major waves on the political scene.

Golbery had been a key strategist on several fronts. He had helped launch the policy of gradual redemocratization in the Geisel presidency, and continued to be its most articulate advocate among Figueiredo's advisers. His political strategy was to press on with *abertura*, ensuring future control by building the electoral strength of the government party through the party reorganization, changes in electoral rules, and traditional patronage mechanisms.

The prime opposition to Golbery's strategy came from the military hardliners, led by General Octávio Aguiar de Medeiros, chief of the SNI.[45] Threatened by *abertura*, these officers favored authoritarian solutions, which for some justified the use of terrorist measures. In the battle over RioCentro, the hardliners confronted Golbery's call for an open investigation with their need to protect their clandestine network, and won.

In economic policy, Golbery clashed with Delfim over austerity measures, claiming that the latter's recessionary policy would hurt the government party in the 1982 elections. By the time Golbery resigned, Delfim had already pushed through measures Golbery wanted delayed, for example, an increase in payroll deductions for social security.

Golbery was immediately replaced as head of the civilian staff by João Leitão de Abreu, who had held the same post in the Médici government (1969–74). Leitão de Abreu was a lawyer from Rio Grande do Sul, and was said to have close links to the rightist military. While many thought this boded ill for *abertura*, the momentum behind redemocratization remained, and there was no significant social base for a return to the repression of the early 1970s. Nonetheless, with Golbery's departure the most powerful pro-*abertura* voice around the President was gone.

The immediate effect of this change was to give Delfim Neto a freer hand to pursue the recessionary policy, on which he had embarked after the failure of his December 1979 policy of pre-announced devaluations and indexing adjustments. Golbery's departure also gave the rightist military more direct access to the President. Finally, such a major shift suggested that Figueiredo was at best an indecisive President. Golbery quit because he could see his influence ebbing, not because Figueiredo decided to change course. While the President continued to proclaim his commitment to elections, his governmental changes tended to favor men closely identified with the Médici, the most authoritarian President since 1964.

Despite Golbery's resignation, the RioCentro incident and its aftermath may have actually facilitated *abertura*, by placing the rightist military more on the defensive. Terrorism stopped, as if to confirm that the rightists had decided to lie low. Finally, the pro-*abertura* military were stimulated to reaffirm even more emphatically their "faith in democracy."

Economic Troubles and a Presidential Heart Attack

Contrary to Delfim Neto's hopes, inflation raced ahead of schedule in 1980, leading to negative interest rates, extensive dissaving, and an overvalued cruzeiro. As a result, inflation increased further and the balance of payments became even more negative. Late 1980 brought a radical switch in policy. Brazil had been following a high growth policy (rapid monetary expansion, high public investment, etc.), but the balance of payments deficit was forcing the country toward the arms of the IMF, always a grave political liability. With inflation at 120 percent and foreign bankers increasingly skeptical, Delfim moved back to a more realistic exchange rate, without pre-set schedules of devaluation or monetary correction.

On the labor front, there was a change from the strike pattern of recent years. The ABC metallurgical workers found the going notably tougher than in 1980. In February a regional military court convicted Lula and ten other union leaders on charges of violating the National Security Law by leading the 1980 strike. The 1981 wage adjustment produced no strike, in part because the union was being run by government appointees, but also because a severe industrial recession had begun, bringing massive layoffs in the area. Furthermore, many leaders of previous strike actions had been fired, since the existing laws provided little protection against employer retaliation. Gaining such protection had been a key demand in recent strikes. Thus union tactics changed, and smaller strikes against layoffs took the place of the mass actions of previous years. (See Keck article in this volume). In April 1982 the Supreme Military Tribunal overturned the National Security Law convictions of Lula and his ten comrades, but this and other trials were part of an ongoing war of nerves by the hardliners against what they saw as a dangerous new labor movement.

September 1981 brought an event that might well have stopped the *abertura*: a presidential heart attack. On September 18 a seriously stricken President Figueiredo entered the hospital and the military ministers met within a day to announce that Vice President Aureliano Chaves would assume the presidency. Opposition fears that the military might run roughshod over the civilian Vice President, as happened with Pedro Aleixo at the time of President Costa e Silva's stroke in 1969, were allayed when Army Minister Walter Pires de Albuquerque declared his support for Chaves, the first civilian to hold the presidency since 1964. Chaves continued to exercise the presidential duties for almost eight weeks, until Figueiredo resumed office in early November. The government had continued to function (although Chaves was excluded from decision-making on military or security matters) and there had been no major shifts in policy. In sum, liberalization, such as it was, had sufficient momentum to survive this scare.

The 1982 Elections: Who Won?

The year 1982 was dominated by the prospect of the elections, scheduled for November. For the first time since 1965, governors were to be elected directly. Because local elections had been postponed from 1980 until 1982, the voter was being asked to choose at every level but the presidential. As the campaign approached, it became clear that the party reorganization alone was not enough to guarantee a PDS victory on a large enough scale to guarantee government control over the electoral college which would elect the next President. In November 1981 the government thus pushed through Congress the "November Package," which prohibited electoral coalitions and required that voters vote a straight ticket.

The campaign was extremely spirited, especially in the urban areas. Both the government party (PDS) and the PMDB mounted highly sophisticated campaigns, while the smaller parties such as the PT, PDT, and PTB had to rely on volunteer canvassing and street corner bullhorns. Like every other federal election in urban areas since 1974, this one frequently turned into a plebiscite on government policies, except in those locales where the PDS could capitalize effectively on voter interest in some local or state issue. Just prior to the elections, the government issued yet another rule change, instituting a ballot on which the names of candidates had to be written in, instead of checked off on a printed list. This measure was aimed at hurting opposition parties, since presumably only the PDS had sufficient local organization to ensure that its voters would learn to fill out their ballots correctly.

One issue the government worked hard to keep out of the campaign was Brazil's foreign debt. Mexico's dramatic debt crisis in August 1982 inevitably cast a shadow over Brazil, whose technocrats were reassuring their countrymen that Brazil's debt situation was far healthier than Mexico's. In fact, Delfim Neto and company were already negotiating with the IMF—a fact that they kept secret until after the elections to avoid providing ideal ammunition for the opposition.

The electoral results largely vindicated Golbery's original strategy.[46] Although the opposition won 59 percent of the total popular vote, it failed to gain a majority in the Congress (taking the two houses together) or in the electoral college which was to choose Figueiredo's successor. In the lower house of the Congress the opposition (combining all four parties—PMDB, PDT, PTB, and PT) outnumbered the PDS by 240 to 235, but in the Senate the PDS enjoyed an advantage of 46 to the opposition's 23. In the electoral college the PDS retained a majority of 359 to 321.

Several points about these results deserve note. First, the government party had lost its absolute majority in the lower house of Congress. That meant that if the opposition voted together it could block any government legislation. Second, even to retain its relative strength in Congress and the Electoral College, the government had to rely heavily on the less-populous and less-developed states, primarily in the Northeast and lightly settled Far West where the government's incumbent political machines could produce the votes. Exceptions to this were the PDS victories in Santa Catarina and Rio Grande do Sul. Thus the government margin was hardly reassuring, particularly in the light of past unreliability of government party votes.

The opposition had won the governorship of nine of the most populous and industrialized states, including such key states as São Paulo, Rio de Janeiro, Minas Gerais, and Paraná. Leonel Brizola, at one time anathema to the military, won the governorship of the state of Rio, bringing with him as Vice Governor Darcy Ribeiro, who as a Goulart staffer had called for a popular uprising in the face of the 1964 coup. Elsewhere the opposition gubernatorial winners were primarily centrist PMDB politicians, such as Franco Montoro in São Paulo and Tancredo Neves in Minas Gerais.

The most interesting new development in party politics was the PT. Although it won fewer votes than its enthusiasts were expecting by election eve, the PT was the most authentic left-wing party in postwar Brazil. It succeeded in gaining registration in all Brazilian states, and built up a network of local volunteers who could be valuable in future electoral battles. The character of the PT's ideology was shown by the fact that the Communist (PCB) strategists scorned it and threw their support to the PMDB, in line with long-standing PCB tactics in Brazil. The real question about the PT was whether it could grow quickly enough to survive under the electoral rules, which require that by the 1986 elections a party be able to win 5 percent of the vote in nine states and 3 percent nationally in order to maintain its legal registration. The party's best performance, not surprisingly, was in São Paulo, not only in the capital but throughout the state. PT enthusiasts were sobered, however, by the fact that Jânio Quadros, the ever enigmatic but quite discredited former President, received more votes for governor than did Lula.[47]

As for the PMDB, it proved stronger and more cohesive than some skeptics predicted. Many had thought the PMDB would eventually split, with the right forming a centrist Partido Liberal, and the left forming a Partido Socialista. Yet during 1982 the electoral logic continued to encourage a single opposition party,[48] a logic that was strengthened by the merger of the PMDB and the PP after the 1981 "November Package" of electoral law reforms. PDS victories in the two states which had been expected to go to the PMDB—Pernambuco and Rio Grande do Sul—occurred for differing reasons. In Rio Grande do Sul, Brizola's PDT ran a strong candidate, which split the opposition vote—precisely the tactic Golbery had hoped to promote. In Pernambuco, on the other hand, the PDS had a strong candidate who defeated a relatively united opposition.

While the opposition had now won executive control of key states, along with their considerable patronage power, the government retained the federal executive, which had gained strength from the steady centralization of power and resources since 1964. The opposition governors quickly found themselves needing support (especially financial support) from the federal government. Their circumstances could hardly have been worse. In March 1983 the Figueiredo government had already been forced to subordinate all economic policy to servicing the foreign debt. The resulting austerity policies meant that the opposition governors found themselves without enough money to meet urgent needs. By early 1984 their popularity had plummeted, as their citizens now faulted the PMDB and PDT governors for the austerity which national policy had imposed.

What were the political prospects for the future? Despite repeated attempts to produce a new and viable constitutional structure (1967, 1969, and 1977, for example), the post-1964 military governments had always focused on how to retain power while cultivating some semblance of electoral legitimacy. The PDS was supposed to be the party on which the government was now relying, yet it was hurt by its lack of input into executive decision making, even on decisions where the party was directly concerned. As a result, its leaders were identified with presidential policies, but did not help make them. They were "of" the government, but not "in" it. When the succession question produced a lack of clear leadership from the center, the party was unprepared to act in a unified manner that could guarantee its future in a different context.

Neither the Figueiredo team nor the PDS leadership seemed to have formulated any viable strategy for the longer run. Their goal had been to keep control of the 1985 presidential election. But they soon lost it in the face of a nationwide campaign favoring direct election of the President. Huge rallies (over a million in Rio and São Paulo) dramatized the frustration of a public impatient with abertura's slow pace. The measure required a constitutional amendment, which meant a two-thirds' vote of the Congress. The vote fell short in April, but the ayes included 55 PDS deputies, whose vote, despite heavy pressure from the President and party leaders, showed surprising government weakness.

Meanwhile, Minas Gerais Governor Tancredo Neves was rapidly gaining favor as the PMDB presidential candidate. He combined old-style political savvy and forward-thinking rhetoric that well suited the transition. He was also on good terms with the military. No less important, he could also appeal to PDS electors, without whose support he could not win, since the PMDB was in a minority.

On the government side, President Figueiredo and his advisers seemed incapable of deciding on a consensus candidate and making that choice prevail within the PDS. The eventual nominee was former São Paulo Governor Paulo Maluf, whose high-powered electioneering and oft-times arrogant manner alienated key PDS leaders and opened the way for Tancredo to win the PDS defectors he needed. The latter joined with the PMDB to create the *Aliança Democrática*, the coalition that supported Tancredo for President and (now former) PDS Senator José Sarney for Vice President. Sarney, a traditional Northeastern politician who first entered public life in the 1950s, had been incensed by Maluf's tactics in seeking the nomination. The Tancredo-Sarney ticket swept the electoral college by 480 to 180.

Brazil now seemed poised for a return to full democracy, i.e., direct election of the next President, restoration of full powers (especially over the budget) for the Congress, the writing of a new constitution, and so on. Tancredo, who had by now assumed heroic proportions in the public mind, would preside over this last stage in the transition from authoritarian rule.

The story had a different ending. On the eve of his inauguration, Tancredo was rushed to the hospital with a severe abdominal ailment that (with complications) soon cost him his life. A shocked and sobered José Sarney suddenly found himself being sworn in as acting President, and soon as full-fledged President. Now the President was a conservative politician who until only a few months earlier had helped lead the party which was a creature of the military government. Brazil's transition to democracy had taken yet another unexpected turn.

Brazil After Two Decades: How Different?

The Brazil of 1985 was not what it had been in 1964. There had been enormous economic change, the turnover of an entire generation of politicians, and a population more than half of which had been born in the last two decades. These vast changes were not always appreciated by politicians and political analysts—whether civilian or military—who often thought of a "return" to democracy in terms of pre-1964 political patterns.[49]

First, the growing "middle class" continued to wield significant weight in the economically developed Center-South. It included, among others, professionals (doctors, lawyers, engineers, etc.), managers, and bureaucrats. Although not of a size proportional to that of the middle class in the industrialized countries, the Brazilian middle class had great voting power in the urban areas. This middle sector was highly sensitive to the media (especially television), had a tradition of political activism, and saw itself as having a stake in redemocratization. Its members were also highly vulnerable to the pressures of inflation. The recession (and wage policy, with its deliberate squeezing of pay differentials) had since 1979 hit them with a sharp loss in real income and made them excellent targets for opposition political appeals.

Second, in the economically more developed states the working-class electorate—especially in the shantytown peripheries of the major cities—voted heavily PMDB. This support for the opposition was related not only to economic trends, but also to new forces which had emerged at the grassroots level. Among the major new forces were the popular mobilization of the Catholic lay action groups (*comunidades eclesiais de base*, or CEBs), and the reinvigorated *sociedades de amigos do Bairro*, or SABs, as well as the impressive grassroots organizing among union members.[50] (See the Mainwaring article in this volume.) In pre-1964 Brazil spontaneous political mobilization was only beginning, and proved highly vulnerable to repression and manipulation in the wake of the 1964 coup.

Seen from the vantage point of 1985, there had been enormous economic growth, greatly expanding the organized urban sectors, especially labor. (See Keck article in this volume.) Unions had shown their power in the series of strikes beginning in 1978. Meanwhile, the repression practiced by the post-1964 governments had created a deep suspicion of all established policies. Radical Christian doctrines, based on the theology of liberation, reinforced this mistrust of political institutions. While the PT had attempted to court the CEBs, many CEB and some union organizers strongly opposed any party identification.

Another major actor was the Brazilian business community.[51] Although Brazilian business had prospered mightily during the "miracle," many businessmen had begun to feel that Brazilian capitalism required a system of more open bargaining. Many pro-*abertura* industrialists thought redemocratization offered the only means to defend themselves against the combined threat of the burgeoning State and the mutinationals. In political terms, however, these businessmen were vulnerable vis-à-vis the technocrats who controlled credit policy (including individual loan requests), import controls, and government contracts.[52]

What about the elite as a whole?[53] The Brazilian elite had long prided itself on its ability to maintain a political consensus by "conciliation." Even if the establishment had often resorted to manipulation and repression, the "conciliation" myth has retained a hold over Brazilians of all classes. The new grassroots movements were challenging this myth, confronting the "conciliators" with workers who rejected manipulation of the old kind.

As for the most important elite, the military, one point was reasonably certain: the military as a whole would oppose any attempt to fix individual responsibility for past repression, especially torture. Any suggestion of trying "war criminals," as happened after the fall of the Greek military dictatorship or as was announced by the Alfonsín government in Argentina, was forbidden territory for any civilian government. But in fact, throughout the political opening, the extent to which all significant elements of the opposition have taken care to avoid provoking the rightist military is impressive.

The greatest danger lay in the possibility of public disorder, especially if the left could be blamed. Left terrorism would galvanize the military, overriding their commitment to liberalization. Some hardline military (and like-minded civilians) were fully capable of faking or provoking public incidents (the RioCentro bombs were apparently part of such a move). Their aim would be to start a violent chain reaction, and thereby panic moderate officers into closing the system again.

Looking Back: Why Did Liberalization Succeed?

No observer of contemporary Latin American politics could fail to be impressed by Brazil's relatively smooth transition away from a military-dominated regime. When President Geisel pledged to redemocratize in 1974, there was deep skepticism. Yet the pro-democratic military prevailed, despite repeated challenges from the "hard line." No less important, the military moderates found the civilian opposition, although neither side would have admitted it, cooperative in the tense process of inching toward less arbitrary rule. Even onetime guerrilla leaders have endorsed peaceful politics. No significant element on the left urged armed action. It was a time of intense political debate, but within the framework created by the *abertura.*

Seen in this light, the post-1979 strategy of the government's most clear-sighted political strategist, Golbery, had been successful. The amnesty satisfied a major claim of those hit by the repression. The phase-out of the compulsory two-party structure in 1979 allowed a wide spectrum of political opinion to organize. The result was a greatly increased "space" in which the elite could debate and mobilize. The discontented were offered an expanded, but still elite-dominated, democratic "game," which they have embraced, while still denouncing the remaining arbitrary powers.

Unlike Argentine, Chilean, or Uruguyan counterparts, the Brazilian authoritarians maintained their allegiance to an electoral process. However distorted or insincere their claim, they *did* allow at least a segment of the civilian opposition to remain active, especially in Congress. The Congressional opposition was therefore able to maintain a thread of democratic legitimacy, even if their votes were buried or disallowed. As a result, the MDB managed in the early 1970s to make a legitimate claim to represent the opposition, as was dramatically demonstrated in the 1974 elections.

But what really motivated the moderate military to take a chance on liberalization? Was it a moral commitment to democracy? Or was it a calculation that Brazilian society could no longer be ruled by the authoritarian structure created in 1968–69?

Close study of the Brazilian government's action since 1974 cannot but lead to the conclusion that Geisel, Golbery, and like-minded officers acted out of a personal belief that Brazil *should* move toward a more democratic regime. The question of how to carry it out after 1981 was left to Figueiredo and a new generation of army officers.

A further question is why the opposition cooperated with the government's political game. Watching the MDB (and then the PMDB) responding to the constant manipulations of electoral law leads one to ask why they kept trying. The point becomes clearest in the case of those young politicians who had no experience with the pre-1964 system and who entered political life to face the arbitrary authority of AI-5 and the SNI. Although the military governments talked of "provocations," the truth is that the opposition was extraordinarily moderate. Even with successive *cassações,* the MDB effectively urged moderation on its more radical members. The most dramatic evidence of this opposition attitude was the 1982 election campaign. Given the issues, it was impressive that the opposition was able to keep the campaign within clear-cut limits. It showed a keen awareness of the dangers of any radical attack.

In the end, liberalization was the product of an intense dialectical relationship between the government and the opposition. The military who favored *abertura* had to proceed cautiously, for fear of arousing the hardliners. Their overtures to the opposition were designed to draw out the "responsible" elements, thereby showing there were moderates ready to cooperate with the government. At the same time, the opposition constantly pressed the government to end its arbitrary excesses, thereby reminding the military that their rule lacked legitimacy. Meanwhile, the opposition moderates had to remind the radicals that they would play into the hands of the hardliners if they pushed too hard. This intricate political relationship functioned successfully because there was a consensus among both military and civilians in favor of a return to an (almost) open political system. What that system would produce for the average Brazilian was, of course, another question.

Brazilian Authoritarianism: Links to the Past

In my paper for the Yale Workshop of 1971 I compared post-1964 Brazil with the *Estado Nôvo* of 1937–45.[54] Can new insights be gained from renewing that comparison? Perhaps, although the parallel grows strained as the years pass. The *Estado Nôvo* lasted only seven years, while the post-1964 military-dominated regime lasted twenty-one years. Here are some further suggested comparisons:

Role of Political Parties: In 1937, Brazil lacked genuinely national parties. The Congress was indefinitely recessed and party politics were banned. When Getúlio lifted the lid in early 1945, the parties that emerged were new. Although they included many familiar faces, they were not an extension of pre-1937 parties. The political scene was therefore far more fluid in 1945 than in 1979. With three new major parties in 1945—UDN, PSD, and PTB, along with the newly legalized PCB—it was not clear how the voting would run.

With the post-1964 military governments the situation was quite different. Unlike the *Estado Nôvo*, the "revolutionaries" of 1964 never abolished the Congress or the state legislatures, nor eliminated all political parties. Instead, they effectively used the electoral system, constantly manipulating it to produce a government majority.

The government's decision in 1965 to abolish the pre-existing parties and institute a two-party system was the most important political act by any post-1964 government. It guaranteed that elections would always give the opposition a chance to capitalize on feeling opposed to the government's repressive and anti-egalitarian policies. Ironically, the compulsory two-party structure ensured opposition unity. The government party only won majorities by flagrant intimidation, as in 1970, or by manipulation of electoral rules, as with the *pacôte de abril* in 1977. General Golbery saw this point early on. He masterminded the 1979 party reform, legalizing a multi-party structure. Yet the government's difficulties in manipulating even this system had already become painfully obvious in the elections of November 1982.

Ideology: In 1971 I noted the similarity in anti-communist rationales underlying both the *Estado Nôvo* and the post-1964 regime.[55] In this there has been no major change. National Security ideology has, however, penetrated more deeply into the officer ranks than its less elaborately defined equivalent in the *Estado Nôvo*, and in addition, the ideology has a huge bureaucratic machine, the SNI, to guarantee its survival.[56]

The strength of belief in the National Security doctrine and the power of the SNI have created a situation quite different from that of the *Estado Nôvo* in 1945. Although the left has been allowed to surface and organize, and even campaign—under a variety of banners—intelligence dossiers are still being compiled. The capacity is still there to crack down as the government did against strikers in São Paulo in 1980, and hardline officers are quite prepared to use it.

Degrees of resistance by civil society: My 1971 paper was written when the Médici government was riding high, having defeated the guerrillas and apparently successful in repressing the opposition. The 1970 elections had given the government a huge majority, and the key institutions of civil society were under heavy surveillance. The only institution able to assert itself against the military government was the Church.

The subsequent decade brought an extraordinary and unforseen blossoming of resistance from long-established but previously conservative groups, such as the Bar Association and the Church hierarchy. There were also new groups, such as the CEBs and newly formed neighborhood associations. The *Estado Nôvo* never faced a comparable level of resistance by civil society. In part it was because Brazilian society was not as highly articulated—many fewer industrial workers had experience in industrial labor relations, for example. It was also because many of the elements mobilizing on the left in 1945, for example the PCB, had decided that they would do better with Getúlio in power than out. They therefore campaigned to keep Getúlio (*Queremistas*), but with an end to the *Estado Nôvo* and the convoking of a Constituent Assembly. In the 1970s and 1980s the new groups made no secret of the fact that they wanted the military-dominated government out.

NOTES

1. Alfred Stepan (ed.), *Authoritarian Brazil: Origins, Policies and Future* (New Haven: Yale Univ. Press, 1973). The most systematic listing of the wide range of

authoritarian measures is Lúcia Klein and Marcus Figueiredo, *Legitimidade e coacão no Brasil pós-64* (Rio de Janeiro: Forense-Universitária, 1978).

2. Juan Linz in Stepan, *Authoritarian Brazil*, 235.

3. The space available does not allow a comprehensive account of the decade between 1974 and 1984, even of political events. Economic policy making and the economic record are discussed only intermittently. There is also relatively little attention paid to important changes in civil society, such as the bar associations, ecclesiastical base communities (CEBs), and the neighborhood associations (SABs). For an excellent overview of many of these sectors, see Bernardo Sorj and Maria Hermínia Tavares de Almeida (eds.), *Sociedade e política no Brasil pós-64* (São Paulo: Editôra Brasiliense, 1983).

4. A fine historical interpretation of national politics from 1964 to 1977 may be found in Peter Flynn, *Brazil: A Political Analysis* (Boulder: Westview Press, 1978), 308–522.

5. The principal secondary source on the wildcat strikes of 1968 remains Francisco C. Weffort, "Participação e conflito industrial: Contagem e Osasco, 1968," *Caderno 5* (São Paulo: CEBRAP, 1972). For the flavor of the student protests, see Luíz Henrique Romagnoli and Tânia Gonçalves, *A volta da UNE: de Ibiúna a Salvador* (São Paulo: Ed. Alfa e Omega, 1979).

6. The most comprehensive study of the evolution of the national security doctrine as it was put into practice by the post-1964 military governments is Maria Helena Moreira Alves, *State and Opposition in Military Brazil*, (Austin: Univ. of Texas Press, 1985). For a valuable analysis of how the national security ideology emerged in Latin America, see José Comblin, *The Church and the National Security State* (Maryknoll, N.Y.: Orbis Books, 1979). The author is a Belgian theologian who taught in Brazil and was expelled in 1972.

7. Carlos Chagas, *113 Dias de Angústia: Impedimento e Morte de um presidente* (Pôrto Alegre: L & MP, 1979) is a first-hand account of the political crisis created by Costa e Silva's stroke. Chagas was the presidential press secretary. His book originally appeared in installments in *O Estado de São Paulo* in Jan.–Feb. 1970. A book version (Rio de Janeiro: Agencia Jornalística Image, 1970) with some added documents was published the same year, but Justice Minister Buzaid ordered it removed from the bookstores and prohibited it from circulation, alleging that it infringed "national security." Only nine years later did it reappear in a new edition. For the account of his presidency by Costa e Silva's chief military aide, see Jayme Portella de Mello, *A Revolução e o governo Costa e Silva* (Rio de Janeiro: Guavira Editores, 1979).

8. The most important analysis of the Brazilian military in politics is Alfred Stepan, *The Military in Politics: Changing Patterns in Brazil* (Princeton: Princeton Univ. Press, 1971). A brief discussion of the December 1968 crisis is given on pp. 259–66. A later Brazilian analysis, which stresses the interaction of civilian opposition groups and the military is Eliezer R. de Oliveira, *As Forças armadas: política e ideologia no Brasil, 1964–1969* (Petrópolis: Vozes, 1976). Clues to military thinking in this period can be found in the collected newspaper columns of Carlos Castello Branco, *Os militares no poder*, Vol. II: O Ato 5 (Rio: Nova Fronteira, 1978); and Fernando Pedreira, *Brasil política* (São Paulo: DIFEL, 1975). Amid the sparse bibliography on the Médici presidency is Hélio Silva e Maria Cecília Ribas Carneiro, *Emílio Médici: o combate ás guerrilhas, 1969–74* (São Paulo: Grupo de Comunicação Três, 1983).

9. The best single source of analysis of this electoral process in the 1970s is

Bolivar Lamounier (ed.), *Vôto de desconfiança: eleições e mudança política no Brasil: 1970–1979* (Petrópolis: Editôra Vozes, 1980).

10. The evidence on torture was gathered and published by several international human rights groups. See, for example, Amnesty International, *Report on Allegations of Torture in Brazil* (London: Amnesty International Publications, 1972), which listed 1,076 torture victims as of September 1972. One of the most tragic cases was Frei Tito de Alencar Lima, a young Dominican monk who, tortured savagely and later exiled, finally committed suicide in France. His story is told in Pedro Celso Uchoa Cavalcanti et al., *Memórias do Exílio: Brasil, 1964–19??* (Lisbon: Editôra Arcádia, 1976), 347–69. An interesting byproduct of the guerrilla era were memoirs by ex-guerrillas that are important historical sources, but are also highly readable. Two examples are Fernando Gabeira, *O que é isso, companheiro?* (Rio de Janeiro: CODECRI, 1979); and Alex Polari, *Em busca do tesouro* (Rio de Janeiro: CODECRI, 1982). Both Gabeira and Polari were in the guerrilla groups that kidnapped foreign diplomats. Both were captured, tortured, and eventually went into exile.

11. There is useful information on Geisel's background and his ministers in Fernando Jorge, *As Diretrizes governamentais do President Ernesto Geisel* (São Paulo: Edição do Autor, 1976). A highly favorable review of Geisel's first two years in the presidency is given in Adirson de Barros, *Março: Geisel e a revolução brasileira* (Rio de Janeiro: Editôra Artenova, 1976). Three years into the Geisel presidency Reis Velloso published a defense of government economic policies in João Paulo dos Reis Velloso, *Brasil: A solução positiva* (São Paulo: Abril-Tec, 1978).

12. The consequences of the Geisel government's policy is made clear in the chapter in this volume by Edmar Bacha and Pedro Malan.

13. A few months later the liberalization attempt was put into historical perspective by the leading journalist Carlos Castello Branco, who described how Professor Samuel Huntington, a Harvard political scientist, had been asked by Geisel and Golbery to draft a memo on "gradual decompression." *Jornal do Brasil*, Sept. 4, 1974.

14. In October 1973 a Congressionally sponsored institute had held hearings on possible strategies for liberalization. Wanderley Guilherme dos Santos, a leading political scientist, presented a paper surveying the state of democracy around the world. His paper and testimony, including questions from Congressmen, are included in Wanderley Guilherme dos Santos, *Poder e política: crônica do autoritarismo brasileiro* (Rio de Janeiro: Forense-Universitária, 1978), which contains also a series of Santos's newspaper columns from 1974. Santos was one of the first intellectuals to spell out, in language palatable to at least some of the military, a rationale for redemocratization. A valuable journalistic account of the political struggle of the 1974–80 period is Bernardo Kucinski, *Abertura, a história de uma crise* (São Paulo: Brasil Debates, 1982). The author was skeptical of the castellistas' real motives. For background on one of the most important actors in the struggle, see Getúlio Bittencourt and Paulo Sérgio Markum, *O cardeal do povo: D. Paulo Evaristo Arns* (São Paulo: Ed. Alfa-Omega, 1979). For an extremely useful index to key speeches, statements, and newspaper articles on the liberalization, see Marcus Faria Figueiredo and José Antônio Borges Cheibub, "A abertura política de 1973 a 1981: quem disse o quê, quando-inventário de um debate," *Boletim informativo e bibliográfico de ciências sociais* (pub. by the Associação Nacional de Pós-Graduação e Pesquisa em Ciências Sociais), No. 14 (1982), 29–61.

15. A favorite target of the censors was the weekly *Opinião*, which routinely had

more than half its copy vetoed by the federal censors in Brasília. Nonetheless, the publication survived from 1972 to 1977, when it closed in protest against government pressure. Documentation of the struggle is given in J. A. Pinheiro Machado, *Opinião x censura: momentos da luta de um jornal pela liberdade* (Pôrto Alegre: L & PM Editores, 1978) General information on censorship, rather chaotically organized, can be found in Paolo Marconi, *A Censura política na imprensa brasileira: 1968–1978* (São Paulo: Global, 1980).

16. The Brazilian Bar Association (OAB) came to play a major role in pressing for political liberalization. One of the best sources on the lawyers' role is the proceedings of biannual national conventions of the Ordem dos Advogados do Brasil. Starting in 1974 each convention had as a theme one of the fundamental issues on which the bar association was crusading: 1974 (human rights), 1978 (rule of law), 1980 (liberty), and 1982 (social justice). Each volume was published with the title *Conferência Nacional da Ordem dos Advogados do Brasil.* A succinct account of the lawyers' movement is given in James A. Gardner, *Legal Imperialism: American Lawyers and Foreign Aid in Latin America* (Madison: Univ. of Wisconsin Press, 1980), 109–25.

17. *Anais de V Conferência Nacional da Ordem dos Advogados do Brasil*: Rio de Janeiro: 11 a 16 de Agosto de 1974 (Rio de Janeiro: n.p., n.d.), 101.

18. Morris related his experiences in "In the Presence of Mine Enemies: Faith and Torture in Brazil," *Harper's Magazine* (October 1974), 57–70.

This account is in part based on interviews with officials in the U.S. embassy in Brasilia and the U.S. consulate in Recife in 1975.

19. For a detailed analysis of the November 1974 elections see Bolivar Lamounier and Fernando Henrique Cardoso (eds.), *Os partidos e as eleições no Brasil* (Rio de Janeiro: Paz e Terra, 1975) and *Revista Brasileira de Estudos Políticos*, 43 (July 1976), which was entirely devoted to the elections. My data are taken from the latter. For an amusing and irreverent view of the elections, see Sebastião Nery, *As 167 derrotas que abalaram o Brasil* (Rio de Janeiro: Francisco Alves Editôra, 1975).

20. The flavor of the MDB campaign can be found in collections of speeches by two of its leaders: Franco Montoro, *Da "democracia" que temos para a democracia que queremos* (Rio de Janeiro: Paz e Terra, 1974), and Marcos Freire, *Oposição no Brasil, hoje* (Rio de Janeiro: Paz e Terra, 1974). Montoro was an MDB Senator from São Paulo, and Freire an MDB Senator from Pernambuco.

21. For a first-hand account of how censorship worked at *O Estado de São Paulo*, see the interview with the paper's editor-in-chief, Oliveiros Ferreira, in Carlos Rangel, *1978: a hora de enterrar os ossos* (Rio de Janeiro: Tipo Editor, 1979), 92–99.

22. In the 1970s the Brazilian Catholic Church emerged as one of the most innovative and controversial in the world. For background on that process, see Ralph Della Cava, "Catholicism and Society in Twentieth-Century Brazil," *Latin American Research Review*, XI (No. 2, 1976), 7–50; and Thomas Bruneau, *The Political Transformation of the Brazilian Catholic Church* (Cambridge: Cambridge Univ. Press, 1974). Bruneau has continued the story to 1978 in *The Church in Brazil: The Politics of Religion* (Austin: Univ. of Texas Press, 1982), which includes studies of a number of ecclesiastical base communities. The most successful attempt to place the contemporary Brazilian church in perspective is Scott Mainwaring, *The Catholic Church and Politics in Brazil, 1916–1985* (Stanford: Stanford Univ. Press, 1986).

23. Among the many accounts of Herzog's death is Hamilton Almeida Filho, *A Sangue-quente: a morte do jornalista Vladimir Herzog* (São Paulo: Alfa-Omega, 1978), a book version of an issue of *EX*-magazine in 1975.

24. The television speech, along with his other speeches, is reprinted in Alencar Furtado, *Salgando a terra* (Rio de Janeiro: Paz e Terra, 1977).

25. Brazil–U.S. relations during the Carter presidency are given an entire chapter in Robert Wesson, *The United States and Brazil: Limits of Influence* (New York: Praeger, 1981). The wider context of U.S. human rights policy toward Brazil is given in Lars Schoultz, *Human Rights and United States Policy Toward Latin America* (Princeton: Princeton Univ. Press, 1981). Economic issues are stressed in two articles by Albert Fishlow: "Flying Down to Rio: U.S.-Brazil Relations," *Foreign Affairs*, 57:2 (Winter 1978/79), 387–405; and "The United States and Brazil: The Case of the Missing Relationship," *Foreign Affairs*, 60:4 (Spring 1982), 904–23.

26. A convenient summary of U.S.-Brazilian friction over the nuclear question may be found in Wesson, *The United States and Brazil*, 75–89. An early analysis and one of the best, is Norman Gall, "Atoms for Brazil, Dangers for All," *Foreign Policy*, 23 (Summer 1976), 155–201. For a useful background to Brazilian nuclear policy, see Wolf Grabendorff, "Bedingungsfaktoren und Strukturen der Nuklearpolitik Brasiliens" (Mimeo: Stiftung Wissenschaft und Politik, Ebenhausen, West Germany: Dec. 1979). For a critique of official policy by Brazil's leading nuclear physicist, see José Goldemberg, *Energia nuclear no Brasil* (São Paulo: Editôra Hucitec 1978). The most widely read domestic critic of Brazil's nuclear policy was Kurt Rudolf Mirow, *Loucura nuclear* (Rio de Janeiro: Civilização Brasileira, 1979).

27. These struggles within the Geisel government are clearly chronicled in three informative journalistic accounts: André Gustavo Stumpf and Merval Pereira Filho, *A Segunda guerra: sucessão de Geisel* (São Paulo: Brasiliense, 1979); Walder de Goes, *O Brasil do General Geisel* (Rio de Janeiro: Nova Fronteira, 1978); and Getúlio Bittencourt, *A Quinta estrela: como se tenta fazer um presidente no Brasil* (São Paulo: Editôra Ciências Humanas, 1978). One well-informed journalist described Geisel as "keeping his uniformed colleagues at a distance and playing masterfully on their individual weaknesses . . .," Kucinski, *Abertura*, 42.

28. Abreu gave his angry version of how Figueiredo's candidacy was "imposed" in Hugo Abreu, *O Outro lado do poder* (Rio de Janeiro: Nova Fronteira, 1979) and in the posthumously published *Tempo de crise* (Rio de Janeiro: Nova Fronteira, 1980). The later volume included Abreu's recommendations for reorienting Brazil's political, economic, and social policies.

29. For a careful study of the new law, including systematic comparisons with previous legislation on national security, see Ana Valderez A.N. de Alencar, *Segurança nacional: Lei no 6.6520/78—antecedentes, comparações, anotações, histórico*, 2nd ed. (Brasília: Senado Federal, Subsecretaria de Edições Técnicas, 1982).

30. In the late Geisel presidency there were a number of collaborative efforts at diagnosing Brazil's ills—coming primarily from the opposition. One of the most interesting is the collection of interviews with prominent scholars, scientists, and technocrats edited by Celcio Monteiro de Lima, *Brasil: o retrato sem retoque* (Rio de Janeiro: Francisco Alves Editôra, 1978). In early December 1978 the opposition-oriented Centro Brasil Democrático held a three-day plenary meeting to discuss Brazil's political, economic, social, and constitutional plight. The proceedings were published in Centro Brasil Democrático, *Paineis da crise brasileira: Anais do Encontro Nacional Pela Democracia* (Rio de Janeiro: Editôra Avenir, 1979), 4 vols. A similar effort, on more sharply focused themes, took place in São Paulo in mid-1979 and the papers were published in Bolivar Lamounier, Francisco C. Weffort, and

Maria Victória Benevides (eds.), *Direito, cidadania e participação* (São Paulo: T.A. Queiroz, 1981).

31. There has been little public analysis of the SNI and its role in Brazilian politics. One of the few is Ana Lagoa, *SNI: como nasceu, como funciona* (São Paulo: Editôra Brasiliense, 1983), a collection of newspaper feature stories along with some official documents.

32. For a collection of his speeches, primarily those given in the Senate between 1973 and 1979, see Petrônio Portella, *Tempo de congresso, II* (Brasília: Senado Federal, 1980). The author of the "apresentação" of this posthumously published volume was, appropriately, General Golbery.

33. The problem of inflation is stressed in "Panorama," the opening article in *Conjunctura económica*, Feb. 1979, the issue with the retrospective survey of the Brazilian economy during 1978. For an excellent review of principal economic problems as of late 1979, see *Revista de ANPEC*, III (no. 4, Outubro 1980), which includes papers given at the annual convention of Brazilian graduate economics faculties. For a concise description of the economic problems facing the new government in 1979, see Werner Baer, *The Brazilian Economy: Growth and Development*, 2nd ed. (New York: Praeger, 1983), chap. 6. The highly conservative U.S. financial weekly *Barron's* ran an editorial in Sept. 1978, entitled "Distant Early Warning: Brazil's 'Economic Miracle' Has Lost Its Lustre." The editorial was another attack on Brazil's great apostasy—indexation.

34. The "new unionism" in Brazil stimulated an outpouring of analysis. Important articles by Amaury de Souza, Bolivar Lamounier, Maria Hermínia Tavares de Almeida, and Lüíz Werneck Vianna appeared in *Dados*, vol. 24 (no. 2, 1981). A useful view from a long-time U.S. observer is Thomas G. Sanders, "Brazil's Labor Unions," *American Universities Field Staff Reports*, 1981/No. 48 (South America). The flavor of the intense interest within Brazil is evident in Ricardo Antunes (ed.), *Cadernos de Debate 7: Por um novo sindicalismo* (São Paulo: Editora Brasiliense, 1980). One of the most important research contributions is John Humphrey, *Capitalist Control and Workers' Struggle in the Brazilian Auto Industry* (Princeton: Princeton Univ. Press, 1982), which is based on a careful study of actual shop floor conditions.

35. Overnight Lula became one of Brazil's most famous personalities. Background on his career and views is given in Mário Morel, *Lula, o metalúrgico* (Rio de Janeiro: Nova Fronteira, 1981).

36. In July 1980 General Golbery delivered a lecture to the Escola Superior de Guerra in which he analyzed the political scene and explained how the government could retain the initiative if it showed patience and negotiating skill. This lecture was Golbery's most complete public statement on the Geisel government's political strategy. Golbery do Couto e Silva, *Conjuntura política national, o poder executivo & geopolítica do Brasil* (Brasília: Editôra Universidade do Brasilia, 1981), 3–35.

37. Useful information on the founding of the PT is given in Mario Pedrosa, *Sobre o PT* (São Paulo: CHED Editora, 1980).

38. The political significance of Portella's death was emphasized in the press, as in the full-page coverage of Jan. 8, 1980, in the *Jornal do Brasil.*

39. Conferência Nacional dos Bispos do Brasil, *Situação do clero no Brasil* (São Paulo: Edições Paulinas, 1981), 11.

40. These conflicts drew much publicity. See, for example, D. José Maria Pires, *Do centro para a margem* (João Pessoa: Editôra Acaua, 1978) and Rivaldo Chinem, *Sentença: padres e posseiros do Araguaia* (Rio de Janeiro: Editôra Paz e Terra, 1983).

41. For a detailed analysis of the "maxi" devaluation, see the article by Edy Luiz Kogut and José Júlio Senna in *Jornal do Brasil*, Dec. 19, 1979. The new measures were widely debated by economists. See, for example, the interview with José Serra, an economist from the University of Campinas, in *Folha de São Paulo*, Dec. 12, 1979. Serra attacked the new policy as a "shock treatment for a very weak patient." Change in economic policy was one among other policy changes announced by President Figueiredo in a major speech on Dec. 7, 1979.

42. For the Labor Minister's speeches spelling out his thinking from 1979 to 1981, see Murillo Macêdo, *Trabalho na democracia: a nova fisionomia do processo político brasileiro* (Brasília: Ministério do Trabalho, 1981).

43. Out of the large literature on wage policy I have relied on the following: Departamento Intersindical de Estatística e Estudos Socio-Econômicos (DIEESE), *Dez Anos da política salarial* (São Paulo, 1975); Edmar Lisboa Bacha, *Política económica e distribuição de renda* (Rio: Paz e Terra, 1978), 25–52; and Lívio W. R. de Carvalho, "Princípios e aplicação da política salarial pós-1964," Universidade do Brasília, Departamento de Economia, Textos para discussão No. 9 (June 1973). The degree to which individual workers benefited by rising through the wage scale is stressed in Samuel A. Morley, *Labor Markets and Inequitable Growth: The Case of Authoritarian Capitalism in Brazil* (Cambridge: Cambridge Univ. Press, 1982). Morley quite rightly argues that the larger question of poverty in Brazil cannot be understood without analyzing the effects of rapid growth in a capitalist economy which has extensive surplus labor.

44. For evaluations of Golbery's resignation by two of Brazil's better informed political journalists, see Mino Carta's column in the *Folha de São Paulo*, Aug. 12, 1981, and Elio Gaspari's in the *Jornal do Brasil*, Aug. 21, 1981.

45. Golbery was struggling against Medeiros as part of his fight against the "Médici group," which he feared might regain the presidency. See the columns of Carlos Chagas in *O Estado de São Paulo*, Dec. 6, 1981, and July 6, 1982.

46. For the evaluation by a veteran U.S. student of Brazilian elections, see Ronald M. Schneider, *1982 Brazilian Elections Project Final Report: Results and Ramifications* (Washington, D.C.: Georgetown Univ. Center for Strategic and International Studies, 1982). A very useful overview of elections and the party system from 1946 through the 1982 elections is given in chapter four of Robert Wesson and David Fleischer, *Brazil in Transition* (New York: Praeger, 1983).

47. For an analysis of the 1982 vote in São Paulo, see Bolivar Lamounier and Judith Muszynski, "São Paulo, 1982: a victória do (P)MDB," (São Paulo: IDESP Texto No. 2, 1983).

48. In mid-1981 the party began publishing a magazine, *Revista do PMDB*, containing party manifestos and articles on principal economic and social problems.

49. There is a vast literature on the political process in the post-1964 era and the changes it brought in voting alignments. Especially helpful are Fábio Wanderley Reis, "O eleitorado, os partidos e o regime autoritário brasileiro," in Sorj and Almeida (eds.), *Sociedade e Política*, 62–86; Bolivar Lamounier, "O discurso e o processo (da distensão às opções do regime brasileiro)," in Henrique Rattner (ed.), *Brasil 1990: caminhos alternativos do desenvolvimento* (São Paulo: Brasiliense, 1979), 88–120; Bolivar Lamounier and José Eduardo Faria (eds.), *O futuro da abertura: um debate* (São Paulo: Cortez, 1981); and Olavo Brasil de Lima Junior, "Processo eleitoral e transição política no Brasil" (paper prepared for conference on "Opportunities and Limits of the Peripheral Society: The Case of Brazil," in Novo Friburgo, July 1983).

50. An excellent starting point on these groups can be found in Paul Singer and Vinícius Caldeíra Brant (eds.), *São Paulo: o povo em movimento* (Petrópolis: Editôra Vozes, 1981); Ruth C. L. Cardoso, "Movimentos Sociais Urbanos: Balanço Crítico," in Sorj and Almeida (eds.), *Sociedade e Política*, 215–39; José Alvaro Moisés et al., *Cidade, povo e poder* (Rio de Janeiro: Paz e Terra, 1982); and José Alvaro Moisés et al., *Alternatives populares da democracia: Brasil, anos 80* (Petrópolis: Vozes, 1982). An official Church view on the CEBs is given in Conferência Nacional dos Bispos do Brasil, *Comunidades eclesiais de base no Brasil: experiências e perspectivas* (São Paulo: Edições Paulinas, 1979). A popularized explanation by a famous survivor of government repression is Frei Betto, *O que é comunidade eclesial de base* (São Paulo: Brasiliense, 1981). A succinct overview by a long-time observer of the Brazilian Church is Thomas G. Sanders, "The Catholic Church in Brazil's Political Transition," *American Universities Field Staff Reports*, South America, 1980, No. 48. For a detailed analysis of CEB's in several regions of Brazil, see Bruneau, *The Church in Brazil*, chap. 8, which cites a host of Brazilian sources.

51. An analysis based on extensive interviews with businessmen is given in Renato Raul Boschi, *Elites, industriais e democracia* (Rio de Janeiro: Edições Graal, 1979), and in Eli Diniz and Renato Raul Boschi, *Empresariado nacional e estado no Brasil* (Rio de Janeiro: Forense-Universitária, 1978).

52. The agonized relationship between business and the military government is analyzed in Luiz Carlos Bresser Pereira, *O colapso de uma aliança de classes: a burguesia e a crise do autoritarismo technoburocrático* (São Paulo: Brasiliense, 1978). The growing state role in the economy was attacked loudly in the early Geisel presidency, an episode well analyzed in Charles Freitas Pessanha, "Estado e economia no Brasil: a campanha contra a estatização, 1974–1976" (Tese de Mestrado, IUPERJ, 1981). For a panel discussion by leading business figures on "As saidas para a crise," see *Brasil em Exame*, May 1983.

53. There is a wealth of data and insight on Brazilian elite attitudes as of the early 1970s in Peter McDonough, *Power and Ideology in Brazil* (Princeton: Princeton Univ. Press, 1981) and McDonough, "Mapping an Authoritarian Power Structure: Brazilian Elites during the Médici Regime," *Latin American Research Review*, vol. XVI (No. 1, 1981), 79–106.

54. Thomas E. Skidmore, "Politics and Economic Policy Making in Authoritarian Brazil, 1937–71," in Alfred Stepan (ed.), *Authoritarian Brazil*, 34–43. The meager historiography on the *Estado Nôvo* has seen some important recent additions. The most important history of the period is Edgar Carone, *O estado nôvo: 1937–1945* (São Paulo: DIFEL, 1976). Carone has published a companion volume of documents on the same period: *A terceira república: 1937–1945* (São Paulo: DIFEL, 1976). There is much valuable information and synthesis in the relevant chapters of Boris Fausto (ed.), *História geral da civilização brasileira*, tomo III: O Brasil republicano, vol. 3: Sociedade e política, 1930–1964 (São Paulo: DIFEL, 1981).

55. For the evolution of the national security doctrine since 1964, see Maria Helena Moreira Alves, *State and Opposition in Military Brazil*. For the *Estado Nôvo* we now have Reynaldo Rompeu de Campos, *Repressão judicial no Estado Nôvo: Esquerda e direita no banco dos reus* (Rio de Janeiro: Achiamé, 1982). The relatively sparse literature on ideology in the *Estado Nôvo* has an important new addition in Lúcia Lippi Oliveira, Monica Pimenta Velloso, and Angela Maria Castro Gomes, *Estado Nôvo: ideologia e poder* (Rio de Janeiro: Zahar Editores, 1982).

56. See Alfred Stepan, *Rethinking Military Politics: Brazil and the Southern Cone* (Princeton: Princeton Univ. Press, 1988).

2

Authoritarian Brazil Revisited: The Impact of Elections on the *Abertura*

BOLIVAR LAMOUNIER

Lovers and madmen have such seething brains,
such shaping fantasies, that apprehend
More than cool reason ever comprehends.
(Shakespeare, *A Midsummer Night's Dream*,
Act V, Scene i).

Authoritarian Brazil is now a minor classic.* It was published for the first time in 1973 and has had numerous printings. It remains one of the major sources for understanding the post-1964 Brazilian political system and, in particular, the harshest period of military domination which began at the end of 1968. Nonetheless, the analyses therein do not seem to have had great predictive power with regard to the liberalization begun in 1973–74, and still less as to the *mechanics* of this process, which led to a rapid enhancement of electoral politics and to the rise, through that route, of a vigorous opposition. In comparative terms, it may even be said that this *abertura*, or "opening up" through elections, is exceptional in that it led to a party system

*A preliminary version of this article was presented at a conference on the prospects for redemocratization in Brazil, Chile, and Argentina at the School of International and Public Affairs, Columbia University, in March 1984. I am grateful to the participants, in particular to Alfred Stepan, Juan Linz, Edmar Bacha, and Manuel Antonio Garretón, for their suggestions and criticisms. I would also like to acknowledge the invaluable help that Amaury de Souza and other IDESP colleagues, particularly Rachel Meneguello, have given me. As usual, however, I am the only one to blame for the deficiencies of the paper in its final form.

differing completely from the one that existed under the previous democratic regime during the 1945–64 period.

The central issue to be discussed in this re-reading will therefore be the institutionalization of an authoritarian regime in Brazil: that is, the hypotheses raised by the different authors regarding the degree of consolidation which this regime had achieved by the beginning of the 1970s; the type of opposition it would eventually face; whether or not the conditions for liberalization existed; and what factors would be likely to set that process in motion. In this return to *Authoritarian Brazil*,[1] then, it is on the cracks in the system, even during the so-called Médici period (a time of authoritarian radicalization), and on the emergence of a vigorous opposition via elections that I wish to focus.

It must be stressed that I have no intention here of presuming to conjure up a defense of hypotheses worked out from a retrospective vantage point. If it is true that some of the analyses included in *Authoritarian Brazil* underestimated the importance of the pluralist traditions of the Brazilian political system and attributed an almost monolithic consistency to the structures erected by the military regime, it is also true that the degree of closure then prevailing and the repressive climate of the time justified this perception. Nonetheless, it seems to me more productive to consider this question as a theoretical discussion which is still relevant, instead of dismissing it by this complacent allusion to the historical context or, as would be more agreeable from a literary point of view, to the "shaping fantasies" to which we are inevitably victims in our triple condition of lovers, madmen, and social scientists.

The central point has to do with the concept of *institutionalization* itself. None of the authors brought together in *Authoritarian Brazil* saw the Brazilian regime as a fully institutionalized political form. All of them agreed that the fundamental aspect of this concept is the existence (or absence) of a dominant *tendency*, wherein the political system is seen by its subjects and by the relevant protagonists as an end in itself—that is, as a form with a vocation to endure. Thus, what matters here is the tendential fit between infrastructure and superstructure: more precisely, between resources and symbols, or between a social (and organizational) matrix, on the one hand, and a viable legalization and legitimation formula, on the other. This fit will never be perfect, given that no large-scale society is highly integrated, but it is almost always possible to perceive it as a question of direction or degree.

Within this framework, then, some of the authors (Stepan, Skidmore, Cardoso, and especially Schmitter) placed more emphasis on the infrastructural context of late dependent industrialization, and on the organizational transformations of the military establishment and of the State. From this point of view, the basic element appeared to be a marked increase in the *resources* of control, from which it could be assumed that the authoritarian elements in the Brazilian political tradition would be enormously reinforced. These authors' approach contrasted sharply with that of Juan Linz, who concentrated on the symbolic or ideological level, the available legitimacy formulas, and the options for institution building with which these were associated. More strictly political and institutional, Linz's approach seems undeniably to have been more effective in highlighting the dilemmas that the Brazilian regime was to face over the short term and the cracks that might lead to a liberalization process. Considering, however, that the other authors also felt obliged or inclined to address the factors which Linz emphasized, it might be useful to review *how* they did so. In this way, I hope to identify certain areas in which their conceptual approach was inadequate when seen from the point of view adopted here, that is, the delineation of a possible scenario of liberalization based on the reactivation of the classical representative mechanisms by means of parties and elections.

The discussion will be divided into four parts, with the first two dedicated to the review of *Authoritarian Brazil* outlined above. The first deals generally with the intellectual and political context, and with the analyses in the book which placed greater emphasis on the new economic model and the technobureaucratic forms of domination introduced or reinforced after 1964. The second discusses the contribution of Juan Linz in greater detail, and argues that his theoretical orientation allows for more development with respect to the return to electoral mechanisms in a process of gradual decompression. In the third part, I have set out to document empirically my ideas on the importance and viability of the electoral road as a vehicle for liberalization. Some global data are presented on the structure of electoral competition since the 1960s, followed by a discussion of the key question, which in my opinion is the impact of the 1974 elections. This question is evidently linked to the deepening of the identification by the mass of voters with the two parties created under the military regime (ARENA and MDB). This deepening eventually confronted those in power with growing evidence of deadlock in the electoral arena, and consequently with the need for successive changes in the institutional framework. Finally, by way of conclusion, I discuss the effects of the 1982 election and the conjuncture which followed it, characterized by an unprecedented aggravation of the economic and presidential succession crises.

I. The Intellectual and Political Context of the Original Analyses

Stepan's workshop met on the Yale campus at *the* moment when Brazil was completing the seventh year of its longest experience of openly authoritarian government, characterized by the direct and corporative occupation of power by the military. The no less evident alliance of the latter with the internationalist branch of the civilian technocracy and with national and international big business suggested an unequivocal break with the patterns of political competition and State organization of the 1945–64 period. On the other hand, the embodiment of these changes in highly bureaucratized structures and procedures, often created in the image of the military institution, and the use of the corporatist labor structure erected in the 1930s and 1940s produced an irresistible temptation to present this new authoritarian period as a return to the *Estado Nôvo* of Getúlio Vargas (1937–45). A restoration or a radically new, equally consistent model: this, as Stepan observes in his Preface, is one of the debates that permeate the book. For Skidmore and Schmitter, the elements of continuity with the past seem more important. Stepan and Cardoso, on the contrary, "though not denying that authoritarian structures have roots in the past" (page viii, Preface) prefer to see them as a product of new transnational processes (particularly the pattern of development which Cardoso calls "associated-dependent" and the reorientation of the role of the military institution toward questions of internal security and national development, which Stepan calls the "new professionalism").

Having set up an ostensibly authoritarian structure such as had never before been seen in the country, and having promoted a spectacular rate of economic growth (the notoriously perverse effects of which are beyond the scope of this essay), this new model seemed immune to fundamental questions about the traditional mechanisms of representation. This is why the writings of the period often suggested that opposition, if and when it arose, would necessarily take the form of one or a few equally monolithic actors, produced by the selfsame logic of concentration set in motion by dependent development or military and technocratic organization, or the "economic miracle." It might come from a nationalist military faction; a serious split within the bourgeoisie, or the desertion of the middle-class sectors once the reduction of the growth rate had made it impossible to assimilate so many interests simultaneously; perhaps a cohesive and militant movement of the "popular sectors." Any of these hypotheses would involve an antibody generated in the entrails of the model itself, whether it were "restored" in all its purity or radically new.[2]

However one may interpret the *abertura* which began in 1973–74, as far as its depth and scope are concerned, there seems to be no doubt that it contradicted these expectations in several important ways. *First*, the *abertura* did *not* originate in a military faction which could be described as rebel, populist, nationalist or in any way hostile to the economic model in force. On the contrary, its core was made up of senior officers, intensely identified with the military hierarchy and organization. Secondly, it was the mounting pressure from the electoral system, more than the initiative or sectoral pressure of any group that produced a significant opposition to the dominant system. The pressure represented by the 1974 elections conflicted, in part, with the perhaps excessively cautious and hesitant project of the so-called Geisel group, but at the same time reinforced it against other military and civilian sectors opposed to any liberalization whatsoever. Finally, the *abertura* was designed from the start to restore the importance of the electoral arena and the role of civilian politicians. Although the size of the opposition vote in the 1974 senatorial results doubtless frightened a part of the military, the return to more competitive elections, as well as to a freer press, was an important ingredient in the decompression plan from the beginning.

In this sense and within these limits, one could say that some of the analyses included in *Authoritarian Brazil* underestimated the importance of liberal-representative traditions, of the electoral process and the party formations, of the symbols associated with them and the resistance which they still potentially represented.[3] The following four factors seem to have reinforced this underestimation:

1) the military clampdowns, with the corresponding shrinkage of the party and electoral arena, set off by the crises occurring at the end of 1965 and 1968;
2) the lack, at that moment, of any research on the difficult adjustment of civilian political forces to the two-party system imposed from above in 1965; one could even speak of the lack of research on and adequate models of the Brazilian party formations and electoral process from the previous regime on;
3) the insufficient or inadequate treatment given to the ideological background of the system, in particular the pluralist components of the Brazilian political formation;
4) certain theoretical assumptions about the relationship between State and Society which tend to go together with the neo-Marxist model of "late industrialization" and of "dependent capitalist development," in spite of the fact that these models can logically dispense with such assumptions, as we shall see below.

The first two factors mentioned can be discussed quickly and directly. Beginning with the second, which is the simplest, it seems necessary to point out that until the early 1970s, there was a lack of empirically defensible models of Brazilian political processes, chiefly with regard to the complexity and the elements of pluralism in the pre-1964 periods. Our party and electoral history was then and perhaps still is buried in stereotyped, often anecdotal descriptions, in which the practices of fraud, personalism, and clientelism are given an obviously unacceptable interpretive weight. Similarly, the interpretation of the processes leading up to the 1961–64 crisis was enveloped in a dense economistic fog, an almost deductive construction in which the crisis appeared as an immediate result of the end of the so-called easy cycle of import substitution. Without any claim to a systematic analysis, it is worth recalling that the basic mapping of the electoral patterns of the 1946–64 period developed by Gláucio Dillon Soares was only published for the first time in 1973; that a more elaborate theorization of the 1964 crisis, pointing to a breakdown of the parliamentary mechanism and the consequent paralysis of decision making, did not appear until 1974, in the doctoral thesis of Wanderley Guilherme dos Santos, still not published in book form; and that the first direct challenge to the predominant stereotypes in the analysis of parties in the pre-1964 period came from the work of Maria do Carmo Campello de Souza in 1976.[4]

More important, evidently, was the influence of the period's real processes: the various military crackdowns and the defeat of the Castelista faction—in short, the predominance of the hard line in the crucial episodes from 1965 to 1968. In these two cases there was more than abundant evidence of intolerance and of a tendency to over-reaction on the part of the armed forces, effectively suggesting that a total break with liberal-representative principles might be in the pipeline. The 1965 episode involved the reaction of the hardline military to the results of the gubernatorial elections in eleven states. The victory of traditional politicians from the former PSD, supported by a center-left coalition and linked to ex-president Juscelino Kubitschek, was seen by the hard line as a dangerous mobilization of corrupt and subversive "anti-revolutionary" forces, and drastic measures were demanded. More than a correction of specific "distortions" which the election had produced in Rio de Janeiro (then Guanabara) and in Minas Gerais, the taking of these measures actually meant a virtually total straitjacket on the moderate intentions of President Castelo Branco: General Costa e Silva was imposed as candidate in the presidential succession, and a new cycle of purges began. Elections for state governor and President of the Republic were made indirect in 1966.

The 1968 crisis, which culminated in the issuing of Institutional Act No. 5, revolved around the refusal of Congress to suspend the immunity of a deputy accused of having made a speech offensive to the honor of the armed forces (in a context also marked by student agitation, by the beginning of urban guerrilla warfare, and by powerful signs of popular opposition). From that point on, with the closing of Congress, the military character of the regime was dramatically accentuated. It was now armed with the unlimited powers of Institutional Act No. 5, and repressive action carried out against clandestine armed groups gave it a totalitarian tendency. The space for a non-militarized or properly political opposition was speedily reduced, while at the same time elections were losing significance as an arena of legitimation: in 1970 there were record levels of abstention and of blank and null votes. From then until the 1972 municipal elections, the MDB (Brazilian Democratic Movement) was almost completely crushed, even reaching the point where the idea of self-dissolution began to be discussed among members of this "tolerated opposition."

Thus the prevailing skepticism and the scant attention paid to parties and elections in the period up until 1974 are understandable. The importance of such mechanisms at that moment was in fact minimal. I do not think, however, that these immediate factors are enough to explain the underestimation of pluralism and of the liberal background to which I referred above. In Brazil there is a whole tradition pointing in this direction. Our sociological literature completely disregards these elements, or recognizes them only as imitative behavior which has long-standing roots in our colonial, slaveowning, and dependent socioeconomic formation. On the other hand, (and here the texts collected in *Authoritarian Brazil* are paradigmatic), authors who are more concerned with the history of institutions and of political ideas make frequent and sometimes emphatic allusions to this pluralist component. Juan Linz refers to an "ambivalent" legacy, both authoritarian and liberal; Skidmore mentions "the Brazilian political elite's faith in the liberal idea" (p. 45); and even Schmitter does not see any evidence that Brazilians are disposed to abandon "their *historical* quest for democracy" (p. 229; my emphasis).[5]

The problem with this point, as we will see below, is that the recognition of this ideological complexity does not always have a consistent effect on the interpretation of the political system. The liberal and pluralist elements appear at times as window dressing, or as an indication of a deeper, though individual attitude, or at others as a symbolic position made necessary by important constraints of the political system. Skidmore, for example, writes that the 1964 revolutionaries "never faltered in their *verbal* commitment" and always conserved "the *pretense* of an eventual return to liberal democracy" (pp. 9 and 45, my emphasis); but he also observes that *even the hard-line members of the military* "have avoided publicly repudiating the Brazilian political elite's faith in the liberal idea" (p. 44). The civilian revolutionaries of 1964, he adds, were historically members of the UDN, anti-populist but also anti-authoritarian, or at least anti-Vargas, and did not want to create another *Estado Nôvo.*

The treatment of these questions is also somewhat contradictory in Schmitter's analysis. The root of the inconsistencies appears to be an excessively wide gap in his description between *ideological* traditions (among which the liberal democratic component is recognized as important) and *social structure* (in his opinion incapable of producing significant resistance to authoritarian consolidation). As already mentioned, around 1971–72 Schmitter also did not believe that the Brazilian regime had been institutionalized as an enduring form of authoritarian domination. He noted major advances in this direction at the material and organizational levels—concentration of resources at the center of the system, and increasing penetration of society as a whole by the means of ideological dissemination and by the intelligence network set up by the military. But he believed, nevertheless, that the decision-making system remained "erratic and capricious," reflecting vacillation and insecurity in its policies. Barely visible obstacles seemed to have prevented "Portugalization." What obstacles were these? Incredible as it may seem, once again they were pluralist values, due to the prevalence of a reasonably liberal world view *even among the military.* It is worth quoting the relevant passage in full:

> Ironically, the military continue to insist on their *salvationist* role and their ultimate commitment to the establishment (under their tutelage) of an open competetive society . . . There is little to suggest that the *golpistas* of 1964 anticipated or desired the outcome they have produced. They seem sincerely to have believed that they were taking the first step toward demolishing, rather than

> perfecting, the Getulian *sistema.* I see no grand design in what they have done during the ensuing six years, merely a series of interrelated reactive responses to emergent crises, each of which further diminished the probability of any return to civilian democratic rule . . . They have gotten where they are institutionally "by the force of things" rather than by specific intent. [pp. 228–29]

If pluralist traditions still have a certain effectiveness even among the military and other wielders of power, they might be expected to be strong in society at large. Schmitter is emphatic in this respect:

> Perhaps the most obvious barrier to a new normalcy [i.e., authoritarian constitutionalization—B.L.] is that many people in Brazil . . . continue to believe in pluralist, competetive democracy. Overtly authoritarian rule, especially its fascist variety, no longer rests on a popular, modern mass ideology, as it did during the 1930s. A manifest and permanent shift away from the symbols of liberal democracy would probably exacerbate factional tendencies within the military, as well as resistance on the part of intellectuals and some bourgeois elements. [p. 226]

Liberal traditions, then, are of considerable weight. They represent the "most obvious" counter to tendencies seen by Schmitter as leading to permanent authoritarian institutionalization. What possibly might there be that these traditions could be strong enough to act in a *positive manner*, that is, in favor of a liberalization, however cautious and gradual? Rigorously speaking, none, to judge by Schmitter's round disqualification of them in other passages as mere wishful thinking.[6] This is something of a puzzle. How can such admittedly widespread democratic values, shared even by many of the military (so much so that they are the "most obvious" barrier to authoritarian institutionalization), be so totally ineffectual? Schmitter's analytical framework precludes even the formulation of such a question. His model, which is based on Barrington Moore, Jr., Gerschenkron, and others, is designed precisely to emphasize the impossibility that the macro-historical trajectory of European and North American liberal evolution be repeated in the Third World. On this scale of comparison, it would evidently be absurd to ignore the enormous structural differences between present-day Brazil and the "early developers" of capitalist industrialization. But this does not mean that such a model is the most appropriate for the discussion which interests us here, as to the difficulty of institutionalizing an authoritarian regime and the viability of a process of decompression based on the gradual reactivation of electoral and pluralist mechanisms.

In addition to this general difficulty, there are two more specific objections to Schmitter's application of this model to the Brazilian case. I am referring here to certain assumptions (reinforced, obviously, by the intellectual and political context) which caused him to impute an exaggerated effectiveness to the State and to conceive the political reactions of society with excessive simplicity.

In the framework referred to above, the bureaucratic consistency and effectiveness of the authoritarian political model are treated holistically, as an almost automatic derivation from the macro-structural context of late industrialization:

> (My) major premise is that there exists a distinctive "authoritarian response to modernization"—a consistent, interdependent, and relatively stable set of political structures and practices that permit existing elites to manage, guide, or manipulate the transformation of economic and social structures at minimal cost to themselves in terms of power, wealth, and status. [p. 205]

For Schmitter, this form of State organization is not a mere legacy from the *Estado Nôvo*. It is in reality an "adequate" institutional and ideological apparatus for the context of late and peripheral industrialization. This context actually tends continuously to create, or rather re-create, that form of State. The Brazilian model appeared to him not only a restoration, but even a purification of the Vargas legacy, which was being purged of its competitive and populist elements, in order to achieve a more effective adaptation to the economic and international conditions of late dependent capitalist industrialization. This purge of the pluralist aspects is what led him to think of a process of "Portugalization" (Salazarist, of course) of Brazil.

The best example of the imputation of an almost monolithic consistency to the new model is that of the very corporatist structure inherited from the *Estado Nôvo*, which regulates labor relations in Brazil. Schmitter and Skidmore are content to say that in the hands of the 1964 revolutionaries it simply returned purified to its original *telos*.[7]

Nonetheless, it seems to me that the major difficulties in Schmitter's analysis flow from the assumptions underlying the other term, i.e. society. Here I refer to his conception that a social structure like Brazil's is incapable of giving rise to resistance to the authoritarian concentration of power. This judgment appears persuasive precisely because his model has previously excluded any concern with the institutional mediations of the political system (especially elections and parties). The effectiveness of pluralist values and of a possible alternative leading towards democracy is thus made to depend strictly on the eventual constitution of collective actors similar to social classes in the Marxist vision: autonomous powers (*puissances*), well defined and clearly counterposed to other powers as suggested in Althusser's analysis of the social basis for power sharing in Montesquieu.[8] Now, the key point of the "late industrialization" model as a contribution to political sociology consists of describing a social matrix which produces exactly the opposite characteristics. Only these autonomous powers, organized and self-conscious, could be the basis for a competitive democracy; late industrialization prevents them from flourishing; *ergo* . . .[9]

II. Additional Notes on the Mechanics of Decompression

At the outset I stated that the objective of this reassessment of *Authoritarian Brazil* is to select elements of contrast for a better understanding of the

Brazilian *abertura*, in its specificity and with its limitations. I am particularly interested in the formulation of a model which can better grasp the possible effectiveness of electoral and representative mechanisms as a vehicle for a process of decompression and eventual democratization. To investigate this possibility, two other sets of more strictly political and institutional hypotheses are necessary.

The first, as we will see below, contains a certain paradox. While on the one hand it implies taking *more seriously* the ideological traditions of the different actors, among which are important liberal or pluralist components, on the other it implies postulating a process which turns *less* on the subjectivity of any actor in particular. In fact, it involves a *calculus* of decompression, that is, an interactive model in which the various actors, whatever their ideologies, calculate the costs of the status quo and of alternative solutions. From this point of view, electoral and competitive mechanisms *may* seem even to frankly illiberal political actors, and even to hardline military officers, to be a rational form of accommodation to highly uncertain situations, particularly, when confronted with the perception that other alternatives (such as Fascism) could present even higher risks.[10]

A second general hypothesis is that the *process* of decompression produces its own effects. Competitive elections can have liberalizing effects even within non-competitive political systems.[11] The existence of an electoral calendar with a minimum of credibility is in itself a source of pressure in this direction. One can also speculate that these liberalizing pressures tend to occur with greater probability not when they are the direct expression of well-defined actors, whose social and numerical force is previously known, but precisely the contrary, when they function as a purely formal, very general game and are thus capable of producing the *quantum* of uncertainty necessary to seduce the different actors to stay at the table.

Linz's contribution to *Authoritarian Brazil* is essential to an understanding of the first hypothesis above, the *calculus* of decompression. The ideological and institutional background, the practices and positions the different actors previously assumed, are incorporated into the core of the interpretation, thereby becoming constraints which then become channels for new processes and interactions. From this standpoint, the Brazilian authoritarian model is even more radically questioned. It is regarded more as an authoritarian *situation* than as a fully formed, or soon-to-be consolidated, authoritarian *regime*.

This theoretical orientation can be mapped, evidently at the cost of a certain impoverishment, by means of four propositions:

1. all processes of political institutionalization presuppose a *legitimacy formula* and have, as a consequence, a strong symbolic component, which is not reducible to the resources, specific policies, or economic performance of the regime in question;
2. these symbols are important insofar as they maintain the allegiance or acquiescence of the population and of the most influential social

groups, but they must also ensure a minimum of unity in the *State itself*, the organizations which constitute it, and, in the Brazilian case, particularly the military;

3. although the military have a great deal of latitude in the use of their coercive powers and in the adoption of specific policies, each important decision creates its own divisions and, as it were, its own jurisprudence, hence acting later on as a factor of limitation; at the symbolic level, above all, the military cannot escape from a certain continuity with the political past of the country, and with the past of their own decisions and their own discourse;
4. finally, the effort of institution-building and the search for a legitimacy formula are not processes confined to national frontiers. They are strongly affected by the great international models (for example, by the prestige of liberal democracy, or of socialism, in contrast to the discrediting of fascism and everything associated with it, such as corporativism, in the minds of the relevant actors); by international opinion, in particular by the large transnational institutions, such as the Catholic Church, the press, the academic community, etc.; and, evidently, by the nature of the international political and economic relations which the Brazilian government is obliged to maintain and cultivate.

The second of these propositions is essential to avoid a poor understanding of the question of *legitimacy*, particularly the tendency to conceive of it as a generic acquiescence, irrelevant to State unity.[12] The great merit of Linz's analysis in this respect was to avoid contrasting State and society in a reified way, and to insist that the choice between different legitimacy and institution-building formulas has immediate and deep repercussions on the unity of the State itself. Once again, it is worth quoting the crucial passage:

> The success of . . . policies based on repression and development can assure some stability in periods of prosperity, but it can never satisfy those who ask questions about legitimacy except perhaps in purely subject political cultures, with traditional rulers in the most narrow Weberian sense of the term. In any society that has developed beyond this stage, as Brazil clearly has, questions about legitimacy will inevitably be asked. They will be asked by intellectuals and those under their influence, by those concerned with religious values, and ultimately by some of those who have to use coercion, like judges and army officers confronted with subversion or public disorder. Only praetorian guards or the lowest ranks of a police force do not ask such questions. Anyone in a position of responsibility, one who must kill or die to defend a regime, must ultimately ask questions about why he should do so and whether he should obey in a crisis situation. [p. 240]

On the basis of these assumptions and of recent Brazilian institutional history, Linz rejected as improbable solutions: legitimation via personalization of power in a charismatic leader; the adoption of a corporatist structure without parties; and the construction of a single party—were it of a fascist-mobilizing or a left populist character. It seemed to him less difficult, but still problematic, to imagine a *Mexicanizing* solution through a hegemonic party. This solution would have as its *sine qua non* condition the transformation of ARENA into a Brazilian PRI, something very difficult given "the tradition favorable to a competitive party system"; the military's reluctance to grant autonomy and transfer resources to a party, even if official and controlled; earlier intimidations and interventions, which had already alienated civilian politicians who would have been potential recruits for a project of this nature; and last but not least, the impossibility of creating for the Brazilian military government, with its technobureaucratic and internationalist economic policies, a populist-nationalist aura with great historical resonance as the Mexican regime has always had.

These considerations throw into relief the dilemmas confronted by the armed forces as they considered the party question and the process of institution building. Another point accentuated by Linz is that the lack of a satisfactory typology of authoritarian regimes made it difficult, in the far away days of 1971–72, to formulate more specific hypotheses on the processes of institutionalization (or of decompression and transformation) of these regimes. Thus, the analysis could not avoid making broad contrasts with the other two main ideal types (liberal democracy and totalitarianism) and with the classical ideologies, which constitute the kernel of the different legitimacy formulas. This, in my opinion, is where Linz's approach permits further development, in terms of an analysis of decompression from an electoral point of view. The question I am raising here is whether this emphasis on the ideological content of the different formulas is not too *substantive*, and therefore inadequate to capture the experimental and incremental aspect often present in this process.

The Brazilian case *circa* 1973–74 suggests indeed that the constraints were enormous, both in the hypothesis of an advance in an authoritarian direction *and* in that of a retreat towards liberal-democratic traditions. This ambiguity can be detected in the *salvationist* emphasis of the military discourse, and in the representation of the armed forces' own role as a temporary intervention to restore conditions for democracy. Just as there were liberal traditions which created obstacles to the advance of authoritarianism, there were also authoritarian traditions (anti-individualistic, favoring a "social" rather than a "political" democracy, and so on) which would hinder a pure and simple return to the pre-1964 situation, or even to the 1967 Castelista Constitution, if any faction of the ruling elite had in fact opted for such a line at the time. The immediate invocation of a *substantive* legitimacy formula, i.e. of a more articulated political and economic doctrine, would certainly tend to multiply the reactions of one side or the other, whereas simply setting the electoral mechanism in motion, under tutelage and within reasonably known limits, would make the strategy incremental, so that little by little the degrees of freedom lost in the immobilism of the Médici era could be recovered.[13]

The basic point of my argument is, therefore, that the importance of electoral and representative mechanisms cannot be adequately assessed without considering more systematically what might be called the *calculus* of decompression. This involves highlighting at least three points of great importance in the Brazilian case. First, as already stated, in the eyes of the ruling group the reactivation or enhancement of the electoral process was fully compatible with the extreme gradualism which it wished to imprint on the liberalization process. Precisely because the electoral mechanism is highly formal, abstract, and in this sense, *uncertain*, it would permit realignment to be cautiously begun without precipitating broader substantive definitions of the government's future intentions. Second, and for the same reasons, the elections would function as a procedural legitimation, that is, as a revitalization of the notion of the *legality* of governmental action, which was in turn indispensable to military cohesion and to what Theda Skocpol calls "support among the regime's own cadres."[14]

Finally, and here we have an important transition from formal or procedural to substantive legitimation, the recovery of the credibility of the electoral calendar could be defined, as it in fact was, as one part of a global strategy of institutional *normalization*. The dysfunctions and the exhaustion of the previous *substantive* appeals created a dangerous symbolic vacuum, which flagrantly revealed the government's growing isolation from society and pushed it towards making new definitions.[15] The somewhat more liberal inclinations of the so-called Geisel group, its link with the former *Castelismo*, and the intention to begin a gradual decompression led it, according to this hypothesis, to fill the vacuum with an idea of normalization located halfway between purely procedural legitimation (which was problematic in any case, given the prevalence of "revolutionary" legality) and a more explicit democratizing appeal, following classical substantive formulas.

It might of course be objected that this abstract elaboration of the benefits of the electoral method is somewhat apologetic, attributing to the strategists of the détente altruistic motives and a level of rationality which is frankly fantastic. This question leads us to examine the return to electoral competition not only *ex parte principis*, but also *ex parte populi*. More precisely, it requires a look at the global parameters of the *structure* of electoral competition at the beginning of the decade, from the standpoint of the changes which occurred between 1974 and 1982.

III. The Structure of Electoral Competition

The arguments outlined in the preceding section suggest that decompression through elections remained viable in Brazil because of the pluralist traditions of the political system and the considerable degree to which the government was able to control the political agenda. I now wish to suggest that this viability was also enhanced by the behavior of the opposition, and that the latter was willing to play this game mainly because of the structure of competition, that is, the distribution of electoral chances, based on the socio-geographic composition of the electorate and the distribution of party preferences. These structural factors are of great importance for understanding how it was possible to continue to take the game reasonably seriously *in spite of the rule of dictatorial legislation* until the end of 1978, and the frequent manipulations based on the government's majority until 1982.

My general hypothesis, then, is that an acceptable level of commitment by the contenders to the institutional rules was created because any changes were then seen as part of a mobile horizon, initially christened détente (*distensão*) and later "opening" (*abertura*) or redemocratization. This is meant to be taken not as a normative proposition, but as a hypothesis about the actual behavior of the actors, particularly those of the opposition, during the period from the elections of 1974 to those of 1982. The analysis of the events after 1982 requires a more complex frame of reference, given the aggravation of the economic crisis, the social tensions, and the succession crisis, all of which produced significant fragmentation in both government and opposition.

On the government side, in the early seventies, as will be seen below, not everyone harbored illusions as to ARENA's electoral potential, in spite of its momentary hegemony. The 1974 elections therefore represent a clear turning point in the projections for the future. By dramatizing the electoral fragility of the party in the big cities, this election certainly reinforced the disposition of some government sectors to slow the pace of change; and to achieve this they capitalized on the fear of a more significant defeat at the hands of the plebiscitary urban vote. It must be pointed out, however, that even this diagnosis does not refer only to the structural bases of the vote. In fact, this perspective of a forthcoming *electoral impasse* was also a perverse product of the rules which the military regime had imposed on the opposition, beginning with the two-party structure itself and with the law on compulsory party loyalty.[16]

Returning to what are referred to above as structural factors, the viability of the electoral alternative in Brazil can be understood, first of all, simply by the size of the electorate and by its increasingly urban distribution (see tables I and II). A more precise indication can be obtained by an inspection of the global election results from the 1966–82 period (see table III). These data allow us to carry out an after-the-fact, albeit persuasive, exercise, which is to verify the electoral chances of a conventional party opposition (as the MDB was and as the PMDB continued to be) even at the most difficult moments.

The term "conventional contender" is here understood as a party which adopts a reasonably broad ideological definition, and which is run fundamentally by professional politicians or by individuals who aspire to that role, thus conforming to normal methods of performing a parliamentary or local governmental role. This definition is intended to emphasize the fact that the sum of votes obtained by this party in each successive election

Table I. Brazilian National Electorate, 1955-1982

	1955	*1960*	*1962*	*1966*	*1970*	*1974*	*1978*	*1982*
A. Number Voting (in thousands)	9,097	12,586	14,747	17,286	22,436	28,982	37,629	48,481
B. Registered Voters (in thousands)	15,243	15,543	18,529	22,387	28,966	35,811	46,030	58,617
C. Total Population (in thousands)	60,202	69,730	73,951	83,175	93,139	102,807	113,481	125,263*
Participation (% A/B)	59.7	81.0	79.6	77.2	77.5	80.9	81.7	82.7
Electorate (% B/C)	25.3	22.3	25.0	26.9	31.1	34.8	40.6	46.8

*Estimate.

Sources: Fundação IBGE, Rio de Janeiro; Tribunal Superior Eleitoral, Brasilia.

(facing a government with a military base which had absolute control over economic policy and considerable means of clientelistic recruitment in the various states) has a direct significance as an indicator of the minimum level of its organizational resources and its legitimacy. It is not influenced by exaggerated gains or losses because of *its own* strategic actions, or because of positions it may have assumed which are too deviant, heroic, or suicidal. Its part of the total vote *is* doubtless affected by some of the restrictions and manipulations of the rules of the game—the *casuísmos*, in contemporary Brazilian parlance—but, it is possible that these have much more effect on the *conversion of votes into seats* than on the distribution of votes itself, on a national scale. Besides, the point at issue here is not the significance of this distribution as a measure of effective power. I am only concerned at the moment with suggesting *the order of magnitude* of the opposition's electoral chances in the different periods of the post-1964 regime. In effect, it can be seen that the ARENA (PDS) vote showed a declining tendency from the

Table II. Total Population and Urban Population, 1950-1980

Year	*Total Population (a)**	*Cities with 20,000 or More Inhabitants: Population (b)**	*(N)*	*% Urban Population (b/a)*	*% Growth of Population of Cities with over 20,000 Inhabitants*
1950	51,940	10,335	96	19.9	80
1960	70,920	21,073	172	29.7	104
1970	92,340	34,207	300	37.0	62
1980	121,151	61,254	482	50.6	79

*Figures expressed in thousands.

Source: Table adapted from Vilmar E. Faria, "Desenvolvimento, Urbanização e Mudanças na Estrutura do Emprego," in *Sociedade e Política no Brasil Pós-64*, ed. Bernardo Sorj and Maria Hermínia Tavares de Almeida (São Paulo: Editora Brasiliense, 1983), 118-163.

Table III. Official Results of the Legislative Elections, 1966–1978, for the Entire Country (in percentages)

	*Senate**				*Chamber of Deputies*				*State Legislatures*			
Year	*Arena*	*MDB*	*Null & Blank Votes*	*Total*	*Arena*	*MDB*	*Null & Blank Votes*	*Total*	*Arena*	*MDB*	*Null & Blank Votes*	*Total*
1966	44.7	34.2	21.2	17,259,598	50.5	28.4	21.0	17,285,556	52.2	29.2	18.6	17,260,382
1970	43.7	28.6	27.7	46,986,492	48.4	21.3	30.3	22,435,521	51.0	22.0	26.8	22,435,521
1974	34.7	50.0	15.1	28,981,110	40.9	37.8	21.3	28,981,015	42.1	38.8	18.9	28,922,618
1978	35.0	46.4	18.6	37,775,212	40.0	39.3	20.7	37,629,180	41.1	39.6	19.3	37,449,488
1982**	36.5	50.0	13.5	48,746,803	36.7	48.2	15.1	48,455,879	36.0	47.2	16.8	48,374,905

*The total votes in 1970 is counted twice, because two-thirds of the seats were up for election. In 1978 the same thing should have happened, but one-third began to be elected indirectly—the so-called "bionics"—due to modifications introduced by the April Package in 1977.

**The PDS votes were included in the column referring to ARENA and those of the opposition (PMDB, PT, PTB and PDT) in the MDB column.

Source: Tribunal Superior Eleitoral.

formation of the two parties on, falling from around 50 percent in 1966 to 36 percent in 1982. With the exception of 1970, when it was drastically affected by the blank and null votes, the MDB's share of the vote (for the opposition in 1982) moved in the opposite direction, from a minimum of 28 percent (for the Chamber of Deputies) to a maximum of 50 percent in 1982.[17]

These data show, therefore, that the distortions and singularities which can be noted in the social context of "late industrialization" do not prevent it from producing a competitive distribution *of votes*, even under extremely adverse conditions for the opposition. The restrictions do not come from the social structure of this electorate, but rather from institutional mediations imposed from the top down, at the moment of conversion of these votes into seats—to say nothing, clearly, of the restrictions on the power of Parliament, the parties, and organized civil society. This point is obvious but needs to be highlighted in numerical terms. A comparison of the data of table III with those of table IV, shows, in fact, that the disproportion between votes and seats is systematically favorable to the government party, reaching the extraordinary disparity of 45.7 percent in the Senate and 23.5 percent in the Chamber of Deputies in 1970.

It must, however, be stressed that three distinct things are under discussion here: *First*, the long-established institutional mediations which are not very controversial in spite of the fact that they systematically favor government parties. This is the case with the proportionally greater weight given to smaller states in the Chamber of Deputies and the equal representation of different states in the Senate.[18] *Second*, direct intervention in the legislative branch, which obviously hit the opposition congressmen hardest: the revoking of mandates and suspensions of political rights based on the *institutional acts*.[19] Finally, the so-called *casuísmos*, or legal and institutional changes which directly or indirectly affected the electoral chances of the opposition. Some of these modifications were also arbitrary acts (i.e. based directly on the institutional acts), but others were approved by the

Table IV. Percentage of the Votes Obtained in the Elections and of Seats Controlled by ARENA (PDS) at the Beginning of Each Legislative Term, in the Senate and in the Chamber of Deputies, 1966-1982

	Senate			*Chamber of Deputies*		
Year	% *Votes*	% *Seats*	% *Difference*	% *Votes*	% *Seats*	% *Difference*
1966	44.7	71.2	(+) 26.5	50.5	67.7	(+) 17.2
1970	43.7	89.4	(+) 45.7	48.4	71.9	(+) 23.5
1974	34.7	69.7	(+) 35.0	40.9	56.0	(+) 15.1
1978	35.0	62.7	(+) 27.7	40.0	55.0	(+) 15.0
1982	36.5	66.7	(+) 30.2	36.7	49.0	(+) 12.3

Original Data Source: Tribunal Superior Eleitoral.

legislature, obviously via the government majority, itself guaranteed by arbitrary interventions effected previously. Examples of this kind of legislation are the figure of the so-called "bionic" senators (elected indirectly), imposed by the April Package of 1977, and the deliberately planned subdivision and creation of new states (Mato Grosso do Sul in 1977, Rondônia in 1981), with the consequent increase in the number of Senators and Congressmen who were submissive to the central government.

The degree to which Institutional Acts Nos. 2–5 diminished the electoral and party spheres is demonstrated in our tables III and IV. The moment for *Mexicanization*, if there had been one, would have been in the period from 1970 to 1972, when theoretically the government could have capitalized on its economic successes, the ARENA victory, and the weakness of the MDB as an opposition party, defeated even by blank and null votes. I say theoretically, and I am not only referring to the already mentioned fact that it would have been unlikely in those circumstances for the Brazilian military to agree to transfer real power to any party. I am referring rather to certain basic data on the chances and the action of the two parties in the period of the "economic miracle" and the crackdown by the Médici government—data of which the government leaders must necessarily have been aware, and which did not leave space for many illusions about a Mexicanization project. Even in the 1966 elections, in the first confrontation of the recently created and ultra-intimidated ARENA and MDB, it was evident how weak the government party was in the large urban centers. In the municipality of Rio de Janeiro, then the State of Guanabara, for example, 15 MDB Congressmen were elected as against 6 from ARENA; in the State of Rio de Janeiro it was 11 against 10; in the State of São Paulo it was 27 against 32.

Referring to the *behavior* of these congressmen, a researcher observes that "by late 1968 the hardliners of the revolutionary circle had apparently decided that they had won a hollow victory in 1966. Congress and some of the state assemblies were being used as forums for the expression of criticism, which many of the sensitive of the military hard line were loath to endure for long."[20] This "expression of criticism" culminated, as is well known, in the crisis of the second half of 1968 and Institutional Act No. 5, which led to the revocation of mandates, including those of some ARENA congressmen. In the case of the MDB, the measures were evidently much more drastic. The same researcher cited above uses the case of São Paulo as an example, where "the 1966 MDB gains were erased. Prior to AI-5, the São Paulo branch of the MDB included 27 federal deputies, 53 state deputies, 71 mayors, 10 councilmen in the capital, and 1,185 councilmen in the interior. By the end of June 1969 the *cassações* under AI-5 and some shifts to ARENA had reduced the party's strength to 12 federal deputies, 20 state deputies, 38 mayors, 7 councilmen in the capital, and about 800 councilmen in the interior." After President Costa e Silva's stroke and subsequent removal, this continued to be the method for taming the parliamentary opposition, as a result of which the 1970 legislative election and the 1972 municipal election became absolute non-events. Nonetheless, it is necessary to note that, "ARENA showed a great deal of weakness in the more modern urban areas," and to insist once again on the high indices of blank or null votes.

It is impossible to understand the presidential succession at the end of 1973, the beginning of liberalization with the new President, General Ernesto Geisel, and particularly the impact of the 1974 elections, without the perspective which comes from these events in 1966 to 1970. The superisolation and political immobilism of the Médici government; the excessively harsh repression of the armed struggle, with the resulting autonomization of the security forces, which in turn led to growing resistance from human rights organizations and severe treatment by the international press; the increasingly sharp criticism of income inequality, which persisted or was even accentuated, in spite of high growth rates—all this was merged in the new government's evaluations. It was facing the alternative of accepting a political "opening", no matter how cautious, or being obliged sooner or later to promote a Fascist-type radicalization. In the first hypothesis, the credibility of the representative mechanisms had to be recovered so that they became one of the most important parameters of the government's global strategy. At the least, it would be necessary to underline the decision to minimize the number of directly coercive interventions, such as revocation of mandates, even if the alternative was to manipulate the rules of the game through *casuísmos*.

This, probably, is one of the meanings of the appeal to the "creative imagination" of the politicians which President Geisel made in his inauguration speech, when he declared the hope that solutions be found which would gradually make "revolutionary" interventions, based on institutional acts, *unnecessary*.[21]

The 1974 election was a good illustration of the hypothesis according to which the *process* of decompression produced its own effects. It is precisely this hypothesis that leads me to disagree with the otherwise acute analysis of Wanderley Guilherme dos Santos, when he states that

> . . . the election results (of 1974) were not a *new and important* factor which compelled the government to reformulate its *political* strategy, though it certainly induced it to modify its expectations, and thus its electoral strategy.[22]

Wanderley Guilherme is right, of course, when he states that *novelty* cannot be established in comparison with the elections of 1966 and 1970, since these were the atypical ones in a historical series going back to the pre-1964 period. He would also have been right in observing that the opposition victory in 1974 was only in the senatorial elections, and only a third of the seats were up for election that year. In other words, the election did not affect the government monopoly of initiative in changes via the legislature and still less the exceptional powers which Institutional Act No. 5, still in full force, concentrated in the hands of the President of the Republic.

It seems to me however that the opposition victory had a substantial effect in at least three ways. *First*, this victory coincided exactly with the rise of a new factor in the context of the two-party experience which began in 1965: a clearly *plebiscitary* meaning to the urban vote in the principal states.[23] *Second* the revival of popular interest in elections and the size of the opposition vote led to an extraordinary deepening of new party identifications, practically eliminating, at the level of the mass of voters, the vestiges of the pre-1964 multi-party system. The change in expectations to which W. G. Santos refers, i.e. the prospect of an impasse in the subsequent elections, is incomprehensible without a reference to this "adoption" of the new two-party structure by public opinion, beginning with the 1974 elections. Thus, the least that could be said of this election is that it established the autonomy of *abertura* as a political process, transforming it into something much less reversible than initially forseen in the governmental strategy. Let us explore these points in more detail.[24]

The immediate intuitive significance of the term "plebiscitary" vote is that of a polarized vote in terms of the simple approval or rejection of a proposition, or of the confidence or lack of it which those in power inspire in the people. The striking clarity of the results, which reflect enormous majorities for or against, and the simplicity of their subjective correlates, are therefore important aspects of the concept, as we will see below. First, however, let us consider two more specific empirical dimensions: a) the generic and prospective nature of the vote, as opposed to contests in which the most important thing is an evaluation of the past record of an incumbent; b) a disproportion between the results and any loyalties or *organized* campaign efforts, not permitting the former to be explained by the latter.

These indications are intended to emphasize the surprising character of the 1974 results. The plebiscitary element derives, in my opinion, from a vast complex of factors, beginning with the imposition of the two-party system (in 1965) *simultaneously with the suppression of direct elections* (beginning in 1966) for mayors of state capitals and for state governors. It is well known that a substantial part of the electorate votes on the basis of the past record of an administration, when it is not out of mere clientelism in the case of local governments, and even in terms of identification with the personality of the candidate, in all elections for the executive. Thus it is evident that if the opposition party controlled the mayorships of the largest capitals, which are strong poles for political diffusion over the whole country, it would not be immune to the erosion of its image and serious risks in direct elections. Protected against this kind of erosion, the MDB grew from the top down, beginning with its big vote in 1974 and moving towards a progressive implantation at the local level, as we will see below. The imposition of the two-party system, the weakness of the opposition party, and the suppression of direct elections combined to make the act of voting plebiscitary, giving the senatorial election a symbolic connotation of a pronouncement about the regime and the situation of the country as a whole.

These hypotheses, however, explain only part of the phenomenon. After all, the 1966 and 1970 elections, and the municipal elections of 1972, had not had this characteristic. The 1970 elections in São Paulo are an interesting case. The MDB managed to have one of its senatorial candidates elected—Franco Montoro (whose alternate was Tito Costa, a candidate linked to the industrial unions of the ABC region)—but it was victory by a very narrow margin. In the voting for the Federal Chamber of Deputies and for the State Assembly, in the municipality of São Paulo, this same electorate which was to give massive victories to the MDB in 1974 and in 1978 gave it, in 1970, *less than half the number of votes obtained by ARENA*. It is thus evident that this change in the function of the electoral process—its transformation into a plebiscite—occurred gradually. Only after some time did the majority of the electorate realize that it no longer voted for President, Governor, and Mayor in the state capitals, precisely the elections which it considered most important. It would therefore appear plausible to suggest that there was a period of "learning" before the vote began to be used in a plebiscitary way.[25]

A massive realignment such as occurred in São Paulo and in various other states between 1970 and 1974 is the result of (and in turn reinforces) strong homogenizing tendencies in the minds of the electors. Thus we can dismiss from the outset the hypothesis that the *organized* mobilization of social sectors produced such a result. In the trade union sector, there was no movement in 1974 capable of having an effect on public opinion, demonstrated by the fact that there were no strikes worthy of note between 1968 and 1978. Even so-called civil society, a term which has begun to be used to describe political activity by business and especially religious and professional associations, did not participate in any important way in the opposition's electoral campaign that year. It is also noteworthy that the propositions of such groups with regard to the politico-institutional process were at that time often timid or insipid.[26]

Another hypothesis frequently raised in this initial period consisted of establishing continuity with the parties in existence before 1964, and seeing the MDB's growth as due to a sort of "rebaptism" of party identifications from that period. This supposition is not supported by the surveys I carried out, nor is it likely, considering the growth of the electorate, and hence the magnitude of the demographic transformations since 1965 and during the authoritarian petrification of the Médici era.[27]

One question which merits closer study is that of party organization. As pointed out above, when we consider the whole country, the only viable generalization is that growth took place from the top down, in other words, that the party machine managed to expand because of the spaces and the stimulus created by its electoral victory. While this description carries regional variations, the case of São Paulo is again illustrative. One of the reasons for the low level of mobilization, not to say apathy, among intellectual militants in 1974 was the choice by the MDB Convention of Orestes Quércia as candidate for the Senate. It was taken for granted that the ARENA candidate, Carvalho Pinto, was unbeatable. Moreover, Quércia's victory in the Convention, defeating the "authentic" candidate Freitas Nobre, placed student and other politicized sectors before an illustrious unknown, a totally colorless candidate, who was unlikely to rid the party of its negative image of "loyal opposition." All that was known of him was that he had been mayor of Campinas and that he was a typical machine politician, responsible for part of the organization of the MDB in the interior of the state. It is impossible to say whether Freitas Nobre would have had more or less success than Quércia against Carvalho Pinto. Considering, however, that the MDB held power in only a small number of local governments and that Quércia received 4.6 million votes against the 1.7 million given to Carvalho Pinto, it can definitely be stated that the decisive factor was the plebiscitary landslide, and not the organization previously set up by the party.[28]

Despite the reservations set out below, the big-city electorate voted, in fact, like a mass in a plebiscite. At least three other factors joined in to produce this result. First, the economic expectations of the population, which had been overly stimulated by the publicity apparatus of the "Brazilian miracle," were beginning to be dashed.[29] Second, the monolithic and invulnerable appearance of the system was breaking down as a result of official promotion of the political *abertura*.[30] The third factor was the important effect of television, which broadcast pre-recorded propaganda, rather than mere photographs, as was to be the case in 1976 and afterwards, thanks to the so-called Falcão Law.

The data presented clearly indicate that the regime's intention *ex parte principis* of restoring the legitimacy of the system via the electoral process collided head on in 1974 with a powerful demonstration *ex parte populi* in the opposite direction. Given the presence of the homogenizing influences discussed above which accelerated this demonstration, and in the absence of strong sectoral pressures to explain it, would it be conceivable in 1974 to forsee the continuation and deepening of such tendencies? If it is true that the regime's dominant group forecast a coming impasse in the electoral arena, what would be the social bases of this impasse, and *what factors would make it improbable that the 1974 electoral patterns would soon be diluted or reversed*? This is an unavoidable question, given—a) the extremely low levels of information and politicization, and the virtual absence of ideological structuration at the level of the masses; b) the absence in Brazil of social cleavages which have historically taken root so far as to be self-conscious and easily identifiable. As is well known, there are no politically and demographically significant ethnic, religious, or regional cleavages.

The answer to these questions can be found, in my opinion, in the deepening identification with the MDB in the large urban centers and in the strong influence which these had on medium-sized towns in subsequent years. The rise on a national scale of contrasting images of the two parties is virtually a byproduct of the 1974 election.[31] The studies carried out since then invariably show that ARENA began to be seen as the party "of the government," "of the rich," or "of the elite," and the MDB as the party of underdogs, "of the poor," "of the opposition." This is perhaps merely another way of mentioning the plebiscitary meaning of this election, but a number of aspects are worth stressing with a view to clarifying the expectations of the political actors with regard to future elections. In the large metropolitan areas, the deepening of party identifications made irreversible the tendency favorable to the opposition, which had been perceptible even at times of maximum repression and government euphoria. Monthly data from the Gallup Institute show that the MDB maintained a significant advantage over ARENA in the distribution of preferences in São Paulo and in Rio de Janeiro during the following years. (See table V.) But this picture becomes clearest in examining the voting percentages for the Senate in the municipality of São Paulo from 1966 to 1978: from a maximum of 59 percent of the votes for the Senate in 1966 (excluding the blank and null votes), the official party fell to 48 percent in 1970, 21 percent in 1974, and 12 percent in 1978. Here we have a vivid picture of the specter of plebiscitary voting. It becomes even more vivid when we break these data down into the eight homogeneous socio-economic areas into which the municipality of São Paulo is divided. (See table VI.) Considered in numerical order from the first to the eighth, these areas are equivalent to a hierarchy of living conditions, from the richest to the poorest. Although there is a certain internal heterogeneity at the level of the individuals or families who live in these areas, the hierarchy is indisputable with regard to the overall infrastructure, available services, quality of housing, and so on. This subdivision therefore represents a stratification of social contexts.[32]

Now, if percentages of ARENA votes in each of these areas are calculated, we can see that they decrease systematically, not only over time, as shown above, but also in terms of this social space. Area by area, this decline can be seen without exception in the four senatorial elections held from 1966 to 1978. Year by year, i.e. in each of the four elections, they decrease from area I to area II, and from this to area III, and thus successively to area VIII, without deviations or inversions. The exact political significance of these tendencies can be appreciated by observing what happened in area VIII, the poorest, in the last two-party election: the ARENA vote for the Senate was reduced to 7.5 percent, as against 92.5 percent for the MDB.[33]

What about the medium-sized and smaller municipalities, which also contain a considerable part of the national electorate? These were almost entirely under the control of ARENA in 1974. The organizational weakness of the MDB often made it easy for the government party, even in the state of São Paulo, to win huge victories in municipal races. Moreover, unlike the state capitals, where there are no elections for mayors, the opposition is not immune to the erosion of support due to holding local power in municipalities of this size. The 1976 municipal elections furnished several examples of this type; suffice it to say that the MDB did not succeed in holding onto city hall in industrial municipalities which were seen as its strongholds, like Juiz de Fora (in Minas Gerais) and Taubaté (in São Paulo).

A study conducted in 1976 in Presidente Prudente, a municipality some 500 kilometers from the city of São Paulo, with 93,000 inhabitants in the

Table V. Political Party Preferences of Voters in Greater São Paulo and Greater Rio de Janeiro, 1975-1978*

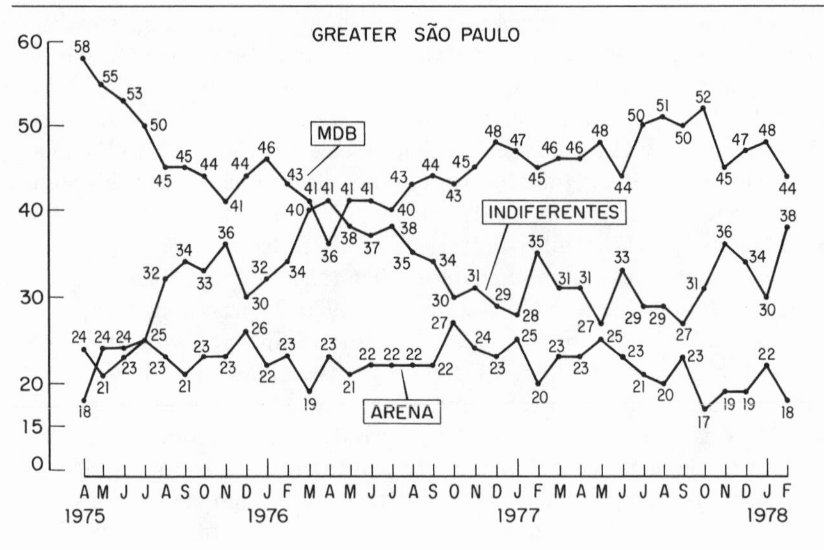

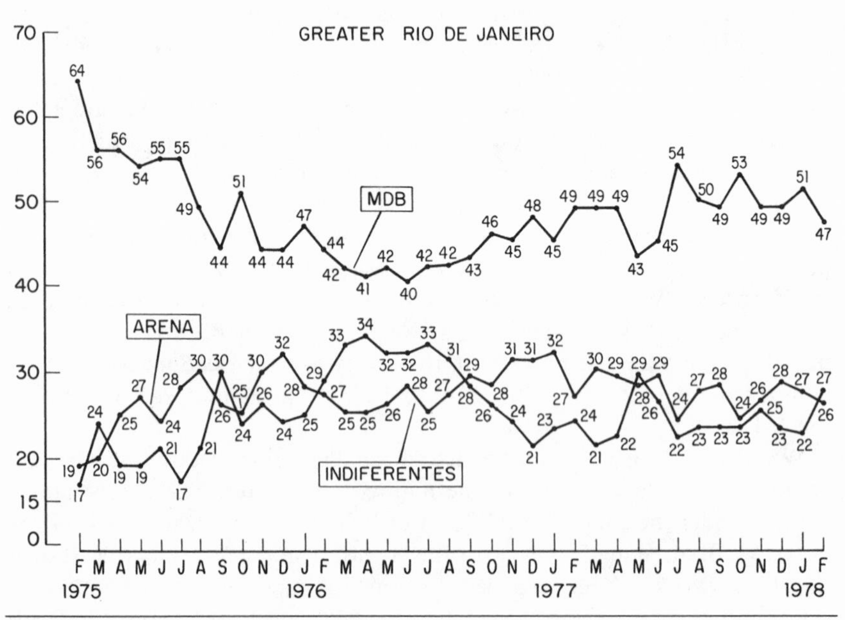

*Source: Gallup Poll, III, no. 26 (February 1978).

Table VI. ARENA Votes in Senatorial Elections, 1966–1978, by Homogeneous Socio-Economic Area of the Municipality of São Paulo (as percentage of total votes cast*)

	*Area***	*1966*	*1970*	*1974*	*1978*
↑ RICHER	I	66.2	57.3	34.2	20.0
	II	63.8	53.7	30.1	17.4
	III	62.0	52.4	27.0	16.0
	IV	60.0	52.2	24.8	15.3
POORER	V	59.8	49.0	22.5	13.0
	VI	52.8	41.5	14.5	9.9
	VII	52.2	39.0	14.5	9.1
	VIII	48.7	35.1	12.3	7.4
↓	Total For Municipality	58.9	47.5	21.2	12.3

*Null and blank votes are excluded. The complement of the percentages given therefore corresponds to the MDB vote. Source: TRE-SP.

**On the division of the municipality into homogeneous areas, see the Secretaria de Economia e Planejamento de São Paulo, *Subdivisão do Municipio de São Paulo em Areas Homogêneas*, Série *Estudos e Pesquisas*, no. 13 (1977).

urban area and with no industrial base, offers some answers. Presidente Prudente in the 1970s may be said to have been an ARENA stronghold. In the 1972 municipal elections, ARENA's advantage over the MDB was 50 to 1 in the race for mayor, one of the largest victories in the state. On the eve of the 1976 elections, in which the MDB was again badly defeated by ARENA, 65 percent of the voters declared themselves Arenistas, as against 15 percent for the MDB and 20 percent with no preference. This identification was a very strong predictor of the votes actually cast, in spite of the habitual supposition that in this kind of locality personalism and clientelism are superimposed over any kind of genuinely party orientation.[34] These indicators provide a background to table VII, which recapitulates the voting for the Senate and for mayor in the município since 1966. The tendency of the opposition to grow, *even at the local level,* is transparent in this historical series. It can be seen that the MDB had already won the senatorial election of 1970 (the candidate was Franco Montoro) although by a slim margin. In 1974, in spite of the absolute control of ARENA over local politics, and also in spite of an ARENA victory in the voting for the Federal Chamber and for the State Assembly, the MDB candidate Orestes Quércia won almost the same advantage over Carvalho Pinto in the municipality as in the rest of the state. This, without doubt, is the effective beginning of oppositionist activity in the municipality, in spite of the fact that the MDB did not win more than 17 percent of the votes for mayor in 1976. In 1978, the sum of votes for the two concomitant senatorial candidates (Franco Montoro and Fernando Henrique Cardoso) gave the MDB nothing less than 78 percent of the votes for the Senate. The turnaround was completed in 1982, with the conquest of city hall by the PMDB. Together, the four opposition parties won 60 percent for the Senate and 68 percent for the mayor, as against 22 percent for the PDS.

Table VII. Voting for the Senate and the Mayor's Office in the Municipality of Presidente Prudente (São Paulo), 1966-1982 (in percentages)

	1966	*1968*	*1970*	*1972*	*1974*	*1976*	*1978*	*1982*[4]	
	Senate	*Mayor*	*Senate*	*Mayor*	*Senate*	*Mayor*	*Senate*	*Senate*	*Mayor*
ARENA	73.5	92.0	39.5	95.2	28.7	78.0	13.0	21.0	22.0
MDB	13.2	4.0	39.7[1]	1.8	64.0[2]	17.5	73.0[3]	60.0	67.7
(Blank & Null)	13.3	4.0	20.8	3.0	7.3	18.8	14.0	19.0	10.3
Ballots Cast (in Thousands)	21,749	29,109	30,079	34,066	36,539	41,733	46,472	58,228	
Registered Voters (in Thousands)	27,845	36,090	38,220	41,101	43,926	49,411	55,851	70,050	

[1]Victory of Franco Montoro in 1970.
[2]Victory of Orestes Quércia in 1974.
[3]Victory of Franco Montoro and Fernando Henrique Cardoso in 1978.
[4]For 1982, the votes of PDS have been aggregated under ARENA and the votes of the four opposition parties under MDB.
Source: TRE—São Paulo.

This case study clearly illustrates the phenomenon of growth from the top down, or from the center to the periphery—in other words, the growing implantation of the opposition party in the spaces opened up by the plebiscitary victory of 1974, whose epicenter was clearly located in the state capitals and in the more industrialized areas. It can be asserted without hesitation that, beginning with the municipal elections of 1976, these tendencies were firmly established in the perceptions of the relevant actors, which explains the high price in legitimacy paid by the Geisel administration in April 1977, in order to establish rules which could ensure it a broad safety margin in the elections planned for 1978. In a broader perspective, it is important to remember that ARENA, just like the dominant parties on the conservative side in the 1945–65 period, became a prisoner of a demographically declining social base. The changes in the social structure responsible for this realignment hindering the regime's plans to obtain legitimacy through elections are fundamentally: a) the growing concentration of the population in large urban centers, accentuating the at least generically "oppositionist" nature of the electorate; b) the composition of this urban electorate in terms of age brackets and educational levels, leading to a similar increase in the weight of the young, relatively unschooled electorate—i.e. precisely those members of the population who have no memories of the events referred to by the military as their justification for having taken power in 1964. The effects of these factors, which were tendentially favorable to the appeal of the opposition, were multiplied by the imposition in 1965 of a two-party structure which facilitated the transposition of these necessarily diffuse cleavages to the political-electoral level.[35]

IV. Conclusions*

Unlike the redemocratization which occurred in Argentina with Alfonsín and in Portugal and Greece in the 1970s, the Brazilian process is characterized by its totally endogenous character and by its gradualism. Even in comparison with Spain, where redemocratization was also an endogenous political process, in that it was not marked by serious military defeats which affected the cohesion and credibility of the armed forces, the Brazilian case is notable for its longevity and its inconclusiveness at least until 1985. This is possibly why the electoral calendar played such a crucial, albeit limited, role. The apparent paradox of this statement requires explanation. Beginning in 1974, the electoral process in Brazil was in reality a test of forces and of legitimacy, and not the symbol and culmination of a transition pact already negotiated on other bases by the relevant actors. It was almost the point of departure of the process. The 1974 results signaled the desire for

*Much of the following analysis appears in "Apontamentos sobre a Questão Democrática Brasileira," in Alain Rouquié, Bolivar Lamounier, and Jorge Schvarzer (eds.), *Como Renascem as Democracias* (São Paulo: Editora Brasiliense, 1985).

change that was forming within society, impelling the organization of various oppositions into one political party (the MDB) and reinforcing the Geisel government's initial disposition to initiate a project of controlled liberalization, even though this met with resistance from the most intransigent sectors of the regime.

This process of *abertura through elections* is somewhat unique. A deeper analysis of the factors which made it viable ought to take as its point of departure the Brazilian ideological and institutional legacy, whose authoritarian character has often been emphasized, but which also contains important liberal components. The long-term legitimation of an authoritarian system is inconceivable to us, let alone a repressive autocracy like the one installed in the Médici era. However, rather than a return to this broader question—the so-called *electoral imperative*—I prefer to concentrate on two points which have a much more immediate impact on the mechanics of the changes taking place between 1974 and 1984:

1) the complexity of the social structure, the degree of urbanization of the country, and the importance of its politico-electoral history make pressure for democratization through the electoral process completely viable, once certain minimal freedoms for the contest are assured.
2) The sponsorship of the "détente" by the core of the dominant system itself meant that for all practical purposes the system could monopolize the *initiative* regarding the politico-institutional changes to be made. Thus the stratification of power could be projected forward in time (even if this required some arbitrary measures and morally dubious expedients), and the government was able to achieve its goal of attenuating the impact of increasingly adverse electoral results.

Between the impossibility of a lasting "Mexicanization," and more dictatorial immobilism, General Geisel opted for a third road, which was "gradual and secure" decompression. This, as suggested above, meant maintaining the stratification of politico-institutional power in force at the time. In other words, the *abertura* via elections was possible largely because the political party forces "reactivated" in 1974 were in reality trying to gain control of a legislature which had been largely deprived of its functions and prerogatives.

It is indispensable to stress the point in the previous paragraph in order to understand, on the one hand, that the *formal* mechanism of representation, i.e. the electoral process, was the field of action "decomposed" by the authoritarian system, thus allowing the opposition to grow; on the other hand, that this opposition, in the beginning, was in reality fighting for space and building its own organization, given that it was fighting, at best, for control of a weakened legislature threatened even by the *legal* instruments of military tutelage.

Thus over these first years, through its parliamentary majority and the unlimited powers that Institutional Act No. 5 granted it, the government conserved an almost absolute control over the agenda of the disputes in the politico-institutional arena. The importance of the movements of so-called civil society—the student and religious movements, and those by professional associations, and finally by trade unions—was not so much that they forced the *beginning* of the *abertura*, but that *little by little they created informal, but effective constraints on the dictatorial exercise of power.* From the angle that interests us here, the relevant consideration is thus the fact that the most decisive levels of power for the question of democracy were not really at stake between 1974 and 1982. These levels are access to the federal executive, via the contest for the presidency of the Republic; the control of economic policy, obviously dependent on the executive branch; the *legal* expressions of military tutelage over the political process, embodied in Institutional Act No. 5 until the end of 1978, and which survived indirectly, for example, through the constitutional provision for the declaration of a "state of emergency."

In this specifically institutional sense, it is the 1982 elections that represent a truly significant landmark. The election of ten opposition governors and the government's loss of its absolute majority in the Chamber of Deputies gave the political system a more or less *diarchical* character (as Juan Linz suggests), and substantially raised the level of confrontation with regard to the presidential succession. Although the parliamentary representation of the PDS and the coercive measures still at the disposal of the Executive were enough to deter the first big opposition offensive for the immediate restoration of direct presidential elections, in April of 1984, there is no doubt that the process of *abertura* reached a new threshold, insofar as the government monopoly over political-institutional initiative was destroyed.

Thus the *abertura* process, in terms of its *initial project*, had reached its limits. In retrospect, the changes which took place between 1974 and 1984 seem to have been based on an implicit negotiation, in the sense that both sides, government and opposition, found space successively to redefine their respective roles, foreseeing advances that would flow from the continuity of the process itself. The opposition benefited insofar as it structured itself as a powerful political-electoral force, and also insofar as in 1979 it obtained from the government the amnesty law, the end of the special powers embodied in Institutional Act No. 5, and the return of party pluralism.

The government, in turn, benefited in various ways. The *costs* of coercion were reduced. It gradually rid itself of the specter of the autonomized apparatuses hidden within the core of the State, responsible for the overdose of repression which had seriously damaged the international image of the country. Thus it capitalized on the political benefits of an atmosphere of progressive *normalization*, as if exchanging the losses in legitimacy produced by assessments of the past for gains based on the growing credibility of its intentions for the future. While it is true that there certainly was an erosion of authoritarian legitimacy, it is also true that this process of change revitalized the government's authority (during the Geisel administration and during the first three years of the Figueiredo administration), by giving it the role of agent of this process of normalization.

The breakdown of the initial project, that of controlled gradualism subject to frequent manipulations of the rules of the game, did not amount *per se* to a clear and complete transition to democracy. It signifies, rather, a first step in the direction of alternation in government, an *explicit* negotiation capable of revitalizing the political project as a whole, without which the regime in power would have been forced to break with its own legality once again, thus installing a new dictatorial period.

In a broader panorama, which links these immediate events to deeper questions about the Brazilian historical formation, it is perhaps possible to speak of a *perverse polyarchy*, that is, a society which does not accept monolithic authoritarian domination, but which also does not have the tradition of pluralist and independent political organization of the State which is typical of the real liberal polyarchies.

To suppose that this basic pattern has altered because of the *abertura* or of the transition with Tancredo Neves in 1985 would certainly be unsound. Nonetheless, there has been a significant change in the way these questions are understood. A few decades ago, the thesis of social "amorphism" was always accompanied by a picture of the State as a cohesive, active, and creative force. A careful researcher would have no difficulty in locating frankly anthropomorphic representations, in which the State appeared as a tutelary, generous, and altruistic spirit. The bureaucratic, legal, and military concentration of power after 1964, exactly because it reached levels never before seen in the country—though even this did not prevent it from plunging into an economic and social crisis of vast proportions—is now calling the attention of analysts to the cracks and the signs of lack of control that are inherent in it. General Geisel's political coordinator, General Golbery do Couto e Silva, used the *black hole* metaphor a great deal, presumably to refer to a power which was infinitely concentrated and hence made useless. Wanderley Guilherme dos Santos described the growing fortification of bureaucratic interests in the State productive sector as a process of *balkanization*. Using a more home-grown expression, ex-minister Melo Franco summed up the situation at the end of General Figueiredo's term in office: in Brazil today, he said, there is power, but there is no government.

NOTES

1. All of the references to *Authoritarian Brazil* in this article have been taken from the hardcover edition published by Yale University Press, 1973.

2. I am not referring here to the texts in *Authoritarian Brazil*, but rather to the whole climate of discussion, which can be found in innumerable sources, including journalistic ones. The nationalist military alternative is explicitly discussed and rejected by Stepan (pp. 62–65) and by Fernando Henrique Cardoso (pp. 166–67). Juan Linz discusses populist or left nationalism, not as a potential rupture, but on the contrary, as one of the hypotheses for *consolidation* of the Brazilian authoritarian experiment (p. 253). Schmitter saw the politicization of the military as a plausible reason for such a break.

3. Only Schmitter included data on the post-1964 elections in his essay. It is symptomatic that these data appear in the subsections entitled "A Single, No-Party System" and "Political Demobilization and Apathy" (pp. 209–14). The only positive indicator he found in the electoral arena was the increase in the voting population, which reached 24% in 1970, and in the formally registered electorate, which almost doubled in absolute terms between 1960 and 1970. The references in the volume published in 1978 by Guy Hermet, et al., *Elections Without Choice* (New York: John Wiley & Sons) to the persistence of "formally liberal and pluralist structures," which are cancelled out on the one hand by the authoritarian situation and on the other by "clientelist controls" (p. 15) are equally skeptical, and lead to the conclusion that even the 1974 and 1976 Brazilian elections had "few direct consequences" (p. 31). See also the article by Alain Rouquié in the same volume (pp. 59 ff).

4. See Gláucio A.D. Soares, *Sociedade e Política no Brasil* (São Paulo: Difel, 1973); Wanderley Guilherme dos Santos, "The Calculus of Conflict: Impasse in Brazilian Politics" (Ph.D. dissertation, Stanford University, 1974); Maria do Carmo Campello de Souza, *Estado e Partidos Políticos no Brasil, 1930–1964* (São Paulo: Editora Alfa-Omega, 1976). See also the bibliographical essay by Bolivar Lamounier and Maria D'Alva Kinzo, "Partidos Políticos, Representação e Processo Eleitoral no Brasil, 1945–1978," *Dados*, 19, 1978.

5. In the same vein, Stepan observes that "an important faction of the military had hoped to eventually allow the inauguration of liberal political forms" (p. 58), after the takeover in 1964. Fernando Henrique Cardoso describes the "project" of President Castelo Branco as one which is "politically and economically liberal, though its liberalism was qualified, of course, by the circumstances of an underdeveloped country" (p. 157). Further on he affirms that "the Revolution aspired to some sort of legitimacy that would ultimately reflect itself in the rule of law, or lawful state." (p. 168) As far as prospects for the future are concerned, it is interesting to note that Fernando Henrique Cardoso was skeptical about the "reconciliation" model proposed at the time by ex-Minister Roberto Campos (pp. 172–73), but did not reject it entirely. A more profound democratic project, which allowed ". . . the reconstitution of popular representative organizations," seemed "a remote possibility" to Cardoso, but, unless I am mistaken, an *abertura* that only increased participation by "the bourgeoisie and the middle classes" did not. On this last point, see p. 175, note 26.

6. Cf. references to the "libertarian ideals of the middle classes" (p. 190) and the hypothesis of a "reconciliatory political outcome" (p. 132, note 88). In order to stress once again that my aim is to discuss Schmitter's *model*, rather than any defects in his analysis itself, it may be permissible to recall that in the same context and also based on the model of "late industrialization," I myself have manifested a similar disbelief. Referring to the middle-level civil servants in the Brazilian State, who then seemed crucial to the legitimation process, I wrote the following in 1974: "Without excluding the possibility that these intermediate groups may sometimes allow themselves loftier concerns, it would not be malevolent to assume that they will normally be most interested in their status and in their own immediate interests. The latter are unlikely to be furthered by any broader political liberalization process. . . . They legitimate the existing political order . . . precisely because it is—and to the extent that it remains—closed and non-mobilizing. (Their) power interests mainly act as reducers rather than increasers of the subjective ordering of civil society." Cf. my article "Ideologia em Regimes Autoritários: Uma Crítica a Juan J. Linz," São Paulo, *Estudos Cebrap*, no. 7, p. 87. The intra-elite conflict, even at the height of the

"economic miracle," was something that the majority of observers certainly underestimated. An important empirical mapping of this question, based on interviews conducted at the end of the Médici government, can be found in Peter McDonough, "Os Limites da Legitimidade Autoritária no Brasil," Rio de Janeiro, *Dados*, 20, 1979. McDonough concludes that although bureaucratic authoritarianism can be considered a "statistical norm" or a "periodical condition," it is *not*, in Brazil, an "*ideological* norm, i.e., a set of beliefs about how intra-elite behavior should be conducted" (p. 114).

7. Other examples could be cited. Schmitter's preoccupation with the fascist-leaning tenor of the first attempts at ideological domestication in "morality and civics" courses has certainly proved exaggerated. The adoption of automatic mechanisms of economic management (monetary correction and mini-devaluations of the exchange rate, among others) may have reduced the level of conflict, as Skidmore foresaw, but it certainly did not impede the "crystallization of an opposition" (p. 31).

8. Louis Althusser, *Montesquieu, la politique et l'histoire* (Paris: Presses Universitaires de France, 1959), chap. 5.

9. The pertinent passage is worth quoting in its entirety, as it is one of the most persuasive and complete examples of *this line* of argument. "Certain characteristics of Brazil's delayed economic and social transformation"—writes Schmitter—(have) "conspired to make it highly unlikely that it would replicate . . . the liberal-pluralist . . . route to political modernity. On the one hand, preindustrial urbanization, proportionately low factory employment, industrialization by import substitution, dualistic stagnation in the rural sector, and heavy dependence on foreign capital and technology seem to have obfuscated some of the major lines of interest and attitudinal cleavage that provided the political dynamism for earlier developers. Above all, these different contextual factors tend to fragment and debilitate class consciousness or even corporate group awareness, making it difficult to establish and sustain aggressive and autonomous movements, parties, or associations. On the other hand, lengthly formal political independence, a prematurely large bureaucracy, sustained high rates of inflation and economic expansion have operated to strengthen the capacity and autonomy of pre-existing state institutions. These factors enabled the state institutions to respond to emerging or latent group protest extending the franchise 'precociously,' by coopting promising (and threatening) individual leaders 'preemptively,' by conceding welfare benefits and governmental social protection 'prematurely,' by promoting widespread corruption and selective favoritism, and, of course, by exercising sporadic but effective repression of more intransigent opponents. . . . Under these conditions, one would not anticipate strong resistance from bourgeois and middle-sector groups steeped in the traditions of political liberalism and angered at having 'lost' something they never had: autonomous participation or hegemony over the political order. Nor would one expect working-class organizations seriously to challenge authority groups with militant demands for greater equality of access or socialization of the means of production. Instead they will be preoccupied with retaining what has already been 'benevolently' granted them from above" (pp. 184–86).

10. A pluralist system does not require that all actors be ideologically pluralist. If they share a basic "calculus" which permits the rule of relatively neutral norms, or if the political situation is such as to force the non-pluralists to accepting the game, the representative mechanism can be consolidated through a process of learning. This point is made persuasively by Dankwart Rustow in his well-known "Transitions to Democracy," *Comparative Politics*, 2:3 (April 1980); and by Giuseppe Di Palma in

"Elections in Societies with an Authoritarian Past," paper presented at the conference *Elections and Democracy in Central America* sponsored by the Inter-American Human Rights Institute, San José, Costa Rica, June 1983.

11. In this sense, it seems to me that the concept of competitive election proposed in Hermet et al., *Elections Without Choice*, is restrictive. See citation in note 3 above.

12. Theda Skocpol, for example, thinks that the problem of legitimacy can be dismissed by saying that "what matters most is always the support or acquiescence not of the popular majority of society, but of the politically powerful and mobilized groups, invariably including the regime's own cadres." It seems to me that her references to "society-centered theories of politics and government" are equally careless, insofar as they include in this type of theorization everything which concerns the concept of representation. Representation in its original Hobbesian sense, as formal authorization, and even the historical transition from the imperative mandate to the independent mandate are inseparable from the constitution of the modern State as an "actor" which strives to become unitary and autonomous. References to Skocpol's work are from her article "Bringing the State Back In," published in *Items*, newsletter of the SSRC, vol. 36, nos. 1–2, New York, June 1982, and from the section "The Potential Autonomy of the State," in *States and Social Revolutions* (Cambridge: Cambridge Univ. Press, 1979), 24–33. On representation, see Hanna F. Pitkin, *The Concept of Representation* (Berkeley & Los Angeles: Univ. of California Press, 1972).

13. Suffice it to remember that a position to the right of the Geisel government became clear within the military as late as October 1977, in the crisis which led to the dismissal of Minister of Defense Sylvio Frota. The rigidity of the ideological, institutional, and military framework in the early days of the *abertura* is an important element in the interpretation proposed by Alkimar Ribeiro Moura and myself in "Política Econômica e Abertura Política no Brasil, 1973–1983," São Paulo, *Textos IDESP*, no. 4, 1984.

14. One of the difficulties in the Brazilian conjuncture after the 1982 elections, as Juan Linz pointed out so well, is that the opposition leaders and parties "have to develop a strategy knowing their assets, as does the government." In 1974, on the contrary, the contenders had to submit their respective hypotheses to the test, and both sides proved vastly mistaken: the government did not have as much support and the opposition was not as electorally weak as expected, or at least as most observers thought. See Juan Linz, "The Transition from an Authoritarian Regime to Democracy in Spain: Some Thoughts for Brazilians," mimeo, Yale University, March 1983.

15. This point should be considered in the light of the *substantive* appeals made by the military governments since 1964. From this point of view, there were two phases of the regime. The first, widely documented in the various contributions to *Authoritarian Brazil*, essentially consisted of legitimation *against* the situation prevailing in 1961–64: against populism, Jangoism, corruption and subversion, and especially against disorder in the administrative and financial management of the public sector. In the second phase, legitimation depended strictly on economic growth, the so-called "Brazilian miracle," which might be called the deification of the GNP.

16. The party loyalty requirement was first introduced by Constitutional Amendment No. 1, in 1969 (which was for all intents and purposes equivalent to a new Constitution, promulgated by the Military Junta). It was undoubtedly a response to

the crisis at the end of 1968, when many ARENA congressmen joined with those of the MDB to turn down the government's request for permission to prosecute an opposition deputy. This new rule (article 152, sole paragraph) established the following: "Whosoever, through his attitudes or his vote, opposes the positions legitimately arrived at by organs of party leadership or leaves the party on whose ticket he was elected, will lose his mandate in the Federal Senate, in the Chamber of Deputies, in the Legislative Assemblies or in the Municipal Councils. The loss of mandate will be decreed by the Electoral Court, upon the party's request, with full rights of defense." There is no doubt whatsoever that this clause was useful to the opposition during the following years, helping to maintain party cohesion and keeping under control the tendency of the so-called *fisiológicos* to support positions of the government majority.

17. In the 1982 gubernatorial elections, the opposition parties together won 52.5% of the vote, against 37.3% obtained by the PDS, and 10.2% blank and null. Nine PMDB governors were elected (including Franco Montoro in São Paulo and Tancredo Neves in Minas Gerais); the PDT candidate was elected governor of Rio de Janeiro (Leonel Brizola); and the PDS won in the remaining 12 states, with the (first) governor of Rondônia being appointed. It can be said that the government party's main victory was in Rio Grande do Sul, where ex-Minister of Social Security Jair Soares was elected because of the division of the opposition electorate between the candidates of the PMDB (Pedro Simon) and the PDT (Alceu Collares).

18. The distribution of seats among the states in the Chamber of Deputies is based, with a few modifications, on criteria established by the 1932 Electoral Code, later revised in article 58 of the 1946 Constitution. Although it has the same directional effect as the arbitrary manipulations, this distribution of seats and the Senate are long institutionalized mechanisms which are subject to little controversy. On these points see Gláucio A.D. Soares, "El Sistema Electoral y la Representación de los Grupos Sociales en Brasil," *Revista Latinoamericana de Ciencia Política*, vol. II, no. 1, 1971; Maria do Carmo Campello de Souza, *Estado e Partidos Políticos no Brasil, 1930–1964*, chap. V, pp. 124–36; and Maria D'Alva Gil Kinzo, *Representação Política e Sistema Eleitoral no Brasil* (São Paulo: Editora Símbolo, 1980), chap. IV, pp. 95–107.

19. On the revocation of mandates and the suspensions of political rights, see Marcus F. Figueiredo, "Política de Coação no Brasil Pós-64," in Lúcia G. Klein and Marcus F. Figueiredo, *Legitimidade e Coação no Brasil Pós-64* (Rio de Janeiro: Forense-Universitária, 1979).

20. Bruce Raymond Drury, "Creating Support for an Authoritarian Regime: The Case of Brazil, 1964–1970" (Ph.D. dissertation, University of Florida, 1973), 18, 193–200.

21. There is no lack of evidence that the initial steps of the *abertura* were deliberate. The simple presence of Armando Falcão as Minister of Justice was interpreted by the press in these terms, as a result of contacts with the governmental team before it took power. Indications that press censorship would be suspended and Geisel's call for "creative imagination" can also be cited. Even in the area of electoral legislation, it should be recalled that the "Etelvino Lins" law, of August 15, 1974, one of the rare instances of parliamentary initiative stimulated by the Executive, contained within it the guarantee of equal and free access of political parties to radio and television during future electoral campaigns, along with measures that were intended to control pressure on poor voters in rural areas. See on this point and on the regulation of access to radio and television the work of Celina R. Duarte, "A Lei Falcão: Antecedentes e Impacto," in Bolivar Lamounier (ed.), *Voto de Desconfi-*

ança: Eleições e Mudança Política no Brasil, 1970–1979 (Rio de Janeiro: Editora Vozes, 1980); and by the same author, "Imprensa e Redemocratização no Brasil," *Dados*, 26:2, 1983.

22. Wanderley Guilherme dos Santos, *Poder e Política: Crônica do Autoritarismo Brasileiro* (Rio de Janeiro: Forense-Universitária, 1978), 95–96.

23. Plebiscitary, evidently, against the regime, when the initial expectation of the Geisel government was to obtain a show of support in its favor. This is what Cruz and Martins suggest on the basis of an extensive survey of the period: "Certain of victory, Geisel invested heavily in these elections, which were to play a crucial role in the activation of his project: with popular support for the "achievements of the Revolution" confirmed in the voting booths, the next year could be dedicated to institutionalizing the regime and to the awaited reforms. But for that to happen, these elections could not be like the others (1970 and 1972), robbed of their legitimacy by the prevalence of censorship and by the violence of intimidation measures. . . . It was necessary to have the opposition become wholeheartedly involved in the contest, and given the guarantees which had been provided, to induce it to accept the evidence of its future defeat with good grace." See Sebastião Velasco e Cruz and Carlos Estevam Martins, "De Castello a Figueiredo: uma Incursão na Pré-História da Abertura," in Bernardo Sorj and Maria Hermínia Tavares de Almeida (eds.), *Sociedade e Política no Brasil Pós-64* (São Paulo: Editora Brasiliense, 1983), 49. The element of surprise and the resounding decisiveness of the 1974 results certainly deserve a more detailed analysis. Suffice it to remember that the opposition won 16 out of 22 Senate races, and that only one (Bahia) of the ten states with more than one million voters gave a victory to the government party. The largest opposition margin (73% of the valid votes) was in São Paulo, which is the state with the largest number of voters (7,100,000 in 1974).

24. The following discussion is based on a series of public opinion polls and studies done on the electoral process since 1974. I personally carried out pre-electoral surveys in 1974 (municipality of São Paulo), 1976 (municipality of Presidente Prudente, state of São Paulo), and again in the municipality of São Paulo in 1978 and 1982. Some of these studies were done in close collaboration with researchers from other states, especially the study of local elections in 1976, published in the volume edited by Fábio W. Reis (see below). This collaboration eventually led to the formation of a Working Group on Parties and Elections as part of the National Association for Research and Graduate Studies (Associação Nacional de Pesquisa e Pós-Graduação). Studies on the 1982 election, coordinated by this group, covered seven capitals, including three in the Northeast (Fortaleza, Recife, and Salvador). The methodology adopted was that of a multi-stage probabilistic sample. The intention was to represent only the electorate of the municipality under the study in each case. See Bolivar Lamounier and Fernando Henrique Cardoso (eds.), *Os Partidos e as Eleições no Brasil* (Rio de Janeiro: Paz e Terra, 1975). On subsequent elections, see Fábio Wanderley Reis (ed.), *Os Partidos e o Regime* (São Paulo: Editora Símbolo, 1978); and Bolivar Lamounier (ed.), *Voto de Desconfiança* (Rio de Janeiro: Editora Vozes, 1980). The 1982 results in the municipality of São Paulo are analyzed by Bolivar Lamounier and Maria Judith Muszynski in the monograph "São Paulo, 1982: a Vitória do (P)MDB" (São Paulo, IDESP, *Textos*, no. 2, 1983).

25. The Gallup polls and the surveys which serve as the basis for the academic studies cited in note 24 contain abundant evidence of the electorate's preference for direct elections. The greater importance attributed to elections for the Executive is a

point often repeated in the literature since the 1950s. For the recent period, see the anthropological study by Teresa Pires do Rio Caldeira, "Para que serve o voto? As eleições e o cotidiano na periferia de São Paulo," in Lamounier (ed.), *Voto de Desconfiança.*

26. On union activity during this period, see Amaury de Souza and Bolivar Lamounier, "Governo e Sindicatos no Brasil: a Perspectiva dos Anos 80," *Dados*, 24:2, 1981; on the public discussions about the *abertura*, see the extremely useful survey by Marcus F. Figueiredo and José Antônio B. Cheibub, "A Abertura Política de 1973 à 1981: Inventário de um Debate," Rio de Janeiro, *BIB* (Boletim Informativo e Bibliográfico de Ciências Sociais) 14.

27. In the study of Presidente Prudente during the 1976 municipal elections, for example, only 25% of the voters declared a preference for one of the parties of the previous period (33% remembered their parents' preferences). Considering the supporters of the former PSD, UDN, PSP, and PTB, we noted that this identification did not determine their options between the new parties, ARENA and the MDB. In 1979, on the eve of the reform which restored a multi-party system, the Gallup Institute asked a national sample the following question: *"If the old political parties which existed before 1964 were to return, for which one would you have the most sympathy?"* In these terms, around 70% of those interviewed declared an option. As could be expected, the preference for the old PTB (Partido Trabalhista Brasileiro, or Brazilian Labor Party) was greater in the state capitals and large cities, where it reached 42%, in comparison with the national average of 33%. Also according to Gallup, this PTB option was chosen by 44% of those who now identified themselves with the MDB, in comparison with 24% of current ARENA supporters.

28. In the state of São Paulo, in 1972 the MDB controlled only 58 out of 551 mayors' offices (excluding the municipalities designated as national security areas and the hydro-mineral fields, whose mayors were appointed) and 809 of the 4,930 municipal councillors. Gallup data indicate that the ARENA candidate, Carvalho Pinto, was favored by 72% of the voters until the month of September, as against 6% for the opposition candidate Orestes Quércia. Two days before the elections these proportions had been reversed: 69% for Quércia, and 23% for Carvalho Pinto. These data also show that the Carvalho Pinto candidacy was clearly identified with the past (age, experience, record in positions which he had occupied) while the opposite occurred with the Quércia candidacy.

29. The exacerbation of expectations, together with the maintenance or even the increase of income inequality during the "miracle," is one of the points underlined by Schmitter in his contribution to *Authoritarian Brazil* (see pp. 197–206). It is interesting to note, however, that the *real* decrease in growth rates was not yet very sharp. On the contrary, the Geisel government opted for a strategy of sustaining growth; the Gross Domestic Product grew by about 9.5% in 1974, and only fell below 5% in 1978, the last year of his term. The inflation rate was 35% in 1974. Treatment of internal macro-economic disequilibria was inconsistent, alternating between measures of monetary and credit expansion and contraction, a typical stop-go pattern. See on this topic the paper by Bolivar Lamounier and Alkimar Ribeiro Moura, cited in note 13.

30. This break, as Fábio Wanderley Reis indicates, is shown in various events which mark the passage from Médici to Geisel: "The critiques of various aspects of the previous government's policy by the new holders of governmental positions, involving even accusations of corruption; and especially, economic revisionism, undermining the politically legitimating effect of the previous triumphalism and

emphasizing the problem of distribution of the benefits of development . . . ; the official promotion of the theme of political *abertura*. . . ." See "As Eleições em Minas Gerais" in the volume *Os Partidos e as Eleições no Brasil*, op. cit, pp. 149–50.

31. We would do well to remember that certain details of the electoral mechanism facilitate this bipolar identification. This is the case, for example, with the compulsory linking of votes for federal and state congressmen (voters must choose candidates from the same party) and the acceptance as valid of any vote cast for a party slate alone, without any indication as to specific candidates. In 1974, 38% of the MDB's votes in the municipality of São Paulo for the Federal Chamber of Deputies were for the ticket, as opposed to only 15% of ARENA's votes.

32. We should also add a purely demographic note on this subdivision of the municipality of São Paulo into eight homogeneous socio-economic areas. The available information shows that the percentage share of areas VI to VIII (the last three) in the total number of registered and active voters, had increased substantially over the last sixteen years. From 40% in 1966, these three poorer areas, which include all the periphery created by recent urban expansion, came to represent no less than 55% of the total number of active voters in 1982. For a more detailed analysis, see *Voto de Desconfiança* and "São Paulo, 1982: a vitõria do (P)MDB," op. cit.

33. In the state of São Paulo, there is a particularly interesting case in the industrial area of the ABC, in the metropolitan area of the capital. The opposition vote was higher there than in the capital itself, as Maria Teresa Sadek showed in her doctoral thesis, "Concentração Industrial e Estrutura Partidária: o Processo Eleitoral no ABC, 1966–1982" (São Paulo, USP, 1984).

34. See Bolivar Lamounier, "O Crescimento da Oposição num Reduto Arenista," in Fabio Wanderley Reis (ed.), *Os Partidos e o Regime*. See also the studies by Olavo Brasil de Lima, Jr. (on Niterói in the state of Rio de Janeiro), Fábio Wanderley Reis (on Juiz de Fora, MG), and Hélgio Trinidade and Judson de Cew (on Caxias, Rio Grande do Sul).

35. The low levels of inter-correlation among opinions, and between these and party preferences, are a constant in the surveys carried out since 1974 (see bibliography in note 24). The low level of sedimentation of social cleavages was already discussed above, with reference to the work of Schmitter and the "late industrialization" model. See, however, two historical studies which investigate the question of political organizations in more detail: Glaúcio A. D. Soares, *Sociedade e Política no Brasil*, and Simon Schwartzman, *Bases do Autoritarismo Brasileiro* (Rio de Janeiro: Editora Campus, 1982).

PART II

The Political Economy of *Abertura*

3

A Tale of Two Presidents: The Political Economy of Crisis Management

ALBERT FISHLOW

In March 1974, when Ernesto Geisel assumed the Brazilian presidency, the euphoria of the economic "miracle" of the previous five years still reigned. Dissident voices were few and discredited. Despite the sharp rise in petroleum prices a few months earlier, perspectives for continuing prosperity were bright. Spectacular growth at rates in excess of 10 percent aroused visions of *grandeza*, the attainment of Brazilian destiny on a world stage. Geisel himself was the best among the generals: a proven technocrat not only capable of managing the economy but also persuaded of the need for a process of guided political liberalization to assure a lasting social tranquility.

In March 1985, João Batista Figueiredo left office in near disgrace. During his term, inflation had accelerated from 40 percent to well over 200. Per capita income declined between 1979 and 1984 by some 10 percent. A staggering external debt had become the unfortunate symbol of Brazilian pre-eminence in the Third World. Table 1 details some aspects of the economic deterioration. In addition, political uncertainty reigned in the midst of transition to not only a civilian but even an opposition government, made possible only by desertions from the official party in the electoral college.

Few could have predicted, or did, such a dramatic reversal. One obvious factor was the counterpart change in the world economy since 1973. The oil shock of 1973–74 had been barely accommodated when new rises in the oil price in 1979 and severe recession in the industrialized countries accompanied by record interest rates created further disruption. Developing countries had called for a New International Economic Order in the early 1970s.

Table 1. Brazilian Economic Performance: 1971-1984

A. Real Rates of Growth of Production

	Gross Domestic Product	*Industry*	*Agriculture*
1971	12.0	12.0	11.3
1972	11.1	13.0	4.1
1973	13.6	16.3	3.6
1974	9.7	9.2	8.2
1975	5.4	5.9	4.8
1976	9.7	12.4	2.9
1977	5.7	3.9	11.8
1978	5.0	7.2	−2.6
1979	6.4	6.4	5.0
1980	7.2	7.9	6.3
1981	−1.6	−5.5	6.4
1982	0.9	0.6	−2.5
1983	−3.2	−6.8	2.2
1984	4.5	6.0	3.2

B. Rates[a] of Growth of Money and Prices

	Monetary Base	*Money Supply M1*	*Augmented[b] Money Supply M4*	*Wholesale Prices, Internal Supply*	
				Total	*Foodstuffs*
1971	36.3	32.3	n.a.	21.4	30.2
1972	18.5	38.3	n.a.	15.9	16.1
1973	47.1	47.0	n.a.	15.5	12.4
1974	32.9	33.5	40.6	35.4	37.4
1975	36.4	42.8	57.5	29.3	33.0
1976	49.8	37.2	55.3	44.9	50.1
1977	50.7	37.5	40.1	35.5	37.5
1978	44.9	42.2	53.2	43.0	51.9
1979	84.4	73.6	65.1	80.1	84.8
1980	56.9	70.2	69.1	121.3	130.8
1981	78.0	87.2	141.7	94.3	85.9
1982	87.3	65.0	105.7	97.7	98.9
1983	96.3	95.0	150.2	234.0	299.5
1984	243.8	203.5	291.9	230.3	223.7

[a]December to December of indicated year
[b]Includes time deposits and federal government debt

Little did they anticipate what a new, and adverse, order would eventually unfold. For the first time in the postwar period, generalized income decline owing to balance of payments difficulties afflicted virtually all the middle-income countries of Latin America in the early 1980s.

The official explanation for the Brazilian reversal gives predominant weight to this disruption of the global economy over the last decade. Brazil, as the largest third world importer of oil, was clearly impacted by the rises in the oil price. Brazil, as the largest developing country debtor, was also

Table 1. (Continued)

C. Balance of Payments and Debt (billion dollars)

	Exports[a]	*Imports*[a] *Total*	*Fuel*	*Net Interest*	*Current Account*	*Net Capital Inflows*	*Surplus*	*Level of Reserves*	*Level of Debt*[b]
1971	2.9	3.2	.4	.3	−1.3	1.8	.5	1.7	6.6
1972	4.0	4.2	.5	.4	−1.5	3.5	2.4	4.2	9.5
1973	6.2	6.2	.8	.5	−1.7	3.5	2.2	6.4	12.6
1974	8.0	12.6	3.0	.7	−7.1	6.3	−.9	5.3	17.2
1975	8.7	12.2	3.1	1.5	−6.7	5.9	−1.0	4.0	21.2
1976	10.1	12.4	3.8	1.8	−6.0	6.9	1.2	6.5	26.0
1977	12.1	12.0	4.1	2.1	−4.0	5.3	6	7.3	32.0
1978	12.7	13.7	4.5	2.7	−6.0	9.4	3.9	11.9	43.5
1979	15.2	18.1	6.8	4.2	−10.0	7.7	−3.2	9.7	49.9
1980	20.1	23.0	10.2	6.3	−12.4	9.7	−3.4	6.9	53.8
1981	23.3	22.1	11.3	9.2	−11.0	12.8	.6	7.5	61.4
1982	20.2	19.4	10.5	11.4	−16.3	7.9	−8.8	4.0	69.7
1983	21.9	15.4	8.6	9.6	−6.8	1.5	−6.0	4.6	81.3
1984	27.0	13.9	7.3	10.2	0	−1.2	−0.8	12.0	91.1

[a]FOB

[b]Registered; excludes inter-bank liabilities and short-term debt.

Sources: *Conjuntura Econômica* and Central Bank of Brazil, *Monthly Bulletins.*

shaken by the rapid run up in the nominal and real interest rate after 1979. Finance Minister Ernane Galvêas stated this view in his deposition before Congress justifying the decision in late 1982 to seek assistance from the International Monetary Fund: "Without any doubt, the energy crisis initiated in 1973–74, with the explosion of oil prices, was the most important factor in the interruption of the accelerated economic development experienced in the previous decades. . . . Between 1968 and '73, the principal economic and social indicators show that the country had solved the most important limitations to the modernization of the economy. . . . The country was recovering rapidly from the first petroleum crisis, when new price increases of that raw material affected us in 1979 and 1980, now further aggravated by the financial shock."[1]

Such an interpretation is too simplistic. It ignores the magnification of external shocks by the style of Brazilian adjustment undertaken. The deterioration of the terms of trade in 1973–74 directly reduced Brazilian real income by perhaps 3 to 4 percent. Other countries faced far worse, prominently the resource-poor East Asian NICs, and yet they now stand comparatively in far better shape.

Equally simplistic, however, are exercises that abstract from the reality of Brazilian political economy in their outward-oriented recipes for efficient adjustment.[2] They emphasize the failure to allow the price system greater sway, and especially the lack of an aggressive exchange rate policy to stimulate exports. These demonstrations of the inadequacy of the Brazilian response pay minimal attention, however, to important structural characteristics of the economy that impede the effectiveness of such a strategy. Of equal importance, they neglect the impact of political considerations in the decisions actually taken.

My chapter in *Authoritarian Brazil* almost fifteen years ago tried to take into account both constraints. For that reason, it anticipated difficulties with the Brazilian model even at the height of its success: "I argue that the return to higher rates of growth in more recent years is based in part upon lagged cyclical adjustment to previous import-substitution oriented industrial development. As such, the present expansion cannot be simply extrapolated, nor its potential imbalances ignored, despite the important strides in economic-policy execution since 1964."[3] Then I foresaw the principal growth problem as the difficulty of generating the higher rates of saving required to sustain expansion once excess capacity had been utilized. And I anticipated that "commitment to solutions that worsen still further the income distribution . . . may not be feasible any longer."[4]

Both observations were accurate. What was missing, however, was the oil shock, and thus an external environment in which foreign exchange limitations would again become the central concern of policy-makers. That led me to the wrong conclusion about the likely small place for foreign savings in sustaining growth: "Foreign savings will also expose the economy to an important source of instability . . . Economic strategists of the present government are aware of the aggravating role earlier played by debt service requirements."[5] That caution was discarded in the acceptance of massive capital inflows in the 1970s.

In the final analysis, I also underestimated the political resistance to recession, and how dominating the *grandeza* theme would prove. I therefore implicitly overestimated the capacity of the new technocracy to respond to new realities. Instead, by the early 1980s, frantic inconsistency had became its hallmark in face of increasingly difficult circumstances.

Brazil's economic evolution over the past decade thus poses one central and overriding question: Why didn't Brazil accommodate better to the deterioration in the external environment? The answer is to be found in a blend of political and economic considerations, and not by either exclusively. This tale of two Presidents is pre-eminently a story of the interaction of political aspirations and economic constraints. It also has a surprising, and happy, political ending, the re-establishment of civilian and more representative governance. That outcome, in the midst of economic decline, casts doubt upon linear views of the relationship between economic performance and political change.

The Geisel economic policymakers, despite their formal obeisance to continuity, faced immediate challenges that required new responses. The new administration inherited from the outgoing Médici government a miracle that was already showing disturbing signs of mortality. Economic growth during the miracle had benefited from accumulated excess capacity. Relatively low fixed investment rates, and domestic savings, had therefore been required. Compared with an incremental capital-output ratio of 2.67 in the period 1965–70, the average from 1971 to 1973 had been only 1.75. To sustain a continuing annual growth rate of 10 percent at historical, pre-miracle investment levels thus implied an increase in the ratio of saving to income of about 4 percentage points, and possibly more, considering the structural changes in the economy that were required.[6] Such an increase in saving requirements conflicted with the incentives to consumption of consumer durables that had been a prominent feature of the miracle years.

Nor were the impressive international reserves at the end of 1973 an accurate measure of the strength of the balance of payments. Real import elasticities had been of the order of two since the late 1960s: growth at a rate of 10 percent a year implied expansion of real imports at 20 percent. Export growth in the same interval was only about half as fast, and only with especially rapid growth in 1972. The difference was made up principally by improved terms of trade. Between 1969 and 1972 Brazilian export prices rose by 7 percent a year; import prices, by 1 percent.[7]

These favorable circumstances kept demands for external financing within reasonable bounds. The current account deficit amounted to about only 2 percent of total product. Capital actually came in larger quantities—this was the first flush of interest in Brazil by the Euro-currency market—and reserves were accumulated. But the underlying potential resource imbalance was a lurking constraint to ambitious continuing growth targets. In the short term, the tendency toward excess import was further exacerbated by the accumulating shortages of inputs in the domestic market and by an exchange rate that was overvalued because its adjustment was determined by the reported—rather than the real and—higher—rate of inflation.

There was a third fragility. After several years of declining inflation, to which high rates of productivity increase and lower real wage increases had contributed, Brazil was subject to reversal of the trend. Demand was strong, fueled by real liquidity increases that had absorbed large nominal increases in financial assets. Now, that expansionary policy was confronting bottlenecks. The response was increased reliance on administrative controls to keep the calculated index at the levels forecast by the departing administration.

To these accumulated economic problems was added the new one of a sharp rise in the oil price in October 1973 in the aftermath of the Yom Kippur War. For Brazil, this was a very serious matter, dependent as the country was on imported oil for some 80 percent of its energy requirements. The transportation system had been predicated upon cheap fuel: the truck, not rail or water, had been the means of penetrating the interior and tying together coastal markets. Consumer durable demand had made the automobile sector the largest among developing countries, and the industry was at the center of Brazilian industrialization strategy. Fuel was therefore a critical input, and one not easily substituted in the short term. The inevitable consequence was a profound threat to economic growth emanating from much larger outlays on oil imports that would reduce imports of equipment and intermediate inputs.

If this agenda were not challenging enough, the new government had a compelling and difficult political project to which it was committed, that of institutionalizing the Revolution of 1964. High growth rates might temporarily distract from the excesses of political authoritarianism and repression, but for Geisel and his close collaborator and adviser, General Golberry, they were not the total answer. The continuity of the changes initiated in 1964, and thus the hopes for a secure and powerful Brazil, resided in a gradual and guided return to a constitutional regime also capable of maintaining order.

Responding with varying intensity to these economic initial conditions and to the administration's political objectives, the Geisel economic policy unfolded in three distinct phases. First, in 1974, came the effort to cool down the overheated economy through the application of orthodox monetary and fiscal policy. These good intentions—in part because they were only marginally effective—soon gave way by 1975 to a more aggressive medium-term development strategy designed to accomplish the dual objective of sustaining high rates of growth while promoting adjustment to the oil shock. To achieve the requisite degrees of freedom for political maneuver, economic growth was seen as an imperative. Economic success was central to the conscious military emphasis upon security and development, and the generals' image of Brazil as an emerging power. In the words of the Second Development Plan: "The government is conscious of the difficulty of maintaining rates of growth of the order of 10%, beginning in 1975, in face principally, of full capacity attained in the industrial sector, of problems related with the energy crisis and scarcity of raw materials, and of their effects on the balance of payments. . . . Nevertheless, the option of preservation of accelerated growth was preferred as the basic policy."[8]

This bold strategy was soon overlaid by a stop-go macro-economic policy designed to keep inflation within bounds and reflecting increasing preoccupation with threatening external disequilibrium. Capital inflows became a central means of reconciling the growth and stabilization objectives. It was this uneasy combination of government investment, monetary stringency, and increasing external indebtedness that dominated from mid-1976 on.

The results, as Table 1 shows, were not entirely unfavorable. Growth averaged 7 percent between 1974 and 1978, and hence somewhat above the postwar trend. Indeed, the Brazilian accomplishment, in a world economy where industrial country growth had fallen to almost half its 1962–73 level, was widely praised. Writing at the end of the 1970s, William Cline concluded that "a provocative lesson from the Brazilian experience is that on balance an ambitious set of growth policies designed to offset external shocks through new import substitution may be preferable to a passive response that accepts sluggish growth rates."[9]

In each of these phases, political considerations and responses of domestic groups played an important role. Policies were not forged in a technocratic vacuum, as they increasingly had been after the imposition of the Fifth Institutional Act in December 1969. Indeed, the conscious objective of the Geisel government was to channel popular reaction, through its program of relaxation of repression, *distensão*. To understand this interaction, and the force of underlying economic constraints, requires a closer look at each of the three periods and their differing priorities and successes.

The initial policy thrust of the Geisel administration was to treat the excess demand inherited from the miracle years. It was a plan put in motion without full regard for the oil shock—although that reinforced the case for slowing down as a means of reducing imports—and without undue concern about implications for trend growth. There was by now great confidence in the apparent Brazilian vocation for rapid economic expansion. The new Finance Minister Mário Henrique Simonsen came to his position determined to set the economy back on course primarily by fine tuning. Writing of the period later, he reveals his dominant concern: "Strong inflationary pressures were built up in 1973 when M [the money supply] expanded 47%, when output advanced far beyond its trend, and when OPEC quadrupled the oil prices. Yet the Médici Government was strongly committed to a 12% a year inflation rate ceiling. With massive price controls and subsidies the general price rise increase was repressed at 15.7%."[10]

Through most of 1974, despite expansionist rhetoric to the contrary, monetary and fiscal restraint was therefore mildly contractionary. The rate of increase of the money supply was reduced to 33 percent, almost all concentrated in the last two months of the year when policy had become more expansionary; and the cash flow of the Treasury recorded a large, and previously unplanned, surplus, of about 0.5 percent of total product.[11] Such measures were in response to evidences of overheating and accelerating inflationary tendencies. They were a modest corrective that did not comprise a strategy to deal with the attainment of high growth in the new international economic environment, while maintaining internal and external equilibrium.

Even this restraint was soon abandoned. By October, there were already signs of greater monetary ease—perhaps in anticipation of the November elections,—and a wage bonus of 10 percent and changes in the wage formula to compensate for the acceleration of inflation soon followed. Deficits in the Treasury accounts likewise gave evidence of a more expansive fiscal policy.

Orthodoxy was condemned to failure on two counts, one economic, the other political. Fiscal and monetary restraint was unable to make much headway in the short term against accelerating inflation, but did provoke a slowdown in industrial activity as well as lead to a major financial failure (the Halles Group). Given the substantial liquidity in the economy, and a commitment to "corrective inflation" that deregulated administered prices, inflation in 1974 could hardly fail to accelerate. Indeed, a simple monetarist model predicts a rate of inflation in excess of 40 percent, without any regard for increases in external prices, and with the restrictive credit policy actually followed during that year.[12] In reality, to the force of past liquidity were added generalized rises in dollar import prices, part of the oil price explosion. Even where moderated by subsidies, they not only were passed along to domestic prices but sometimes anticipated. Orthodox restraint was thus additionally handicapped.

At best, the policy would have required patience and further reductions in growth to have yielded some moderation in price increases. Neither were now politically palatable. The great confidence in ARENA's unchallenged hegemony that had made the *distensão* experiment possible in the first instance was badly shaken by a major electoral defeat in 1974. Moments of political uncertainty, and they abounded in early 1975, were not ones to impose unpopular and conventional austerity policies. Rather, there was great temptation to demonstrate a singular Brazilian capacity to overcome challenges other countries, and other regimes, could not manage. Brazil was to be an "isle of tranquility" in the midst of international economic turbulence.

By early 1975, the die was therefore cast for resumed expansion. The choice was feasible for two reasons. First, higher inflation was tolerable domestically because of widespread indexing. And, second, a weaker external balance of payments would not deter Brazilian growth now that new conditions of more liberal international finance prevailed owing to petrodollar recycling.

One essential component of the post-1964 reforms had been the creation of an extensive indexing system for financial assets to guarantee against inflationary erosion. The new government emphasized positive returns to private saving: now long-term assets could compete with more liquid instruments and money. The counterpart was an increase in the real resources directed to long-term investment. In 1968, the automatic adjustment of the exchange rate to differential inflation in Brazil and the rest of the world further guaranteed against inflation-induced distortion of relative prices. Finally, the wage formula, as modified to rid it of some of the deliberate bias in its earlier application, seemed to guarantee against erosion of real wages owing to higher inflation, especially so long as price rises remained moderate (the higher the inflation rate, with a fixed period of adjustment, the lower the average real wage).

The doubling of the inflation rate between 1973 and 1974, from 20 to 40 percent, was therefore not terribly preoccupying in these institutional circumstances. This was the more so when price increases were accelerating in the industrial countries to double digit levels. Brazilians prided themselves on their greater ability to adapt to such an inflationary environment without paying a high real cost. Indeed there was even boastful talk of exporting their proven techniques of monetary correction to the United States.

The unfavorable balance of payments was potentially more limiting. But Brazil had discovered during 1974 that the old pre-oil crisis financial rules no longer held. As authorities had made conditions of capital inflow substantially easier by reducing minimum maturities and domestic taxes, there was no shortage of either domestic borrowers or international lenders. In 1974, a trade deficit of $6.2 billion was financed with use of less than $1 billion in reserves; recycling of petro-dollars had begun in earnest, and Brazil was not only an eligible recipient but an attractive target.

Of that total deficit, only $2 billion can be attributed to higher prices of oil. A further $1.5 billion is explicable by higher prices of other imports. The rest is due to an unprecedented surge in real imports: the realized import elasticity rose to almost 3. Various restrictions on imports were adopted through the year, but without effect. Much of the buying was in anticipation, self-fulfilling, of later rationing or higher import cost. The sudden increase in capital inflows brought the gross debt level to more than $17 billion, with exports less than half as great.

Few voiced concern. Instead foreign financing increasingly became an instrument of choice, satisfying multiple objectives. First, it dampened domestic inflationary pressures inherited from 1973. It did so not only by augmenting the supply of imported inputs in 1974 but also by allowing the government the luxury of not raising domestic prices of imports, and especially fuel, to the full extent that might have been anticipated. There was no imperative to cut back on consumption of energy or other imports if the greater cost could be covered by borrowing. Only gasoline prices went up substantially, and even they, well below the percentage increase in the world price of oil.[13] Thus Brazil partially insulated itself against imported inflation, and avoided still greater acceleration from the recorded 1973 levels.

Second, abundant and relatively cheap imports helped to sustain high rates of fixed investment by providing access to needed equipment and intermediate inputs. Although the rate of industrial growth declined throughout the calendar year, the principal sectors affected were automobiles and other consumer goods. The producer goods sectors retained their dynamism, averting more serious deceleration in aggregate industrial growth, and assuring a basis for subsequent resumption of rapid expansion.

Third, foreign saving resolved the dilemma of inadequate finance for Brazilian high growth rates now that excess capacity had been used up. Ambitious growth targets could be compatible with continuing increases in consumption, and they did not require large rises in domestic saving. On the contrary, thought could be given to correcting the lagging consumption standards of the poor by implementing a more liberal wage policy.

A full commitment to external debt as a way to facilitate balance of payments adjustment and finance growth progressively became the basis of the Geisel development strategy. The Second National Development Plan, approved in December 1974, actually took a somewhat different initial stance. Its strategy of rapid growth, led by import substitution in the intermediate and capital goods sectors, but with due attention to exports, only partially addressed the oil crisis and implicitly underestimated its magnitude. The Plan was primarily one of accommodation to a new stage of industrial development that had been prepared independently of the new international environment. Greater attention to domestic sources of energy was readily added without doing violence to the initial framework. Dependence on external finance was not so easily cobbled on. The balance of payments received as much attention as the policy of price control, and it was concluded only that debt was to be held to prudent limits.[14]

The new style of industrial expansion continued to be advertised as favoring the Brazilian private sector in deliberate preference to foreign enterprises. It would strengthen private capacity to participate in large-scale ventures, and where that was impossible, the Plan envisaged State enterprises as the national counterpart. Such discrimination in the basic sectors conformed to the national security doctrine of the military and also built upon a strong and vocal national presence in the production of capital goods. Furthermore, the Plan placed an increased emphasis upon autonomous technological capacity appropriate to an emerging power.

The strategy had clear roots in a Brazilian structuralism that the accomplishments of the miracle had partially obscured, but had far from eradicated. Those who believed that post-1964 Brazil had been converted to the magic of the market or to outward-oriented growth were quite mistaken. The origins of the miracle, after all, had precisely grown out of commitment to expansive rather than restrictive monetary policy despite continuing inflation, and had derived its success from increasing internal demand, led by the public sector, to utilize excess capacity. For all the increase that exports had achieved after 1968, and they had grown faster than world trade, subsidies (and high international prices for primary products) were an important contributing factor. A new emphasis on the internal market would also provide an opportunity to convert the critics of the unequalizing income distribution consequences of Brazilian expansion in the 1960s.

As adapted to the new energy and international economic environment, it became a plan for all seasons. Later, in the hands of Planning Minister Reis Velloso, it was even to take on a guise totally absent from the original: a "strategy of progressive deceleration" in contrast to the option of recession as response to the less favorable external context. Contrast that with the pledge "to grow markedly, in the next five years, at rates comparable to those of the last years."[15]

What counted was not the Plan but the policies of public investment and import substitution it inspired. It is useful to contrast its heterodox emphasis upon economic growth and State controls with a more conventional policy of market adjustment to the rise in oil prices. An orthodox response would have placed principal weight upon realignment of relative prices and reduction of real incomes in conformity with the deterioration of the external terms of trade. That implied higher domestic prices of energy, comparable to the international increase, in order to signal direct, and indirect, substitution in use. In the second instance, standard adjustment required real devaluation to encourage production of exportables and import substitutes, and to reduce domestic absorption by reducing real income and curtailing expenditure. Some domestic inflation was inevitable with the higher tradable prices caused by devaluation, but restrictive monetary and fiscal policy could contain it to tolerable levels by curbing aggregate demand and promoting domestic saving. Thirdly, temporary adjustment lags could be eased by relying on external finance to tide over the balance of payments until the realignment of domestic production had been successfully accomplished.

The structuralist economic perspective then dominant in Brazil took issue with such a program in five important respects. For one, reliance on devaluation presumed significant price responsiveness of imports and exports. Many Brazilian policy-makers had limited confidence in the absorptive capacity of the world market, particularly during international recession: after all, imports during the miracle had grown even more rapidly than exports, financed by newly available Euro-dollar loans. In the second place, structuralists saw domestic elasticities of substitution as quite low, with relative price changes having limited impact upon input mix. Oil demand was widely regarded as wholly inelastic. Price changes would lead to higher costs and not reallocation of production. Third, indexing would convert initial price changes into generalized inflation through widely ramified nominal responses of wages and other inputs. Fourth, a restrictive monetary policy to dampen excess demand would curb industrial production by rationing working capital and thereby prejudice rather than facilitate the reallocation of domestic productive capacity. Finally, the higher interest rates produced by tighter credit would not induce more savings but they could discourage real investment, once again complicating medium-term adjustment.

In sum, technology, behavioral responses, and Brazilian institutional arrangements rendered the market route one of dubious validity. More direct intervention, and more securely tied to the import than the export side of the balance of payments, was a natural preference. Policy relied on governmental incentives rather than prices alone, and indeed sometimes at the expense of prices. Such a course suited the activist inclinations of the new Geisel administration, and responded to a reasonably accurate portrayal of objective Brazilian economic conditions. (The Appendix to this essay sets out the technical rationale for rejecting a more conventional approach.)

Yet the conceptual framework underlying the heterodox Brazilian response to the oil crisis was flawed in two important dimensions. Ironically, one weakness was the apparent coherence in resolving the short-term balance of payments problem at the same time that Brazil was responding to longer-term requirements to alter its style of development and to deepen its industrial structure. In so doing, the Geisel administration erred in its excessive weight upon import substitution as the source of relief from the foreign exchange constraint. There was a second contradiction in the strategy. It foresaw both a strong public sector and constructive relations with the private national sector. The reality was to be otherwise. Public sector expansion entailed increasing deficit finance and came to rely on external resources. The State became larger but also economically weaker. At the same time, it impinged upon private terrain, requiring further subsidies (and incurring larger deficits) in order to mollify objections of national entrepreneurs.

Import substitution was too import-intensive to work in the short run as an effective policy to ameliorate the balance of payments. As formulated in the Development Plan, moreover, with very large-scale projects requiring significant advance investment, this generic feature of the import substitution approach was further exaggerated. Table 2 presents the import coefficients of some of the priority intermediate sectors as well as of aggregate indexes for capital goods, petroleum, and total imports. As can be seen, there were significant declines in the ratios of imports to domestic production in the case of some intermediate goods, but not uniformly. For aluminum, fertilizers and the petrochemicals, where demand was growing, absolute imports also continued to grow. Apparent successful substitution in iron and steel comes only after very large import increases in 1974 and 1975.

In capital goods the story was different. No substitution at all occurs until 1977, and a return to higher participation of imports follows in the next year. In addition the volume of capital goods imports rose sharply after 1973 with the acceleration in investment. While sectoral indices show a declining role of imports, in the aggregate it is not until 1977 that the import ratio (in quantity terms) falls below its 1973 value. And for petroleum, it is apparent that neither import substitution nor conservation had occurred: it took just about as many barrels of imported oil per unit of output in 1979 as 1973.

Table 2. Imports Relative to Domestic Production

	1973	*1974*	*1975*	*1976*	*1977*	*1978*	*1979*	*1980*	*1981*
Sectoral:									
Intermediate Products									
Paper	.22	.25	.12	.13	.13	.10	.11	.08	.08
Cellulose	.16	.20	.10	.05	.05	.04	.03	.02	.01
High-density Polyethelene	.76	.99	.34	.72	.38	.45	.15	.03	.02
PVC	.13	.63	.21	.45	.33	.35	.47	.08	.03
Rolled Steel	.25	.63	.33	.15	.09	.06	.03	.03	.05
Fertilizers[a] (NPK)	2.68	1.98	1.86	1.34	1.48	1.30	1.34	1.17	.85
Aluminum	.58	1.05	.68	.58	.62	.45	.37	.26	.14
Capital Goods Made to Order	.66	.64	.65	.64	.46	.55	.37	.49	.40
Total[b]	100	123	111	100	88	88	90	84	74
Petroleum[b]	100	93	93	94	88	93	97	78	77
Capital Goods[b]	100	125	144	98	70	67	64	65	57

[a]Imports of inputs for national production excluded

[b]Indexes of quantity of imports divided by index of gross product; 1973-100.

Sources: Sectoral: Calculated from Brazilian producers' estimates of quantities produced and imported as reported in *Exame*, May, 1983. Total: Import quantity indexes and gross product from *Conjuntura Econômica*.

Import substitution between 1974 and 1979 contributed only 10 percent of the total demand increase for Brazilian industry; export expansion was comparable in importance. Only in metallurgy and machinery did import substitution amount to more than 20 percent. Relative to the previous period, in which liberalization had led to an increase in import participation, and hence a negative contribution, there was a significant turnaround. But the limited degree of openness of the economy precluded large generalized gains from import substitution even from a 1974 base.[16]

Import substitution can only work in the short run to alleviate balance of payments requirements when there is significant excess capacity to be exploited. Brazilian reality was otherwise. Brazil entered 1974 with the highest level of capacity utilization in the entire postwar period. In addition the Plan called for moving into entirely industrial sectors, and not only expanding domestic share in traditional ones. Import substitution could only be pursued at the initial expense of larger import needs and hence greater external vulnerability.

Seen from the vantage of longer-term strategy, the policy also had its limits. Specific projects were not vetted by careful analysis of benefits and costs, and certainly not by market forces. A simple appeal to import savings seemed to suffice. There were no calculations of rates of return on the massive investments involved in the program, only absolute dollars saved.[17] Extreme examples of the consequences of such an optic are the Alcohol and Nuclear programs, rationalized for their abilities to stanch the vast flow of resources required to import oil, as well as their contributions to national technology. Even in its first, more limited anhydrous phase, and with high oil prices, the social returns from gasahol were negative.[18] From almost its inception, the Nuclear program was a highly questionable venture, explicable by pretentious Brazilian claims as an emerging power. There were also expensive infrastructure undertakings drawn from an extravagant wish list, like the *Ferrovia do Aço*, later to be abandoned.

This industrialization strategy, nominally in the name of an increased role for the private sector, wound up with a larger public role than had been anticipated. The State had to channel resources to private firms, as it did through rapid enlargement of the National Development Bank. It had to subsidize to stimulate priority investment. But because of the scale of the projects and the very reluctance to offer credit subsidies when private equity participation was limited, the government frequently opted for more direct intervention. State enterprises became the standard form of productive engagement in the new sectors, frequently in association with foreign capital. Investment of the largest State enterprises grew from an average of 4 percent of gross product in the period 1970–73 to 5.4 percent in the period 1974–78. Equally significantly, it went from 17 percent of total investment to 23 percent.[19]

This more prominent role helped provoke a debate over *estatização* that raged between 1975 and 1977. Increased centralization of authority, and rising dependence of private sector fortunes upon public decisions over which entrepreneurs had minimal control, became distasteful. That the "miracle" years were evidently past made criticism come easier now, even when the trends had been long-standing. *Visão*'s publisher, Henry Maksoud, was a relentless opponent of the Development Plan's justification of governmental initiative when there were "empty spaces" that had to be occupied. Minister Reis Velloso was forced to defend his policies at length, to the point of apparent exasperation. New limitations were imposed on State enterprises, and new promises of restraint were exacted.[20]

An increasing public role was implicit in the adjustment strategy Brazil had embarked upon. It was not a simple matter of choice, or altered legislation. So long as the objective was large-scale productive capacity in the intermediate goods sector, and rapid growth, there would be a balance of payments problem as well as internal disequilibrium. Imports remained necessary, albeit increasingly restricted, while the domestic market absorbed exportables. The State stimulated demand while implementing its ambitious plans, but confronted an inability to finance them. Higher taxes were resisted. On the contrary, the private sector sought increasing transfers to compensate for price controls that were utilized to prevent inflation from surging out of control. Credit subsidies through limited monetary correction became a chosen instrument. Fiscal incentives, including generous allowances to exporters of manufactured products, were another. The Brazilian problem, at root, was a weak, not a strong, State.

Foreign resources increasingly helped to reconcile some of the inconsistencies. Private banks favored loans to State enterprises. The latter became agencies of the State, not merely in undertaking new productive activities, but also in financing the public sector and balance of payments deficits. The new capital markets, with their bias in favor of the public sector, were a welcome savior, especially with low and even negative real rates of interest. Brazilian adjustment became based on growth-led external debt.

The process could not continue with its initial intensity. Soaring growth in 1976—industry expanded at miracle rates of more than 12 percent—and accelerating inflation—48 percent versus the 30 percent of 1975—were clear signals of domestic disequilibrium. A current account deficit of $6 billion had to be financed, of which net interest payments already represented almost a third. The dangers of excessive growth loomed large.

At the same time, the campaign against *estatização* took on political significance. An important base of governmental support was weakened. More than that, an influential, conservative body of opinion for the first time openly contested the claim of the authoritarian regime to be the promoter of the national interest. The further repercussions would be seen in 1978, when some prominent São Paulo entrepreneurs went further in declaring themselves in favor of return to civilian governance.

Caution was indicated, even for a government still firmly able to manipulate the rules in its behalf. The aggressive economic strategy had to be partially abandoned in favor of greater attention to macro-economic policy and broader support from national industrialists.

The hallmark of the policies followed from mid-1976 to the end of the Geisel government was mild restraint. Luiz Bresser Pereira captured the spirit of the period well: "Though the monetary authorities adopted a basically neoclassic and monetarist discourse in theory, their practice was more moderate, combining basically Keynesian monetary and fiscal instruments of macroeconomic policy with instruments of administrative control such as price controls through the Interministerial Price Council, control of exchange rates through the minidevaluation policy (which had its unsuccessful start in 1967), control over interest and rents through indexation, and wage control."[21]

The most impressive policy achievement was a progressive rise in real market interest rates. Through 1977 and 1978 Treasury bills (LTNs) yielded more than 10 percent in comparison with monetary correction. Borrowers could pay in excess of twice that rate.[22] Such higher rates were central to two objectives. The first aim was to discourage private demand by making financial assets more attractive than expenditure, and by causing borrowing to be increasingly costly. Central Bank resolutions operated directly by allocating credit away from housing and consumer durables. The second goal was to encourage domestic borrowers to contract external loans. This would at the same time help to alleviate interest-inspired inflationary pressures while closing the balance of payments deficit through capital inflow.

Yet higher interest rates did not fully signify more effective macroeconomic policy. Monetary expansion continued to exceed the targets set in the Monetary Budget. Capital inflows financed not only current account deficits but also reserve acquisition. The latter spilled over to increases in the monetary base and commercial bank deposits. Even when the borrowing was sterilized by issue of internal government debt, its near money status meant large increases in real liquidity, as the series on M4 for 1977 and 1978 in Table 1 testifies.

And higher rates contributed to increasing segmentation of capital markets. Demands for subsidized credit mounted. Although the favorable 1975–76 BNDES treatment of priority industrial sectors was cut back in 1977, agriculture and exporters benefited. It is estimated that total subsidies amounted to more than 5 percent of gross domestic product in 1977 and 1978. These were financed not by explicit fiscal transfers but by Central Bank credit. In other words, positive real interest rates became a destabilizing, expansionary force on the monetary base even before there was a large internal debt.

The indirect financing of these subsidies was a measure of the increasing fiscal constraint on the Brazilian State. They could no longer be met by explicit transfers as they had been earlier. Meanwhile, there was a mounting need to underwrite public sector firms. Efforts to curb State enterprise expenditures were not wholly effective. In addition, administrative controls on prices of these enterprises, an instrument to curb inflation, showed up as still larger deficits. Shortfalls were covered by foreign borrowing, especially in 1978. And indexed interest payments on internal public debt were a rising, if still modest, share of total obligations.

Restraint, even if imperfect, yielded some results. Growth slowed in 1977 in response to the deflationary policies applied. Larger exports, to which higher coffee prices contributed, helped to yield a bare, but welcome, trade surplus in the same year. On the domestic front, inflation not only stabilized during the year, but decelerated to a 35 percent increase in wholesale prices. The policy of moderation had managed to keep the Brazilian economy from surging out of control. In the easing that inevitably followed, as complaints from the private sector mounted, growth again resumed, with some modest upward pressure on industrial prices in 1978.

Stop-go was an inelegant and ultimately ineffective solution to the problem of incomplete adjustment. Its limits became apparent in 1978, when inflation returned to its 1976 level without pressure from excessive economic growth. Conventional restraint was powerless against adverse agricultural harvests and rising food prices that were replicated in other sectors. The Brazilian system of indexing was now operating to the disadvantage of policymakers. It had helped in promoting deceleration of inflation so long as it had been manipulated to reduce real wages and when supply side shocks exerted a positive force. Now, as civil society was increasingly voicing its demands, real wages could no longer be determined as the residual income share.

Instead, controls proliferated at the expense of priorities. As the absolute level of subsidies increased, sectors and interests sought to defend entrenched positions. Pure market considerations were ever further from being decisive in the allocation of resources. Relative price changes had to be minimized to avoid new inflationary impulses.

The external situation was not much better. Exchange rate policy was constrained to accompany relative inflation, and not much more. Larger devaluations were ruled out by the expectation that they would be soon passed along in prices and wages. Between 1974 and 1977, the real exchange rate moved within a narrow band. Its limited devaluation in 1978 derives primarily from the dollar's decline relative to other currencies. Subsidies to manufactured products, excluding indirect tax exemptions permitted under GATT, did rise from about 20 percent to 40 percent of their value over this period; their effect, however, was partially offset by downward pressures on international prices of Brazilian exports.[23]

There was thus no mechanism to assure increased incentives to penetrate external markets. While exports continued to grow in the 1974–78 period, as they had done earlier, and their composition diversified to include more manufactured goods, a decomposition of the sources of growth shows a marked difference relative to the preceding period. In 1971–74, increased Brazilian competitiveness accounted for close to half of the rise in exports; in 1974–78, less than 20 percent. Indeed, Brazilian participation in world trade remained approximately constant. For manufactured products exclusively, the conclusion is again telling: 71 percent explained by competitive improvement in 1971–74, 43 percent in 1974–78.[24]

A vigorous export expansion was a necessary component of the adjustment strategy. Import restrictions and import substitution were limited in the foreign exchange they could make available to meet the increasing debt service on the accumulated debt. These results were indicative of difficulties ahead. Debt increases on the order of 28 percent a year, the average between 1973 and 1978, had led to ever larger debt-export ratios and postponement of the required balance of payments realignment. At the same time the real resource contribution of the debt worked in the opposite direction. Ever larger return interest payments diminished the surplus of imports over exports available for national use. Domestic savings would have to rise to compensate and to sustain the high rates of investment—about 25 percent of product—that external saving had helped finance. The debt had been productively applied, not entirely in investment to be sure, but without the diversion of large capital flight that occurred elsewhere. Still, debt is a mechanism for transferring adjustment over time, not for avoiding it. Quite apart from later rises in interest rates and industrialized country recession, there was a potential debt problem in Brazil's future.

Brazil during these years was clearly becoming more vulnerable, as its integration into the world economy was progressively more asymmetric: its share of debt far exceeded its share of trade. Were export growth not to manage to keep up, or the rigid control over imports to be breached, or the larger debt to give rise to much more costly interest remittances, or supply conditions to prove less favorable, the potential balance of payments constraint that overhung like a sword of Damocles could easily become operative.

Table 3 makes apparent the nature of the adjustment during the Geisel years. Although the current account was continuously reduced from its 1974 level of $7.1 billion to $4 billion in 1977, the net debt had quadrupled from its value of $6.2 billion at the end of 1973. Oil imports took an impressive and increasing toll as Brazil increased its volume of purchases. Poorer world economic performance handicapped exports relative to their earlier pace. One of the important reasons for Brazil's ability to show progress in rectifying its balance of payments was the rise in coffee prices that in 1977 offset two-thirds of the adverse oil price effect. Indeed, overall, Brazilian terms of trade in 1977 were more favorable than they had been in 1973.

Had Brazil taken other steps to adjust, it might have taken more advantage of this favorable tendency. Table 3 shows the effect of three other policy responses. Export promotion through a 10 percent rise in price of non-coffee exports would have made a modest, but increasing, contribution. Limits on imports, to restrain income elasticity to unity, would have had a more significant impact, particularly between 1974 and 1976, when they could have reduced reliance on debt finance by 40 percent. Slower economic growth, at a constant 5 percent rate, has a smaller effect because it is coupled with the actual large import elasticity in 1974; in this sense, there was justification to the prescription of orthodox recession as a regulator of

Table 3. The Balance of Payments and the First Oil Shock (billion dollars)

	1973	*1974*	*1975*	*1976*	*1977*
Trade Balance	0.0	−4.7	−3.5	−2.3	0.1
Net Interest	−.5	−.7	−1.5	−1.8	−2.1
Current Account	−1.7	−7.1	−6.8	−6.1	−4.0
External Effects					
Oil Prices		−2.2	−2.3	−2.9	−3.1
Recession Export Volume		−.6	−1.5	−1.4	−1.9
Coffee Price		−.1	.1	1.4	2.0
Policy Responses					
Export Promotion		.4	.5	.5	.6
Import Limits		2.4	1.2	0	−1.7
Slower Growth		1.6	.8	−.3	−2.0
Actual Net Debt	6.2	11.9	17.2	19.5	24.7
Policy Adjusted Net Debt					
Export Promotion		11.5	16.3	17.4	21.8
Import Limits		9.5	13.3	14.1	20.7
Slower Growth		10.3	14.5	16.3	23.2

Sources: Actual Balance of Payments data: Table 1
External Effects
Oil Price: Fixed at 1973 nominal value
Export Volume: Deviation in estimated non-coffee export volume caused by slower OECD growth relative to 1968–73 average. Deviation calculated from regression for period 1969–1982 relating Brazilian export growth to OECD growth rate, yielding an elasticity of 2.
Coffee Price: Fixed at 1973 nominal value
Policy Responses
Export Promotion: consequences for non-coffee export growth of a 10 percent real price increase, using a medium-term price elasticity of .6. The latter is a composite of the medium-term value of about .75 for manufactured products, and .5 for nonn-manufactures. See the elasticities in IPEA, *Perspectivas de longo prazo da economia brasileira* (Rio, 1985), ch. 4 and Alberto Roque Musalem, "Subsidy policies and the export of manufactured goods in Brazil," *Brazilian Economic Studies*, vol. 8 (1984), pp. 169 ff.
Import Limits: consequences of a unitary real import elasticity, post-1973, with actual growth rates.
Slower Growth: consequences of 5 percent annual growth 1974–77, using actual 1974 import elasticity in 1974, but unity thereafter.

the balance of payments. Note that both of these last two policies yield higher imports in 1977 than were actually realized. The actual restrictive policies put in place had finally begun to take effect.

These calculations make clear that a steady, and early, combination of efforts, rather than a single grand alternative, was necessary to counter the more hostile external environment. Instead, Brazil had been late to react to a flood of imports in 1974 and had then gambled on an ambitious plan of import substitution. Muddling through could keep the resulting disequilibrium within bounds; it did nothing to correct it. But the policy was effective enough to assure a successful transition to Geisel's chosen successor, João Figueiredo, the chief of the security agency. Geisel successfully outmaneuvered both the rising opposition from the political left and the military right to assure continuity in the process of political *abertura*. So long as disaster was averted, as it was, and the economy continued to grow, as it did, primary attention could be directed to the delicate task of commanding the pace of popular participation. The new administration was left to unravel the economic adjustment problem.

The occasion of the presidential succession in March 1979 defined a new approach to economic policy. That was the style of Brazilian authoritarianism: in the fifteen years after 1964, no finance or planning minister had ever been removed during his term; significant changes in direction were reserved to new administrations. The decision not only to retain Simonsen in the cabinet—Velloso left—but also to upgrade his responsibilities as super economics minister signaled the new policy style. It went beyond validation of another episode of stop-go to slow down accelerating inflation and to protect the balance of payments. Rather, Figueiredo acknowledged the need for a major overhaul of the Brazilian economy to bring it once more under firm control.

The guiding principles were restoration of the role of market forces and enhancement of the effectiveness of policy instruments. In the first instance, Simonsen sought to reduce the destabilizing subsidies on credit: as inflation accelerated, so did the difference between fixed interest charges and the real cost of resources, with the monetary authority picking up the bill. As one element of the reform he proposed explicit fiscal transfers to cover the subsidies rather than their implicit finance through monetary expansion. That would restore the possibility of monetary restraint. In the second place, he sought to exert greater control over public expenditures, including those of the State enterprises. Despite repeated efforts, the latter had escaped the firm discipline of the central government, in part because of their access to external resources. Third, Simonsen accepted a gradual reduction in export subsidies, which had been a source of increasing conflict with the United States, and compensated by an acceleration in mini-devaluations. This shifted more of the burden of export promotion to the exchange rate rather than relying on off-budget tax exemptions and credit subsidies. At the same time, import restrictions were to be modestly liberalized. Finally, Simonsen made clear that slower growth might very well be a temporary price for reordering the Brazilian economy.

This reformist vision encountered opposition on all sides. Private industrialist critics challenged the utility of a looming recession, when their profits were already under some pressure. Workers were already experiencing erosion of real wages from accelerating inflation because indexing was only annual, and saw no attention to their concerns. Private banks did not welcome a Bank of Brazil in direct competition for prime clients rather than as a source of Central Bank subsidized credit to priority sectors. Within the government the new ministers were eager to spend, not to reduce their expenditures and their political power. Poor agricultural harvests reinforced the case for abundant, and subsidized, credit to that sector. The alcohol program was in need of new investment for a second, hydrated phase based on alcohol-powered cars. State enterprises equally resisted controls over their operations. Petrobras, in particular, wished to increase substantially its expenditures for exploration.

President Figueiredo, beset by the rise in OPEC prices in June and the need for a "War Economy," soon yielded to the chorus of complaint. Unlike the Mexican political cycle, which is noteworthy for more restrictive policy at the beginning of the presidential term than at the end, the Brazilian pattern since 1967 had been the opposite. A new administration preferred to show immediately its capacity to manage rapid economic growth and thereby legitimize its mandate. Simonsen's approach seemed to be yielding only stagflation. In a matter of months, in August, Simonsen was dismissed in favor of Antônio Delfim Netto, already in the cabinet as Agriculture Minister, and one of the principal spenders. Delfim promised a supply side approach that would make demand restraint unnecessary. That was sweet music to an administration pledged to validating wider popular political participation.

Delfim's attempt to re-create the earlier miracle soon foundered. But Delfim persisted. In a remarkable reversal, he retreated fully from his heterodox initiatives of 1979 in favor of orthodox austerity in November 1980. After years of insistence that it could not happen, the Brazilian recession finally came. The downturn proved, in part because of the extent of deterioration of the international economy, more durable than had been envisioned. Not until 1984 would Brazil again experience rising per capita incomes. Before then, it would have to accept the unthinkable a second time: after the election of 1982, Brazil entered into a stabilization agreement with the International Monetary Fund—but not entirely gracefully. It took a series of letters of intent before the program took effect, and again during its duration. *Grandeza* was a dream from a distant past.

Delfim, widely acclaimed as the author of the "miracle," was greeted by national euphoria and confidence in his capacity to save the day. In reality, he could have hardly selected a more unpropitious moment to return. The second oil shock had struck. Severe and prolonged recession in the industrialized countries and higher interest rates were shortly to follow. And as we have seen, the Brazilian economy was far from secure in its adjustment to the 1973–74 shock.

Nonetheless. Delfim boldly designed a program to reduce inflation while abetting growth. On the real side, first priority would go to agriculture and energy. The former bore much of the weight, and hopes, of the policy. Rapid growth of agricultural production would diminish the food component of the price index that had been so troublesome in recent years; provide the exports to assure continuing service of the debt; permit energy substitution through the alcohol program; and facilitate more equal income distribution. Energy was a self-evident need, whether for domestic oil production or substitutes like alcohol. Both sectors were assured all the subsidized credit they wanted.

Macro-economic policy started from a theory of cost-push inflation. One of the characteristics of the policies that had been followed since 1976 had been a large increase in real interest rates. Delfim set out to undo their effect through strict controls in September that brought nominal rates sharply down. At the same time, many administered prices were freed during the fall, and a new wage law enacted in November provided semi-annual adjustments as well as relative gains for lower wage workers. While these last actions accelerated inflationary pressures, they were justified as the residual of the previous administration, and brought some reduction of fiscal deficits and labor peace. The increases were prelude to what was a hoped for substantial deceleration that was to occur beginning in 1980.

On the external side, Delfim enacted a maxi-devaluation of 30 percent in December, the first in more than ten years. Export subsidies and prior deposits on imports, no longer necessary after adjustment of the exchange rate, were removed. In addition, alert to the deterioration in the balance of payments, he took new measures to encourage private foreign borrowing to bolster foreign exchange availability.

The final element, and the true novelty in the program, came last. Delfim pre-announced both monetary correction and devaluation during 1980, the former at 45 percent, the latter at 40 percent. Domestic credit was to be limited accordingly. This move was designed to change inflationary expectations: if everyone believed that inflation would be only 45 percent in 1980, after the relative price changes in the fall, then they could be. Heavy doses of controls were to reinforce the message.

The Delfim strategy, as it thus took form, was a mixture of the standard IMF formula of devaluation to stimulate exports and import substitutes; Southern Cone international monetarism, predicated on liberalism and the strict relationship between domestic and international prices, and traditional Brazilian interventionism. Brazilian policy was a source of confusion to the international financial community. At first, when the thrust was toward freeing markets in the Christmas package of December 1979, the bankers applauded; as disequilibrium became rampant in 1980, they rebelled. The bankers had the final word. Their refusal to roll over the debt without a stabilization package eventually led to a more orthodox approach in November 1980.

By that time the results of Delfim's policies were falling short of their ambitious goals, to say the least. Although economic growth proceeded in 1980 at a rate in excess of 7 percent, it was fueled by consumer demand. The investment ratio declined. Financial assets, now yielding much less than the inflation rate, were abandoned in favor of speculative acquisition of physical assets. Meanwhile inflation soared and crossed the three digit threshold for the first time in Brazilian experience. And the current account deficit in 1980, under pressure from further rises in the oil price, attained a record $12.4 billion and required massive finance. The net debt stood at almost $60 billion and three times the level of exports, compared with a 1977 level of little more than two.

Delfim's special blend of heterodoxy failed in 1979–80 for four reasons. For one, it suffered from a very large dose of excess demand. The public sector deficit in 1980, excluding monetary correction, is variously estimated at between 5 and 7 percent of gross domestic product.[25] Although possibly smaller than the 1979 level, the deficit did not find ready finance in the controlled financial markets of 1980, instead transmitting their effects primarily to prices. Unlike the "miracle" years, there was no elastic domestic supply available to confront demand. Capacity was substantially utilized, especially in the sectors growing rapidly.

A second factor was the wage law of November 1979 that conceded semi-annual rather than annual adjustments. Increasing labor unrest, as accelerating inflation eroded real wages, put pressure on the government to devise a new scheme. Delfim, in a bid to secure order, accepted not only more frequent adjustment but a law that redistributed income in favor of those least paid. Their adjustment was to be greater than the inflation index. The apparent higher real wage attendant upon semi-annual correction and special concessions to lower wage workers did not have a substantial independent effect upon accelerating inflation. The World Bank report is unequivocal: "A simple examination of trends in total and per unit labor costs in industry from November 1979 to May 1982 suggests that the formula was not a major contributing factor to inflation."[26] The reason is high turnover at the bottom of the wage hierarchy as well as the lag in the new INPC price index behind general inflation. But what the law did do, because wage correction was based exclusively on past trends, was make impossible a significant deceleration in inflation without leading to a large real wage increase. The much lower target of 45 percent for 1980 was thus doomed from the beginning.

In the third instance, international monetarism could not rely upon a ready supply of imports to discipline domestic price changes as the theory required. Brazil was in the midst of a balance of payments crisis in 1980, despite rapidly rising exports, as a result of higher oil prices and increased interest rates. Imports remained under restraint, as they had for several years. It was an inopportune time to experiment with this new approach—as Argentina and Chile were also to discover.

Finally, this was not the moment to reverse inflationary expectations. Rising import costs, fears of oil shortages, and a demonstrable commitment to expansionary policies all negated the rhetoric of prefixed monetary correction and exchange rate devaluation. Rather, as the disparity between reality and the government targets increased, the relevant question was when the policy would have to change. Expectations, and attendant holding back of exports and financial speculation, centered on that assumption, not the announced inflation objective for the year.

Delfim has been rightly criticized for the errors of this aberrant 1980 policy. Bolivar Lamounier and Alkimar Moura are especially harsh: "The monumental failure of that heterodox experiment of economic policy can, in part, be explained by the attempt to implement a strategy of economic growth, without consideration for the accentuated deterioration in the conditions of the international economy in 1979 and 1980. . . . It cannot be said, however, that there had been a generalized inability, among the government technocrats, to interpret the unequivocal signals of economic difficulty arising from the international economy. The predominant attitude was to try to exorcise such ghosts with the optimistic rhetoric inherited from the years of the Brazilian miracle."[27] Delfim, himself, was later to emphasize the strong export response in 1979-81, and the fact that Brazil in 1981 "had already reestablished equilibrium."[28]

At one level, he was right to ignore the failure of his 1980 heterodoxy. Table 4 presents the balance of payments effects of the second oil shock, the interest rate shock, and the attendant international recession. These show how Brazil was overwhelmed by the adverse turn of the external environment. More than half of the $4 billion deterioration in the current account in 1979 is explained by rising oil prices and interest rates. An additional $1.2 billion derives from interest costs on increased debt accumulated in the single year 1978. Conversely, alternative policy efforts, including holding real imports constant at their 1978 levels, would have made modest inroads in 1979. None taken individually would have been adequate to cancel even the oil price increase.

Table 4. The Balance Payments and the Second Oil Shock (billion dollars)

	1978	*1979*	*1980*	*1981*	*1982*
Trade Balance	−1.0	−2.8	−2.8	1.2	.8
Net Interest	−2.7	−4.2	−6.3	−9.2	−11.4
Current Account	−6.0	−10.0	−12.4	−11.0	−16.3
External Effects					
Oil Price		−1.8	−5.7	−7.1	−6.1
Recession Export Volume		—	−.6	−1.4	−2.4
Interest Rate		−.3	−1.1	−2.5	−5.9
Policy Responses					
Export Promotion		.8	1.0	1.3	1.1
Import Limits		1.6	1.9	−1.2	−2.6
Slower Growth		.6	1.9	−.2	−.6
Actual Net Debt	36.2	46.4	57.7	68.0	83.5
Policy Adjusted Net Debt					
Export Promotion		45.6	55.8	64.5	68.3
Import Limits		44.8	54.0	64.9	82.5
Slower Growth		45.8	55.7	65.9	81.6

Sources: Actual Balance of Payments data: Table 1 External Effects
Oil Price: Fixed at 1978 nominal value
Export Volume: As in Table 2, using deviation from 1974–78 average growth of 3.1 percent
Interest Rate: Using constant 1978 average real (with respect to U.S. GNP deflator) interest rate on net debt of previous year. Net interest payments are in Table 1 and net debt was obtained, inclusive of short-term debt, from Paulo Nogueira Batista, "International Financial Flows to Brazil Since the Late 1960's," mimeo (Rio, May 1985). These net debt values are tabluated below.
Policy Responses
Export Promotion: As in Table 3
Import Limits: Real imports held constant at 1978 levels
Slower Growth: Unitary elasticity effect on imports of product growth at 3 percent.

In 1980, rising oil prices and interest rates made matters considerably worse. The actual worsening of the balance of payments was smaller than would have been anticipated owing to the 58 percent increase in the value of exports since 1978. In the absence of the oil price rise, Brazil would have enjoyed a healthy trade surplus in 1980.

Starting from its much higher debt in 1978, and the continuing large volume of oil imports, Brazil had less flexibility to deal with the second oil shock than the first. Delfim inherited the problem of inadequate adjustment; it was not simply of his own making. That said, however, there was therefore a strong argument for conserving what few degrees of freedom remained. A more cautious policy would have marginally improved balance of payments performance, prevented spreads—and hence interest costs—from rising, and retained domestic credibility. Even if restraint did not avert the next step of declining income, Brazil would have entered such a recession considerably stronger, rather than afflicted by a new overlay of distortions originating from the policies actually followed. In particular, the beneficial effects of the 1979 devaluation were completely wiped out by inadequate continuing correction; by the end of 1980, the exchange rate had appreciated relative to its pre-devaluation level.

Expectations did turn out to be rational. The virtually universal disbelief in the adequacy of the initial heterodox policy was confirmed by its change in November and December of 1980 under increasing pressure from foreign creditors. Unlike some of his compatriots in the Southern Cone, who both believed in their policies and understood that sustaining them when they were not working was the only way to make them work, Delfim was more pragmatic. Yet it is a measure of the strong separation between the tracks of ongoing political liberalization and technocratic economic policy formulation that he yielded to external influence rather than domestic critics.

More orthodox policies of restraint became the order of the day. Monetary expansion was severely limited, provoking a liquidity shortage. Real interest rates rose to levels of 40 to 45 percent. Firms cut back on production and tried to work down their bloated, and increasingly expensive, inventories. Private investment declined. Investments of State enterprises were cut back. These deflationary impulses produced a decline in gross output of 1.6 percent between 1980 and 1981 and a still larger drop in industrial production. Urban unemployment became overt. Brazil entered a period of cumulative falling per capita income that was to be more severe than that of the Great Depression.

The immediate gains were relatively modest. Inflation decelerated from 121 percent in 1980 to 94 percent in 1981. The trade balance moved into modest surplus. The primary effect of the recession was to unloose a new flow of capital from commercial banks, placing Brazil further in debt. Instead of conceding the need for more fundamental changes, and implementing them, Delfim's primary stabilization objective was to retain international creditworthiness and liquidity.

It was, in other words, an unproductive recession, just as the preceding prosperity had been. To avoid going to the IMF, Brazil undertook an even more severe stabilization to persuade international creditors of its sincerity. But in so doing, it lacked a real program.

The balance of payments remained troublesome. The exchange rate, after acceleration of mini-devaluations in the second half of 1981, was not quite back to where it had been in early 1979. Now, with the dollar appreciating, more aggressive offsets were necessary. Export growth, aided by restoration of the subsidy to manufactured products, was respectable in 1981 but not spectacular. Almost as much of the trade improvement was achieved by more rigorous import controls. Meanwhile, much of the new international finance was dangerously short term and obtained through the inter-bank market. During 1982, exchange rate devaluation again fell behind.

On the internal side, tight monetary policy and fiscal restraint were not sustained. The deficit of the consolidated federal public sector actually rose in 1981 from 5 to 6.5 percent of gross product; in 1982 it reached 9.9 percent.[29] While controls over the money supply were apparently effective, suggesting large real declines in liquidity, they conceal an increasing issue of internal debt that was used instead to finance the deficit. In 1981 and 1982, as seen in Table 1, the augmented series of money and quasi-money exceeded inflation. Progressively greater reliance on internal debt, which was to increase from 5 to 15 percent of gross debt between 1980 and 1984, meant rising financial obligations of the federal government since interest rates were much higher than growth of revenues.[30] It also meant fewer policy degrees of freedom as government bonds had to be guaranteed against changes in the exchange rate, and not only domestic inflation.

The hope was that a short, albeit severe, recession would permit Brazil to resume its access to external finance and economic growth. The crucial November election loomed in 1982, and government hopes for controlling the process of *abertura*, and the selection of the next President, turned on a respectable showing. That fact helps to explain why, despite the gathering clouds in financial markets in early 1982, domestic restraint did ease and industrial decline moderated. That motivation was reinforced by the recalcitrance of inflation: it stopped falling in 1982, presenting a much more unfavorable tradeoff than in the previous year.

The government was later to blame the decline in exports in 1982 on spreading international recession, the increase in net interest payments on the high international interest rates, and the closure of financial markets on the Malvinas/Falklands war and Mexican default. As Table 4 shows, Brazil by 1982 was certainly laboring under very adverse external conditions. But the truth of the matter is that the 1981–82 recession was also poorly staged. Brazil waited too late to go to the Fund—formally, until after the election although earlier contacts had been established. By March 1982, net reserves of the Central Bank were already negative. Until virtually the very end, however, the technocrats insisted on their capacities and the fact that Brazil was different from its profligate neighbors. Indeed, before negotiating with the Fund, Brazil put together in October its own spartan plan for presentation to the private banks, a plan calling for minimal finance and exuding confidence: "It is precisely this blending of short- and long-term adjustment which will create the preconditions for the Brazilian economy to find a path of relatively more stable economic growth with smaller imbalances and being threatened neither by growing inflationary pressures nor by the imprevisability of external factors."[31]

It was precisely this blending that had been absent in the previous three years. Policy had been very short-term oriented, and frequently altered. *Ad hoc-ism* was rampant. Solutions were designed for immediate problems, but frequently introduced new distortions that later would inhibit effective policy. The government failed to clear the baggage of credit subsidies and tax incentives inherited from the past, or to establish meaningful priorities. The principal audience and judges were the external creditors. Planning and Finance ministers undertook well orchestrated forays to the exterior to assure and reassure that overly optimistic targets were securely within reach. Meanwhile domestic credibility dissipated. Delfim remained in office because there was not even governmental capacity to define an alternative strategy.

When the Fund program was formalized, it incorporated the limited relief that had been asked of the private banks. Brazil was obligated to achieve a $6 billion trade surplus in 1983, which it did at the expense of another decline in output, this time steeper than that of 1981. Another requirement was a change in wage indexation, limiting correction to 80 percent of the new consumer price index, despite the fact that this measure already had a tendency to lag behind inflation. Real wages in the industrial sector fell significantly during 1983. Contributing was another burst of inflation to the 200 percent level, of which the February 1983 maxi-devaluation of 30 percent was a principal cause.

The Fund Program provoked increasing internal criticism as an inadequate response to Brazil's difficulties. Much of the opposition was directed to the continuation of onerous external interest payments that came to rival the entire import bill in their magnitude. The Fund itself was continuously unhappy with Brazil's performance and lack of compliance with policy targets. Brazil's consequent series of revised letters of intent offset the favorable impact of its excellent trade performance. Creditors never conceded a multi-year rescheduling and reduction of spreads as they had done Mexico. The longer it was delayed, the less sense it made, in view of the government's lame duck status.

Brazil epitomized the limits of the IMF approach. There was impressive improvement of the external accounts. Between 1982 and 1984 the current account deficit of $16 billion was converted into a small surplus. Import reduction, and the rapid growth of exports in 1984, were equally responsible. But the internal stabilization and basis for sound future growth that was also supposed to occur did not. Inflation more than doubled rather than decelerating. High real interest rates, the counterpart of tighter money and large government sales of debt, meanwhile discouraged private investment. Together with controls over public investment, this led to a decline in the ratio of gross capital formation to only 16 percent in 1984, just about its lowest level of the postwar period. The public sector deficit has exceeded its target regularly not merely because of hard to control expenditures and reduced tax receipts, but owing to the rapid growth of internal interest payments.

To critics of the IMF stabilization approach, the stark asymmetry of the results came as no great surprise. Contrary to the IMF's implicit monetarist model, linking external and internal equilibrium, the Brazilian experience confirmed a very different interpretation. Emphasis upon improvement of the external accounts has become an important source of the internal disequilibrium.

The very policies required to permit large trade surpluses and payment of external interest add to inflation and subtract from investment. Thus aggressive devaluation of exchange rates reflects itself sooner or later, and mostly sooner, in domestic inflation because of the ubiquity of indexing. In addition, the public sector must attract ever greater resources from the private sector in order to service the now largely public external debt. To do so on a voluntary basis, interest rates must be kept high and these costs are reflected in prices. In addition, governmental deficits, whether financed by money or internal debt, then replenish nominal demand to sustain the inflation. The State is too weak to accomplish the large transfer needed in a noninflationary way.

The extensive resources that have been transferred externally, amounting to some 5 percent of gross product in 1983 and 1984 and reducing national income proportionally, have come primarily at the expense of investment. Consumption outlays have resisted further compression. Even with the changes in the wage indexing arrangements required by the IMF, there were limits to further declines in standards of living. Saving has not been responsive despite the continuation of high real interest rates; bank certificates of deposit yielded about 25 percent in 1983 and 1984.

These economic circumstances, despite attempts to show that Brazilian performance was better than elsewhere in Latin America, and a modest recovery led by manufactured exports in 1984, contributed to the Figueiredo government's loss of political control. The rules of succession, which had been thought to assure not only continued dominance of the government party, PDS, but also a final military President to guide transition, proved unable to withstand the clear lack of popular acquiescence. It has been left to the New Republic, deprived of Tancredo Neves's leadership, to define a new economic strategy.

Three lessons emerge from this tale of Brazilian adjustment to the oil price shocks of 1973 and 1979. One is the important role of institutional and political limitations in getting policies right. A combination of slower growth, relative price changes, and direct interventions would have been desirable in accommodating to the first shock. Whatever the rigidities of oil consumption in the short term, there was some elasticity in the longer term. Other countries did much better at conservation. Export growth, while rapid, needed still further stimulus in light of the changed international economic environment. Imports had to be reliably available even if more expensive, rather than rigidly controlled and possibly inadequate. Debt was a desirable component of adjustment, but not as a sometimes disorderly residual to the balance of payments. Too much was put off to the future, even disregarding the intensity of the second set of shocks. Financial openness was much greater than trade openness. The asymmetry contributed to a vulnerability that was later to prove costly.

Commitment to rapid growth, and a pervasive system of indexing, biased policy choice toward a massive investment in import substitution sustained by external borrowing. That policy was marred by its own short-term import intensity, not least with respect to oil imports, as well as mounting resistance to expansion of the public sector. Its execution was correspondingly handicapped. Very quickly, macro-economic restraint was required to achieve a tenuous internal and external equilibrium. From that time forward, Brazilian policy increasingly became oriented exclusively to short-term, and external liquidity, problems. Medium-term adjustment suffered. It was steered neither by the market—for controls were increasingly necessary to prevent large imbalances—nor by a feasible government program. Attempts to define a strategy early in the Figueiredo government, successively by Simonsen and Delfim, both failed: the former for lack of support, the latter for its heterodoxy. Thereafter, the only guide, and not entirely coherent, was that imposed by international creditors and the IMF.

The second lesson is the weakness of the Brazilian State in directing the process of adjustment. This goes against the conventional wisdom that sees only the large public participation in the economy, with claims on some 40 percent of resources in the early 1980s according to World Bank calculations.[32] Yet the Brazilian State neither commanded nor effectively cooperated with the private sector. It had no secure control over real resources. Rather the Brazilian public sector experienced progressive fiscal difficulty over this period, diminishing its capacity to guide needed structural adjustment. Credit subsidies and other incentives proliferated, but required an increasing inflationary tax to finance them. Intervention had a diminishing impact; its generalization was a symptom of weakness. State enterprises were a direct instrument, but they foundered on the need to increasingly secure their finances from abroad. Increasingly, parastatals became means of obtaining resources for other activities rather than agents for priority real investment. Weakness was self-fulfilling. Larger and larger efforts were necessary to persuade the private sector in the validity of signals emanating from policymakers. As subsidies multiplied in response to private demands, the *relative* inducements required to undertake the reallocation of resources became blurred, while deficits increased.

The conclusion from the successful East Asian developmental experience in the last two decades is not only that export promotion can yield favorable results in an expanding world economy, but also that strong State intervention can be positive. In the Latin American case, the problem has been that intervention has frequently yielded negative outcomes. This is because the intervention often reflects State weakness rather than strength. State priorities can be diluted and deflected by decisions of private interests who are opposed; when pursued, they become the more costly by reason of the lack of cooperative support.

These conclusions, and the earlier text, stand somewhat at variance with Antonio Barros de Castro's recent provocative reinterpretation of the same peroid.[33] Three of his central theses therefore require brief comment. First is his rehabilitation of the National Development Plan by attributing large foreign exchange savings in 1983 and 1984 to it. Second is his justification of Brazilian state intervention by reference to the international competitiveness of Brazilian industry. Third is his view that future export surpluses, and hence full transfer of interest payments abroad, are a desirable basis for a continuing growth strategy.

Castro's defense of the import substitution adjustment strategy seems to me deficient for three reasons. First, the calculation of gross foreign exchange saving in 1983 and 1984 finesses the question of the investment required to produce them, and the costs in macro-economic disequilibrium thereby provoked. One cannot evaluate a strategy only by reference to its benefits. Second, the calculation itself, attributing *all* changes after 1980 to the earlier planned investment, is as much a caricature as his criticism of conventional techniques that consider only contemporaneous changes in imports. In particular, the largest gains are from national production in petroleum, to which the Geisel government attached less priority than the Figueiredo administration, measured at least by investment in exploration. Third, the insistence upon import substitution in the programmed sectors as the "primordial" source of the surpluses in 1983 and 1984, and secondarily, exports of such sectors, seems to run counter to the evidence. Between 1980 and 1983, and hence even more so until 1984, export growth was a more important source of industrial expansion than import substitution. And export growth from 1982 to 1984 is a greater source of improvement on the merchandise account than non-oil import substitution.[34]

Castro's aversion to neo-liberalism is one I share. Brazilian industry survived the distortion of excessive financial openness because import protection offset the artificial advantage thereby bestowed on imports. Chile, especially, and Argentina were less fortunate. But to acquiesce in intervention of the public sector even when it is misguided and counterproductive is to fail to impose priorities that weak States especially require. Import substitution industrialization worked more effectively in Brazil in the 1950s than elsewhere because the prior Korean War boom had provided an export bonanza. The State was able through its commercial policy indirectly to tax those resources, not otherwise accessible, and redirect them to the expanding industrial sector. In the 1970s the problem was of another kind. There were no traditional exports to tax and discourage; rather all exports had to be encouraged. One way simultaneously to have stimulated production of all tradables, exports and import substitutes alike, would have been through devaluation. Or even excluding that, use of a more balanced set of price incentives. That balance has eluded not only Brazil but other Latin American countries, and understanding why is a central issue of regional political economy. The limits of state intervention, as well as its uses, must be understood.

Castro's advocacy of export surpluses equally seems misplaced. He succumbs to the allure of arithmetic showing its easy, and inevitable, continuation despite higher rates of import growth. That arithmetic misses the internal transfer problem that currently helps constrain sound Brazilian development, and ignores the counterpart application of domestic saving to interest payments rather than investment. More domestic resources can come only at the expense of consumption, and doubtless having an impact on the income distribution. On the other side, export growth, even at modest rates, does not simply occur. New investment and technology are required to maintain continuing competitiveness in a range of products. Imports are essential to both. It is well to note that the East Asian countries did not grow by export surpluses; for the longest time, even while exports were expanding rapidly, they ran deficits and achieved resource inflows despite high levels of domestic saving. Only subsequently have trade surpluses appeared. Brazil is unlikely to prove an exception to this pattern.

For all of these reasons, despite its many insights, I diverge from the central themes of Castro's analysis and have offered another interpretation of this decade of Brazilian development. What remains is a final word on politics and its influence.

Even in Brazil's technocratic and insulated style of decision-making, politics counted. Indeed, at decisive choice points, political objectives helped undermine what might have been more sensible economics. The 1975 decision to abandon restraint, without having integrated it into a more comprehensive response to the oil shock, was clearly motivated by the belief that economic growth was necessary to *distensão*. The 1979 decision to replace Simonsen with Delfim had a similar motivation, in circumstances that were still less appropriate. Finally, the persistent refusal to go to the IMF, even as it had become evident by 1980 that Brazil's external situation was precarious, can be explained by the refusal to admit an inability to cope.

It was, of course, politics of a special kind. Rapid growth was hardly the result of irresponsible populist pressure. Nor, for that matter, was the acceleration of inflation to unheard-of Brazilian levels. Technocrats and the military dominated. Their politics was *their* interpretation of what seemed necessary to maintain society upon *their* chosen path. Indexing, a mechanical formula for adjustment, was their preferred incomes policy, one devoid of participation.

Society's voice mattered much less, and initially not at all. When it began to be heard, its force was critical rather than constructive. The efforts of entrepreneurs were directed against expansion of the State, not in behest of effective cooperation. The Brazilian business community was vocal in its aversion to recession, and supported Delfim's return with enthusiasm. Labor was disinclined to pay again through lower real wages while it remained disenfranchised and marginalized.

There was no coherent opposition economic strategy until quite late, and even then, some aspects had a quixotic flavor. Being out of power, and the prospect for power, for more than two decades is a poor learning experience. More generally, under Brazilian conditions during the Geisel and Figueiredo governments, it was impossible to construct the coalition of interests necessary to an alternative, and more satisfactory, adjustment policy. It would not have been easy to assure a fair distribution of the perceived costs. In the end, in the 1980s, sacrifice was imposed, and highly unequally, rather than shared. Returns to assets increased as wages fell and unemployment grew.

The ultimate irony was that a deteriorating economy hastened a transition to civilian government under possibly the best political circumstances: the selection of a moderate by a fusion of former government supporters and the opposition. The military had failed at what they were supposed to deal with best, reality in the face of adversity. That failure, like that of the civilians whom they had earlier replaced in 1964, was an important asset to their successors. A thriving economy, permitting more orderly retreat, would have left the next government much more beholden to military judgment. Now the threat of renewed intervention is more remote. There is no alternative to real political compromise to resolve the serious problems ahead. This tale of two Presidents offers some lessons on how not to proceed.

APPENDIX

The purpose of this appendix is a more formal comparison of orthodox adjustment with the Brazilian economic strategy adopted in the 1970s. A simple model will suffice. We begin with an orthodox version.

Let Brazil be specialized in a single product, which it consumes and exports. Production requires imported inputs. Further, in the short term, the input proportions of labor and the input are fixed. Then, the domestic price of Brazil's product, P_b, can be written:

$$(1)\ P_b = l_t w + m_t P_m$$

where P_m is the domestic price of the import good and l_t and m_t are the unit requirements of labor and the import good. Each of the prices can be specified in dollars, P^*_b and P^*_m, by converting by the exchange rate expressed as dollars per cruzeiro, and adjusting for domestic subsidies:

(2) $P_b^* = e(1+s_b)P_b$

(3) $P_m^* = e(1+s_m)P_m$

The balance of trade can be written in the short term as:

$$\text{(4a)}\quad T = X\left(\frac{P_b^*}{P_x^*}, Y_w\right)P_b^* - M(Y)P_m^*.$$

Exports depend upon the relative price of the Brazilian product compared with international competitors and upon the level of world production. Imports in the short term are totally price inelastic. In the longer term, there are substitution possibilities between imports and labor so that import requirements depend upon relative wages:

$$\text{(4b)}\quad T = X\left(\frac{P_b^*}{P_x^*}, Y_w\right)P_b^* - M\left(Y, \frac{P_m}{w}\right)P_m^*$$

Domestic output equilibrium can be written:

(5) $Y = E(Y) + T(Y, Y_w, P_x^*, P_b^*, P_m^*)$

Expenditure demands on output and the trade balance depend on domestic output, world income and prices of competitive goods, and world prices of the Brazilian export and import.

This simple system has six endogenous variables: P_b, P_b^*, P_m, T, w, and Y. P_m^*, Y_w, and P_x^* are exogenous. The exchange rate, e, and subsidies, s_b and s_m, are policy determined. Specifying a particular trade balance level adds a sixth equation and completes the model.

Rewriting equation (1) in foreign-exchange terms shows the critical importance of the wage rate measured in foreign exchange:

$$\text{(1)}\quad \frac{P_b^*}{(1+s_b)} = l_t\,(e \cdot w) + \frac{m_t\,P_m^*}{(1+s_m)}$$

Solution of the system can be seen graphically in Chart I, with the dollar wage and output as the axes. The output schedule YY requires a progressively lower real wage, and trade surplus, at higher levels of output to offset increased saving. The schedule corresponding to a zero trade surplus, TT, requires lower wages with increasing output to offset larger imports. Chart I is drawn on the supposition that there is a full employment equilibrium with a zero trade balance.

Now let there be an oil shock. The rise in prices of imports, with real wages constant, increases the real price of Brazilian production and hence reduces their competitiveness. The trade balance as a consequence must deteriorate. This is portrayed in Chart II as a shift downward in the schedules. The higher price of imports requires a lower real wage if Brazil is not only to remain competitive, i.e. maintain P_b^*, but increase its competitiveness to pay for a larger import bill. Since the trade balance enters into

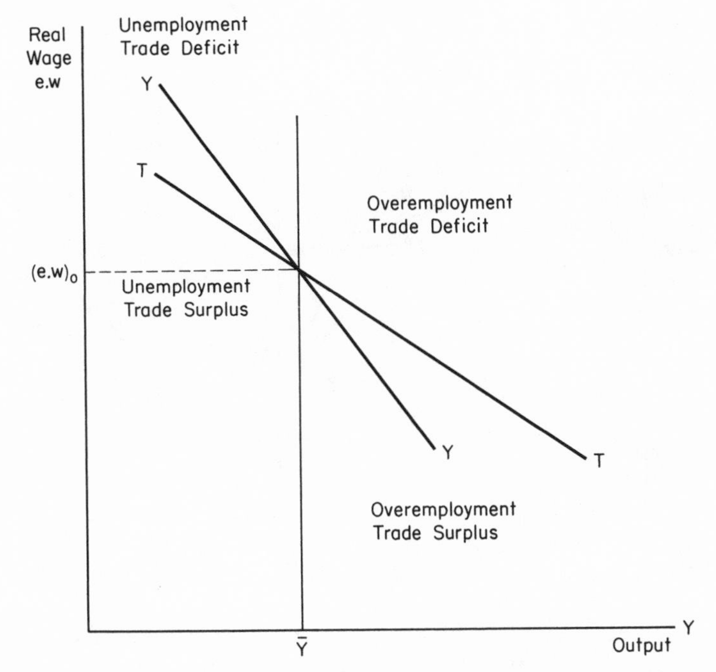

Chart I

YY, it also shifts. The policy implication is clear: devalue. (Or with a fixed exchange rate, reduce wages.) But not necessarily to the full extent of eliminating the trade deficit. For the schedules shift and become more elastic in the long term. Hence, finance is required and justified.

In Chart II we show the initial real wage $(e \cdot w)_0$. The real wage required to satisfy immediate equilibrium is $(e \cdot w)_s$, compared with a smaller decline of $(e \cdot w)_l$ in the long term. Setting the wages at the latter implies a trade deficit. Note that in this model the import surplus made possible by debt finance adds a deflationary impulse that would have to be offset by larger domestic expenditure.

The heart of the matter is the reduction in real wages, assuming resistance to any reduction in profit margins. Here we add in the specific Brazilian circumstance of indexing.[1] This leads to another equation:

(6) $\hat{p} = f(e \cdot w)$

The inflation rate $\hat{p}$ is dependent upon the level of the real wage in dollars. Since indexing is designed to prevent the real wage from falling, by automatic increases to compensate for higher domestic prices, the only way to secure a real wage decline is by accelerating the rate of inflation. With a

[1]For a clear presentation of this relationship, and its relevance to Brazilian adjustment, from which I have borrowed, see Francisco L. Lopes and Eduardo M. Modiano, "Indexação, choque externo e nivel de actividade: notas sobre o caso brasileiro," *Pesquisa e Planejamento Econômico*, vol. 13, no. 1 (1983).

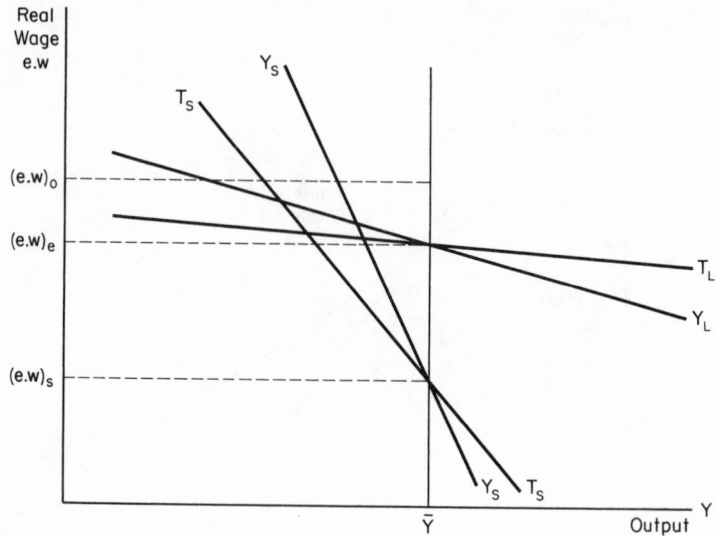

Chart II

fixed indexing period, the average wage is effectively lowered because an initial wage level is eroded more rapidly over the period.[2] This central feature of the Brazilian economy makes inflation even more clearly into the equilibrating variable reconciling inconsistent real claims.

Adding this relationship to the system leads to an augmented graphical presentation in Chart III. Now devaluation increases the inflation rate. Suppose as well that there are no long-range substitution possibilities so that the required devaluation is large. Moving to the required short *and* long term real wage $(e \cdot w)_s$ implies an acceleration of inflation from $\hat{p}_o$ to $\hat{p}_s$. Note that the impact of devaluation upon inflation is greater the larger the rise in import prices that must be offset and the larger the share of import costs in total costs.

The appeal of a debt strategy becomes clearer. Moreover, by subsidizing domestic prices of oil, domestic prices of export products can be kept down without resort to lowering the real wage. Finally, export subsidies can also be used to keep products competitive. Both were in fact extensively used. The critical problem in both cases was the lack of fiscal capacity to finance the subsidies. If the foreign debt were all absorbed by the public sector, and were allocated for this purpose, that would take care of the macroeconomics. But if the foreign debt is also used for massive investment required to

[2]A numerical example may help. Suppose an initial wage of 100 and a linear inflation rate of 10% over a year. Then the final wage is 90 and the average real wage, 95. With an increase of inflation to 20%, the average real wage is now 90. If the nominal wage is now reset to the earlier real peak value of 100, the reduction becomes permanent. An attempt to re-establish the average real wage will lead to further acceleration: under perfect indexing, admitting *no* declines in real wages, inflation tends to explode and nominal devaluations are eroded.

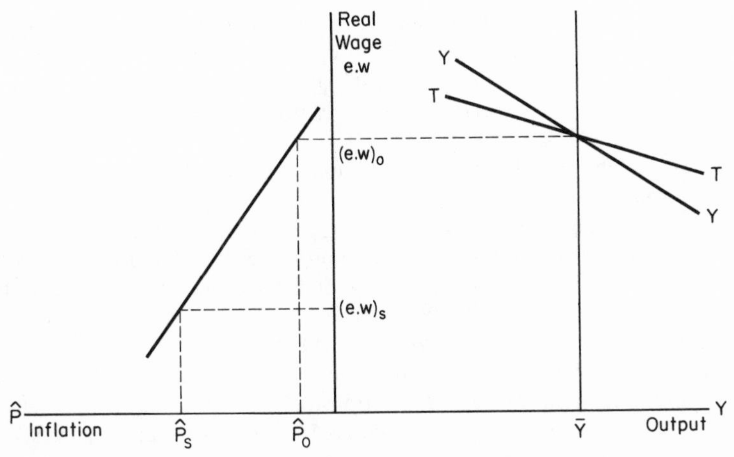

Chart III

reduce the import coefficient, as it was, a fiscal constraint is confronted that limits even the short-term sustainability of a subsidy strategy incompatible with international price ratios.

NOTES

1. Ernane Galvêas, *A Crise Mundial e A Estratégia Brasileira de Ajustamento do Balanço de Pagamentos*, Exposição no Senado Federal, March 23, 1983, pp. 7 ff. My translation.

2. For the arguments in favor of outward-orientation, see, for example, Anne O. Krueger, "Import Substitution versus Export Promotion," *Finance and Development*, vol. 22, no.2(1985). For an analysis in the context of the oil shocks, see Bela Balassa, "The Newly Industrializing Developing Countries after the Oil Crisis," *Weltwirtshaftliches Archiv*, vol 117, no.1(1981) and "Adjustment Policies in Developing Countries: A Reassessment," *World Development*, vol.12, no.9(1984). Jeffrey Sachs has recently compared the performance of the Latin American and Asian countries, attributing differences to the exchange rate policies followed, in "External Debt and Macroeconomic Performance in Latin America and East Asia," *Brookings Papers on Economic Activity*, no.2(1985).

3. Alfred Stepan (ed.), *Authoritarian Brazil* (New Haven, 1973), 70.

4. *Ibid.*, 108.

5. *Ibid.*

6. With an incremental capital-output ratio of 2.67, 10% growth implies a saving rate of 27%; domestic saving had been between 22 and 23% in 1970–72. See World Bank, *Brazil: Economic Memorandum* (Washington, 1984), 251–52.

7. The trade price and quantity indices are those of the *Conjuntura Econômica*.

8. República Federativa do Brasil, *II Plano Nacional de Desenvolvimento* (*1975–1979*) (Rio, 1974), 29. My translation. See also, Minister João Paulo dos Reis Velloso, *Brasil: A Solução Positiva* (Sao Paulo, 1977): "It was concluded that the 'recession strategy' . . . would be *inconvenient* for its effects . . . Inconvenient for a thousand reasons, economic, social and political: mass unemployment, decline in the

standard of living of the workers, rupture of the development process, traumatization of the entrepreneurial structure in formation. And that in a country that is undertaking a necessary political *distensão*" (p. 115). My translation.

9. William Cline, et al., *World Inflation and the Developing Countries* (Washington, D.C., 1981), 134.

10. Mário Henrique Simonsen, "Inflation and Anti-Inflationary Policies in Brazil," *Brazilian Economic Studies*, no.8(1984), p. 7.

11. These monthly data can be found in the *Conjuntura Econômica*.

12. See the results of simulating actual policy in Simonsen, "Inflation," 26.

13. For a discussion of relative prices, see Eduardo Modiano, "Choques Externos e Preços Internos: Dificuldades da Política de Ajuste," in Pérsio Arida (ed.), *Dívida Externa, Recessaõ e Ajuste Estrutural* (Rio, 1982).

14. Brasil, *II PND*, page 129. The position of Carlos Langoni, however, that the strategy of finance implied a belief in the transitory character of the shock, does not follow (*A Crise do Desenvolvimento* (Rio, 1985). Rather, external finance became a source for carrying out the structural realignments that had earlier been deemed necessary. Carlos Lessa, "A Estrategia de Desenvolvimento 1974–1976," unpublished thesis, Faculty of Economics and Administration, UFRJ (Rio, 1978), is right in stressing the pre-oil shock basis of the industrialization strategy.

15. Velloso, *Brasil*, 119. Cf. President Geisel's introduction to the Plan, bound with it.

16. World Bank, *Brasil: Industrial Policies and Manufactured Exports* (Washington, D.C., 1983), 39.

17. Velloso, *Brasil*, 119 for estimates of the foreign exchange saving already achieved.

18. See, not only for the cost-benefit analysis but for an insightful essay in political economy, Michael Barzelay, *The Politicized Market Economy* (Berkeley, 1986).

19. These calculations are reported in Thomas Trebatt, *Brazil's State-Owned Enterprises* (Cambridge, 1983), 130.

20. For some flavor of the argumentation, and a *Visão* interview with Velloso, see its special number of April 19, 1976. To some extent, Velloso's book was a lengthier defense of his views.

21. Luiz Bresser Pereira, *Development and Crisis in Brazil, 1930–1983* (Boulder, 1984), 176.

22. World Bank, *Brazil: An Economic Memorandum*, 299.

23. For the calculations of export subsidies, see Alberto Roque Musalem, "Subsidy Policies and the Export of Manufactured Goods in Brazil," *Brazilian Economic Studies*, no. 8(1984). For trade-weighted exchange rates, and the role of changes in the dollar, see IPEA/INPES, "Perspectivas de longo prazo da economia brasileira" (Rio, 1985), Table 12.1.

24. These decompositions are performed by Maria Helena T. T. Horta, "Sources of Brazilian Export Growth in the 70's," *Economic Studies*, no. 9(1985), pp. 164–65.

25. Brazil National Monetary Council, *Foreign Sector Programme in 1983* (Brasilia, 1982), 8, charts a deficit of almost 7% in 1980 (and even higher in 1979) to show a downward trend. That of the World Bank, *Brazil: Economic Memorandum*, 75, goes the other way.

26. World Bank, *Brazil: Economic Memorandum*, 108. While there is some debate over the transitional effect of the new wage law, evidence suggests that the largest firms had already granted wage adjustments before it came into effect.

27. Bolivar Lamounier and Alkimar Moura, "Política Econômica e Abertura Política no Brasil, 1973–83," mimeo for Vanderbilt Conference, November 1983, p. 27. My translation.

28. A. Delfim Netto, *1973/1983: Dez Anos de Crise e, Apesar de Tudo, Crescimento* (Brasilia, 1983), Exposition in the Federal Senate on May 17, 1983, p. 18.

29. World Bank, *Brazil: Economic Memorandum*, 75.

30. Shares of internal debt calculated from Central Bank of Brazil, *Brazil Economic Program: Internal and External Adjustment*, vol. 7(1985), p. 28. Average, rather than end of year, ratios were obtained, differing from the published ratios.

31. Brazil National Monetary Council, *Programme*, 10.

32. World Bank, *Brazil: Economic Memorandum*, 31.

33. Antônio Barros de Castro and Francisco Eduardo Pires de Souza, *A Economia Brasileira em Marcha Forçada* (Rio, 1985). I do share Castro's view that Carlos Lessa, "A Estratégia," exaggerates the abandonment of the import substitution policies. In part, that is because Lessa's principal interest is the Plan itself.

34. For data on investment in exploration, see IPEA/INPES, "Perspectivas," Table 10.2. Note also Velloso, *Brasil*: "A strategy was sought, therefore, that, at bottom, has the same logic of substitution of imports of petroleum, because it refers principally to raw materials" (p. 116, my translation). He goes on to demonstrate how such self-sufficiency, even with larger imports of petroleum, yields a better balance of payments result.

The decomposition of manufacturing growth is found in "Perspectivas," Table 8.4.b.

4

Brazil's Debt: From the Miracle to the Fund

EDMAR L. BACHA
and PEDRO S. MALAN

This essay provides an interpretation for the rise and fall of Brazilian foreign borrowing in the international credit market from 1968 to 1982.* This period represents a unique historical phase that concluded in 1983 with the agreements reached with the commercial banks and the International Monetary Fund.

This essay starts with an overview of major trends in the world economy from the late 1960s to the early 1980s with special reference to the re-emergence of private international credit markets, and the associated and very rapid accumulation of external debt by developing countries. The reason for this introduction is that Brazil's rapid build-up of foreign debt was part and parcel of a wider world phenomenon which has to be kept in mind when analyzing a specific country experience.

Next, we devote some time to a discussion of the main features of the changing domestic decision-making processes which presided over Brazil's foreign debt accumulation. Three major periods come distinctively to the fore: the years of the so-called economic-miracle of 1968–73; the "adjustment" attempt of 1974–78; and the critical 1979–81 period during which sudden policy shifts were incapable of avoiding the foreign exchange collapse and debt renegotiation of late 1982 and early 1983.

The third section deals with more recent phenomena, especially the international and national implications of the near-closure of private international credit markets to Latin America since the second half of 1982. As

*Edmar Bacha gratefully acknowledges a research grant from CNPq-Brasil.

in section two, we take a global perspective before discussing Brazil's dilemmas and the costs of the options taken. Some tentative lessons from this historical experience still unraveling before us are briefly summarized in the last section. The influence of the debt overhang on the economic future of the country justifies the consideration of this section in a book dedicated to the analysis of the prospects for democracy in Brazil.

In fifteen years, from 1967 to 1982, the world saw the rise and fall of developing countries' foreign borrowing in the private international credit market. The consequences of the 1982 near-collapse of this market will stretch for some years to come, appearing most visibly in the new roles and changing relationships between private banks, multilateral credit organizations, and the central banks of the major countries, now facing the well-known dilemmas of lenders of last resort.

It is reasonably clear that the present crisis of liquidity cannot be solved in the realm of the private international credit system. Less clear, but perhaps more important, is a recognition that there is no purely financial solution to the question of the external debt of non-OECD countries, which reached over $800 billion by the end of 1983.

Skeptical observers may consider these recent developments and their still unfolding consequences as simply another manifestation of "history as usual." After all, it has happened before: for well-known reasons, between the early 1930s and the early 1960s there was a virtual cessation of private international financial flows. The European return to full current account foreign-exchange convertibility for non-residents and banks after 1959, the U.S. government regulations on national banking in the sixties, and the Soviet fears of holding dollar-denominated deposits in the U.S., led to the impressive flowering of the Euro-currency markets, through new and booming trans-national banking activities.

The more advanced developing countries eagerly seized the new opportunities to borrow "with no strings attached," a flexibility which previously they did not enjoy under World Bank project loans or IMF stabilization programs—and much less under the bilateral government to government loans which marked the previous period.

As a result, the total external debt of non-oil developing countries grew from less than $40 billion in 1967 to $97 billion in 1973 to $375 billion in 1980. The impressive rate of increase of 22 percent per year in nominal terms during the 1970s is lower than the real rate of increase of 12 percent during the 1960s, due to the small initial size of the debt.

Debt service flows increased at faster rates than debt itself, reflecting the shift to private borrowing at higher (market) interest rates and shorter maturities. This shift to private sources is one of the outstanding features of the period: in 1969, the first year for which comprehensive debt data are available, 55 percent of the outstanding debt of developing countries was from official sources. The remainder was split between officially guaranteed suppliers' credits and bank debt, the latter representing around one-third of the grand total. In 1973, private borrowing accounted for barely 50 percent of outstanding total debt, but by 1980 to nearly two-thirds of a total nearly four times as large in nominal terms.

There also occurred a significant concentration of debt, especially commercial bank borrowing, among a small group of developing countries. By mid-1982, only ten countries, out of the total membership of 143 at the World Bank, accounted for more than half of the total LDC debt. This trend was sharply accentuated by the private bank's commercial recycling of OPEC's surpluses after the first oil shock.

The sudden, sharp rise in the price of an internationally traded commodity the demand for which was inelastic with respect to price created a global real disequilibrium which in practice required a combination of recession and inflation to be sorted out, plus years of structural adjustment.

Indeed, for the rest of the world, the only long-term options for real adjustment were the reduction in oil imports, the increase in exports for OPEC, or the transfer of real assets to OPEC countries, through OPEC's direct investment in oil importing countries, most likely those with convertible currencies.

The international financial system added a degree of flexibility to this adjustment process. The creation of short-term financial assets in the financial centers of the world allowed for both the satisfaction of OPEC's preference for liquidity and the extension of credit to deficit-ridden countries. This extension of credit, in turn, permitted either a postponement of the adjustment or its distribution over time.

It is worthwhile in this respect to quote from Cooper, who wrote the following passage before the second oil shock and the rise in international interest rates had taken place:

> What happened was that a number of countries took conscious and, I think, rational decisions to ride out the recession. They chose not to experience it in 1974–75, but to borrow abroad instead, to maintain growth and domestic demand, and external debt rose accordingly. They took a gamble that I think was rational and that, indeed, was very helpful from the point of view of the world economy as a whole, because they helped to limit the extent of the downturn. *But it is a gamble that they essentially lost*. The recession was much sharper and much longer than was anticipated at the time, and now these countries face serious decisions as to how much to retrench and how to accomplish it . . ." [emphasis added].[1]

Brazilian experience fits admirably well in this description, dramatically confirmed—and aggravated—by subsequent events. The remainder of this section, therefore, is an attempt to provide a brief overview of the main features of fifteen years of foreign borrowing—from the initial year of the so-called Brazilian economic miracle (1968) to the foreign exchange collapse of 1982.

We will discuss separately three broad periods: (a) the years of the "miracle," 1968 to 1973; (b) the "adjustment" to the first oil shock, 1974 to 1978; and (c) the critical 1979–81 period. In these sections we attempt to place the Brazilian discussion in the evolving international context and to emphasize the nature of the domestic decision-making processes.

The performance of the Brazilian economy from 1968 to 1973 was rather impressive when measured by conventional indicators: (a) an average annual rate of growth of GDP of over 11 percent in real terms; (b) an average rate of inflation of around 20 percent slightly declining over the period and partially neutralized by widespread indexing; and (c) overall surpluses in the balance of payments in every year from 1968 to 1973, with capital inflows running at rates over and above those required to finance Brazil's secular current account deficit, leading to a simultaneous accumulation of reserves (from $199 million at year-end 1967 to $6,417 million at the end of 1973) and gross foreign debt (from $3,344 million in December 1967 to $12,572 million at the end of 1973). See Table 1 for more details.

In many interpretations, an authoritarian and centralized régime and its imputed "rational and pragmatic" economic policies have been presented, both in Brazil and abroad, as the main elements behind these rather impressive achievements, in an attempt to use economic performance as the basic criteria for political legitimacy.[2]

Elsewhere, we have argued that the "economic miracle" depended heavily upon some cyclical phenomena operating endogenously in the Brazilian economy and upon an exceptional and elusively temporary international situation.[3] In our view, the simultaneity of a domestic upswing (after the stabilization crisis of 1964–67) and of very favorable international conditions with respect to trade and finance provide the basis for a clear understanding of the economic boom Brazil experienced from 1968 to 1973.

We will not go over the arguments and evidence in support of our earlier interpretation. Our concern in this section is to put Brazil's rapid build-up of foreign debt in its proper perspective. For that endeavor, one must first recognize the outstanding historical singularity, for Brazil, of the 1968–73 period: it was the only period, in half a century of our history since 1930, when growth and capital accumulation were not foreign-exchange constrained.

In fact, during this six-year period, Brazil experienced growth with unlimited supplies of foreign credit at nearly negative real rates of interest. Using Fishlow's characterization in this volume, foreign debt was an option, and it certainly allowed higher levels and rates of growth of investment without any significant pressure to reduce the growth rates of public and private consumption.

As long as Brazilian exports were growting faster than world trade, as indeed they were (an amazing 24.8 percent per year in dollar terms as against an also amazing 18.3 percent per year for world exports), and much faster than the nominal international rates of interest, even a rate of growth

Table 1. Balance of Payments: Summary Accounts, 1973-1983 (US $ Million)

	1973	*1974*	*1975*	*1976*	*1977*	*1978*	*1979*	*1980*	*1981*	*1982*	*1983*[a]
Trade Balance	+7	−4 690	−3 540	−2 255	+97	−1 024	−2 717	−2 823	−1 202	+780	−6 470
Exports (FOB)	+6 199	+7 951	+8 670	+10 128	+12 120	+12 659	+15 244	−20 132	−23 293	+20 175	+21 899
Imports (FOB)	−6 192	−12 641	−12 210	−12 383	−12 023	−13 683	−17 961	−22 955	−22 091	−19 395	−15 429
Balance of Services	−1 722	−2 433	−3 162	−3 763	−4 134	−5 062	−7 057	−10 152	−13 135	−17 083	−12 748[a]
Non-factor Services	−965	1 533	−1 429	−1 574	−1 576	−1 805	−2 317	−3 120	−2 863	−3 588	−2 435
Factor Services	−757	−900	1 733	2 189	−2 558	−3 257	−4 740	−7 032	−10 272	−13 494	−10 313[a]
(Net Interest Costs)	(−514)	(−652)	(−1 498)	(−1 810)	(−2 104)	(−2 696)	(−4 104)	(−6 311)	(−9 161)	(−11 353)	(−9 555)
Current Account Deficit[1]	−1 715	−7 122	−6 700	−6 013	−4 037	−6 015	−10 478	−12 807	−11 734	−16 311	−6 171
Capital Account	+3 512	+6 254	+6 189	+6 651	+5 269	+10 916	+6 194	+9 679	+12 773	+7 851	+3 372
Direct Investment	+940	+887	+892	+962	+810	+1 071	+2 226	+1 532	+2 326	+2 547	+657[a]
Loans and Financing	+4 495	+6 961	+5 933	+7 761	+8 424	+13 811	+10 924	+10 596	+15 553	+12 515	
Amortizations	−1 672	−1 920	−2 172	−2 992	−4 060	−5 324	−6 356	−5 010	−6 242	−6 952	+2 715
Short Term Capital	−251	+326	+1 536	+920	+96	+1 358	−558	+2 561	+1 135	−259	
Overall Balance[2]	+2 179	−936	−950	+1 192	+630	+4 262	−3 218	−3 472	+625	−8 828	−3 330
Gross External Debt[3]	12 572[b]	17 166[b]	21 171[b]	25 988[b]	32 037[b]	53 600	61 300	68 400	80 000	89 100	97 000
Reserves[4]	6 417	5 252	4 041[c]	6 544	7 256	13 900	10 800	7 900	9 200	6 000	
Net External Debt	6 155[c]	11 914[c]	17 130	19 441[c]	24 781[c]	39 700	49 000	60 000	70 800	83 100	
Gross Domestic Product (US$ billion)	96.3	125.7	145.5	165.4	184.3	207.5	240.8	283.8	304.6	327.4	330.5

[a]Excludes reinvestment
[b]Excludes short-term debt
[c]Excludes commercial banks
[1]Includes unrequited transfers
[2]Includes errors and omissions
[3]Includes authors' estimates for short-term debt after 1977
[4]Includes authors' estimates for commercial bank reserves after 1977
Sources: Central Bank Bulletins, Central Bank of Brazil (1984), and authors' estimates

of imports of nearly 27 percent per year as observed from 1967 to 1973 seemed bearable and implied a substantial real resource transfer to a country with an undoubtedly high absorption capacity.

Nevertheless, a "foreign exchange illusion," vividly described by Hirschman, developed in Brazil.[4] The government apparently took for granted as a new formality what were in fact rather exceptional international developments in both trade and finance. After the domestic recovery of 1968 to 1970, when income grew faster than investment due to generalized excess capacity (associated with the previous boom of 1956–61 and the stabilization crisis of 1963–67), both public and private investment plans were ambitiously designed as if Brazil had definitely solved its secular problem of foreign-exchange constrained growth. Either exports or debt or a combination of both seemed to allow for maintaining a perfectly elastic supply of imports at given prices. An increase of nearly 150 percent in the capacity to import from 1967 to 1973 due to better terms of trade and export volume served to boost expectations.

The synchronized boom in the advanced economies in 1972–73 led to a commodity price explosion which helped to feed the worst inflation the integrated capitalist world economy had ever experienced. But up to 1974, the world economy was booming, Brazil was booming, debt was manageably accumulating, and euphoria developing as never before. The collapse of the Bretton-Woods system of fixed parities and the acceleration of world inflation seemed events of minor importance for the Brazilian planners, at the time exclusively concerned with projecting past trends into future dreams.

The first oil shock in late 1973 and early 1974 painfully showed that growing euphoria had its costs and that a given economic policy, pursued single-mindedly to its limits, would outlive its utility.

The Geisel administration, inaugurated in March 1974, had a perception of the need for structural adjustment. As in Cooper's quote, it chose "rationally" to "ride out the recession and not to experience it in 1974–75 but to borrow abroad instead, to maintain growth and domestic demand". But the Geisel administration took more than a gamble. It decided to continue to pursue the projects which had been initiated in the previous years of growing euphoria and, additionally, to launch an ambitious program of import-substitution in capital goods and basic raw materials. The idea of the need to either increase net exports or transfer real assets was there as the only long-term solution to the real adjustment imposed by the oil shock. The growth of external debt was thought to be a temporary price to be paid for financing this adjustment over time.

Indeed, net medium and long-term foreign debt rose fivefold, from $6,155 million in December 1973 (approximately equal to the value of exports in that year) to $31,616 in December 1978 (two and a half times the value of exports).

With the exception of 1974 and 1975, when there was a loss in international reserves (of around $950 million in each year) for the whole eleven-year period from 1968 to 1978, there was an overfinancing of Brazil's current account deficit through the capital account. In 1978 alone gross capital flows were nearly $11 billion, exceeding the current account deficit and thereby boosting the Central Bank held international reserves by more than $4 billion. OPEC surplus had been reduced to virtually nil in that year, indicating an important fact, not well understood at the time: that the international private credit market was not dependent on OPEC surpluses to extend loans to developing countries.

However, for many optimistic observers, by 1977–78 the balance of payments adjustment in Brazil was nearly completed. The trade deficit of $4.7 billion in 1974 had been eliminated in 1977. The current account deficit was reduced from $7.1 billion in 1974 to $4 billion in 1977. Imports, after rising slightly more than 100 percent in 1974 (to $12.6 billion) were kept in dollar terms in the $12 billion mark for the next three years. Exports rose from $6.2 billion in 1973 to $12.1 billion in 1977. Moreover, the economy was growing at an average annual rate equal to the historical (i.e. postwar) rate of around 7 percent, and inflation had stabilized at the 35 to 40 percent rate.

But could one say that adjustment "Brazilian-style" had been accomplished by 1977–78 and, were it not for the second oil shock and the U.S. "monetarist revolution" of 1979, that Brazil—together with forty other developing (and socialist) countries—would not have reached the debt rescheduling stage in the years from 1980 to 1982?

It is hard to answer this counterfactual question even with the benefit of hindsight, but it is undeniable that the adjustment, as attempted, did increase the vulnerability of the Brazilian economy to further external shocks. Oil dependence was the most apparent one for a country importing more than 80 percent of its consumption. Oil came to represent around one-third of total imports in 1976–78 (as against 10 to 11 percent over the 1968–73 period), despite a virtual stabilization of its nominal price around $12 per barrel between 1974 and 1977—a sharp fall in real terms due to world inflation and the rise in the price of Brazilian exports. Other imports fell accordingly. Capital goods went down from 41 percent of total imports in 1971–72 to 26 percent in 1977–78 and intermediate goods from 39 percent to 33 percent.

However, as serious as the real problem posed by oil dependency was, the increased vulnerability to upward changes in international interest rates which came to regulate nearly 70 percent of long- and medium-term gross Brazilian foreign debt of $43.5 billion by the end of 1978 (of which $29.5 billion consisted of currency loans). As a result, net interest payments came to represent, in 1977–78, nearly half of Brazil's current account deficit, as can be seen in Table 1.

Now this fact has an important implication, namely, that the same current account deficit implies a much lower transfer of real resources from the rest of the world. In other words, the net contribution of foreign resources to finance Brazilian growth was steadily declining over the period.

In order to maintain the economy growing, this reduction in the contribution of foreign resources had to be matched by an increase in internal savings and its translation into additional foreign exchange through expanded exports and import substitution.

This real adjustment could have been postponed or distributed over time but not avoided by building up foreign debt. In a mixed economy it would have required a substantial shift in relative prices—sustained and perceived as such—to induce substitution not only in consumption, but in investment as well. This adjustment was very timidly running its course, together with its slowly working real effects (due to the higher capital ratios and long maturation periods of the big investment projects being simultaneously implemented) when a more vulnerable Brazilian economy suffered two near-fatal blows: the second oil shock which sent oil prices from $12 to $30 per barrel from late 1978 to early 1980; and the sharp rise in international interest rates, which doubled in nominal terms from 1977 to 1979 and increased even further in 1980 under the impulse of a disastrous monetary and fiscal mix in U.S. economic policy. The game was clearly over. The next paragraphs attempt to explain why it took nearly three years, from 1979 to 1982, for Brazil to recognize it fully.

The second oil shock led to an OPEC surplus in a two-year period (1979–80) of nearly the same magnitude, in current dollar terms, as the surpluses observed in the 1974–78 period (around $170 billion). The lower magnitude in real terms was more than compensated for by the higher share of oil consumption in world income, by the sharp rise in international interest rates, and by the difficulties involved in a second round of recycling the OPEC surpluses, due to the increased vulnerability of the already highly indebted developing countries.

Brazil's trade deficit rose from $1 billion in 1978 to $2.7 billion in 1979, and to $2.8 billion in 1980, despite an impressive growth of exports of more than 50 percent in nominal terms, in the two-year period 1979–80. But imports rose by more than 70 percent, due to sharply deteriorating terms of trade. Oil came to represent nearly 45 percent of total imports in 1980. The apparently high level of the Central Bank reserves at year-end 1978 ($11.8 billion) was sharply reduced by the overall balance of payments deficit: $3.2 billion in 1979 and $3.6 billion in 1980.

The level of gross international reserves started to lose meaning after 1979, with the growth of Brazil's then unrecorded short-term borrowing in international credit markets.

According to Nogueira-Batista, Jr., net reserves, defined as gross Central Bank reserves minus its short-term liabilities, in June 1980 were barely one-third of their level of December 1978.[5] The net reserves/imports ratio declined from 71 percent in December 1978 to 15 percent in June 1980, and to 13 percent in September, when a major economic policy decision was taken: to adjust the economy through recession in order to revert the trade balance from deficit to surplus in 1981.

This decision represented a major shift in the previous policy orientation and requires a brief explanation. The Geisel administration was succeeded by Figuereido's presidential term in March 1979. The strong man in economic affairs, Mário Simonsen, was preparing a slowing down of the economy in mid-1979 (as indicated by his never-published Third National Development Plan for the 1979–85 period), when he was replaced in the position by Delfim Netto—until then Minister of Agriculture and a critical voice within the government of Simonsen's planned slowing down of the economy.

Delfim Netto's version of the Third National Development Plan, as submitted ritualistically to the Brazilian Congress in September 1979 contained no figures whatsoever, but made a firm commitment to furthering economic growth. The Cabinet reshuffling was hailed by the entrepreneurial class, who strongly supported the new minister's growthmanship.

Indeed, the rate of growth of real GDP in 1979 was 6.8 percent and nearly 8 percent in 1980. A devaluation of 30 percent in December 1979 helped boost exports from $15.2 billion in 1979 to $20.1 billion in 1980. However, imports rose from $18.0 billion in 1979 to $23.0 billion in 1980, practically repeating the trade deficit of 1979. The current account deficit rose from US$10.5 billion in 1979 to $12.8 billion in 1980, half of it represented by the interest costs on net foreign debt.

Inflation, which was kept in the 35–40 percent range from 1974 to 1978, rose to 77 percent in 1979 and to an all-time high of 110 percent in 1980. By mid-year, the first hints of Brazil's need for rescheduling its foreign debt started to appear in public discussion—the suggestion of the need to IMF advice being voiced by foreign bankers.

The reasons were, as so often in the past, the balance of payment crisis, the rather precarious net reserves position, and very specially, the erosion of credibility in the government's policy after the disastrous decision (as of January 1980) to predetermine the rates of devaluation (at 40 percent) and monetary correction (at 45 percent) for 1980, when all informed observers were expecting a rate of inflation of at least the same magnitude of 1979 (77 percent), probably much more, due to the second oil shock, the December 1979 maxidevaluation, the acceleration of wage readjustments decided upon in late 1979, and the widespread use of backward looking indexing mechanisms.

The policy shift in the direction of a contractionary policy started in the second half of 1980 and was pursued throughout 1981. It resulted in a $4 billion swing in the trade balance (from a deficit of $2.8 billion to a surplus of $1.2 billion), at the cost of the sharpest income drop in statistically documented Brazilian history: some 5 percent fall in real per capita income. Inflation was only moderately reduced from 110 percent in 1980 to 96 percent in 1981, basically due to a fall in both domestic agricultural prices (expressing the combined effect of a good harvest and falling demand) and internationally traded agricultural commodities in world markets.

But the basic problem of the balance of payments remained unsolved. The current account deficit reached $11.7 billion—of which $9.2 billion were represented by interest costs on foreign debt, since the prime rate reached 18.8 percent on average in 1981 and the LIBOR averaged 15.5 percent. On top of that, Brazil was paying a spread of 2⅛ percent. Registered external debt reached, officially, the value of $61.4 billion. This figure refers to medium- and long-term debt. Unrecorded at the time, short-term debt can be estimated at $18.6 billion. Therefore, Brazil's total debt at the end of 1981 was around $80 billion, more than one-fourth of total GDP, and nearly three and a half times the value of 1981 exports ($23.3 billion) of which two-thirds were required to pay for amortization and interests costs, which reached $15.4 billion in 1981.

By the end of 1981 it was clear that the old indicators of debt-servicing capacity were losing their traditional meaning—in Brazil and everywhere. The world recession and stagnant world trade, the amazingly high levels of real interest rates, the associated fall in commodity prices, and the new protectionism in the advanced countries were transforming what were medium-term real adjustment problems into liquidity problems associated with the lack of convertible currencies and inability to pay contractual obligations, in several developing and Eastern European countries. In 1980 six countries had to renegotiate their debts in operations involving $4.4 billion. In 1981, fourteen countries initiated renegotiations with a value of $10.8 billion. The debt issue had become a global question, involving not only finances but also trade patterns and the state of the world economy. The next section deals with recession and financial distress in the world economy and Brazil's prospects under IMF's conditionality.

Historical experience demonstrates that every major recession is accompanied by a major financial crisis. We had both in the early 1980s. The world economy was in deep recession: the rate of growth of real GDP in OECD countries was, on average, less than 1 percent in 1980–82 (as against 5.2 percent from 1960 to 1973, and 2.7 percent from 1974 to 1979). The average rate of growth of world trade was slightly more than 1 percent in real terms in 1980–82 (as against 8.6 percent from 1960 to 1973, and 4.5 percent from 1974 to 1979). The ratio of unemployment as a percentage of civilian labor force stood at more than 10 percent, on average for the OECD (as against 3.1 percent from 1960 to 1972, and 5.1 percent from 1974 to 1979).

The real shocks associated with the second oil price rise in 1979, and the tightening in U.S. monetary policy after October of that year, led to an international propagation of recession which was aggravated by the U.S. indifference to the external consequences of its domestic economic policies. The very high degree of integration of world financial markets forced a general tightening of monetary policies in Western Europe and Japan. There followed a generalized rise in real interest rates, an associated fall in commodity prices (given the higher financial costs of retaining stocks), and three years of world recession and nearly stagnant world trade, in 1980, 1981 and 1982.

International lending continued up to 1981, albeit at declining rates. In current dollars, commercial banks increased their exposure to developing countries to $60 billion in 1978, $55 billion in 1979, $49 billion in 1980, and $48 billion in 1981. The major change, however, was in the increasing importance of short-term debt: from December 1979 to December 1981 the stock of short-term debt of developing countries increased from $68.1 billion to $115 billion, whereas the stock of medium- and long-term debt rose from $131.4 to $175.2 billion.

The well-known herd instinct of the banking community helps to explain not only this continuing extension of credit (at shorter maturities, higher spreads, higher profits and higher risks) but the contraction which slowly began in the second quarter of 1982 and rushed to a near-panic shortly after the Mexican collapse of August 1982.

What happened was not, as some Brazilian policy makers were quickly trying to point out, a temporary interruption in the international lending activities of commercial banks, while they were waiting to watch the water clear. On the contrary, the capital market rupture of the third quarter of 1982 should be considered as a major turning point in international banking history.

Confidence was badly shaken, and the market shrank. A crash in 1982 was averted only because of rescue packages quickly arranged by heavy players such as the U.S. Federal Reserve Board, the Bank for International Settlements, and the International Monetary Fund moving to fill the role of international lenders of last resort. In 1982 some two dozen countries initiated what promises to be a long and painful renegotiation of their foreign debts, involving some $35 billion.

It is true that the major U.S. and European banks did not have much choice. After all, it is estimated that many of them had more than half of their assets in international markets and almost all had at least one-third. As a respected business publication put it "They are locked into their international customers, like it or not and will be increasing their loans to troubled borrowers such as Mexico, Brazil, Chile and Argentina".[6] But this was not true of all the estimated 1,100 to 1,300 banks involved in international lending. Estimates were that this number could be reduced by half by decisions not to send "good money after bad," and especially, not to lend in the interbank market to weak countries' banks disguised financing of their own countries' balance of payments deficits.

The events discussed in the previous subsection created for Brazil and several highly indebted developing countries a near-fatal combination: a simultaneous contraction of world trade and international credit. To countries outside the charmed circle of freely convertible national currencies, this combination usually generates—as in the 1930s—a rather serious liquidity crisis. There simply is not enough cash in convertible currencies to meet immediate contractual obligations and import needs. Liquidity problems, as is known, can easily lead to insolvency and to the need to postpone payments and renegotiate the foreign debt, as indeed happened to Brazil by the end of 1982. This subsection attempts to show why.

In December 1981, the National Monetary Council published its projections for the foreign sector for the year 1982. Rather optimistically, exports were projected to grow from $23.3 billion in 1981 to $28 billion; imports from $22.1 billion to $25 billion. A trade surplus of $3 billion would result, reducing the current account deficit to $10 billion from nearly $12 billion in 1981. Net capital inflows were projected to reach the same level as in 1981, leaving unchanged the level of official reserves, estimated at $7.5 billion.

Brazilian exports, however, after increasing by 32 percent from December 1979 to December 1980, since then steadily declined every month, with its rate of growth measured on an annual basis. In December 1981, when projections were made it was 15.7 percent. In March 1982, 8.9 percent; in June 1982, 2.6 percent—and increasingly negative after that. The annual rate of growth in October 1982 was *minus* 9 percent.

As a result, the $28 billion exports target was progressively reduced, first to $26 billion and then to $25 billion by mid-year. In fact it was $20.2 billion for the whole year. The second year of domestic recession plus stringent import controls reduced imports from one projected $25 billion to $19.4 billion, allowing a modest trade surplus of $780 million.

But the bad news was not only related to trade. The amazingly high levels reached by international interest rates during 1981, especially in the third quarter, when the prime rate reached 20.2 percent and the LIBOR 18.46 percent—exacted their toll. Net interest payments, which had already increased by nearly a half in 1981 (to $9.2 billion), increased further to $11.4 billion in 1982, nearly 70 percent of the current account deficit of $16.3 billion, which represented 5 percent of the estimated Brazilian GDP for 1982.

The extreme vulnerability of the Brazilian economy, which started to appear during the Malvinas episode, was dramatically signaled after the 90-day Mexican moratorium declared on August 23rd, and the frustration which marked the Annual Meeting of the IMF/IBRD in Toronto in early September, when the seriousness of the potential liquidity crisis was not clearly perceived. In fact, the advanced countries failed to take appropriate actions on a global scale, such as a general increase in IMF's quotas or the creation of an emergency fund to deal with the troubled period ahead.

The private international credit market was much more than merely paralyzed as far as loans to most developing countries were concerned. It was not a temporary phenomenon, but a deep-rooted one. The market would not return to its pre-1982 pattern of lending. The contraction had been set in movement, fed by the financial markets' typical behavior patterns of following the herd, in expansion and contraction.

The stage was set—and the need was clear—for the entrance of lenders of last resort. The rescue package designed to allow Mexico to go through its 90-day moratorium was formed in less than 48 hours by the U.S. government and the Federal Reserve, in an operation involving nearly $4 billion.

Brazil would follow shortly after. Concerned with the November 15 elections, the government, while recognizing that it would need the IMF and U.S. government aid, decided to keep this from public discussion.

The decision was ill-conceived, for the situation was untenable. Brazil's gross foreign borrowing requirements for 1982 were estimated at well over $20 billion. Even assuming a sharp fall in international reserves, it was increasingly clear that Brazil would not be able to borrow mainly from commercial banks the nearly $1.4 billion a month required for the last quarter of 1982.

Additionally, the government was clearly overestimating the liquidity of its short-term assets. Included in the value of the government's reserves were financial assets which were becoming increasingly non-liquid, such as the debts to Brazil of other highly indebted developing and socialist countries.

At the same time, the short-term liabilities of the Brazilian monetary authorities were rapidly increasing. Brazilians had to wait for the first Technical Memorandum of Understanding with the IMF, an annex to the Letter of Intent sent to the Fund on January 6, 1983, to know that the Central Bank's net reserve position as of September 31, 1982, was only $1 billion, enough to finance less than three weeks of imports.

Secret negotiations with the IMF and with the U.S. government started in October. The Brazilian government had to ask President Reagan to postpone his trip to Brazil until after the November 15 elections, to avoid undue political interpretations. When he finally arrived in Brasilia, President Reagan was happy—as he declared—"to come to Bolivia" and announced U.S. support for the country's financial problems. In fact, aid had come earlier in the form of a $1.4 billion "bridge loan" from the U.S. Treasury, of U.S. support for another $500 million "bridge loan" from the BIS, of U.S. Treasury officials cajoling commercial banks to arrange another $2.3 billion "bridge loan," and of U.S. support for a Brazilian mid-term loan arrangement with the IMF.

The Brazilian government approved the balance of payments projections for 1983 as early as October 25, 1982. This was partly for reasons of domestic consumption (to show that it was taking the initiative to propose a voluntary adjustment to the balance of payments constraint as envisaged for the rest of 1982 and 1983) and partly for foreign consumption (an attempt to differentiate the Brazilian case from Mexico's and Argentina's by showing its commitment to austerity and rationality).

Published the same day, the projections for 1983 caused great perplexity—and well-founded fears of another year of recession and financial distress. Indeed, for the first time in Brazilian foreign sector programming, the starting point was a somewhat magical and curiously precise number: $10.6 billion. This figure was stated in October as the maximum amount that Brazil would be able to borrow in international capital markets during 1983. Subtracting $7.2 billion of amortization and adding $3.5 billion of direct investment and suppliers' credits one would reach the figure of U.S. $6.9 billion for net capital inflows. The government assumed that this would be the required current account deficit (since there were no reserves to lose). As the deficit on services was projected at $12.9 billion, the trade surplus which came out as a residual was supposed to be as high as $6 billion. Moreover, most of this surplus would have to come from a curtailment of imports—of 15 percent to 20 percent—since exports were supposed to reach a maximum of $22.5 billion (a 10 percent increase in nominal terms). The projected fall in imports from 1980 (when they reached $23 billion) to 1983 (a projected $16.5 billion) was equivalent to 50 percent in real terms.

These projections formed the basis for the renegotiation with the IMF and for the wild persuasion efforts involving the commercial banks and were advertised as manifestations of how serious the Brazilian government was in its commitment to reduce the current account deficit from 5 percent of GNP in 1982 to slightly more than 2 percent in 1983. As expected, the IMF mission required other measures, especially a 50 percent cut in the public sector borrowing requirements as a proportion of GDP, and the usual strict targets for the expansion of domestic credit.

As soon as an agreement was reached with the IMF, by mid-December, the Brazilian authorities arranged for a meeting in New York with representatives of the 125 major banks involved with Brazil's debt. There, on 20 December, Mr. de Laroisière's opening statement supported Brazil's request and announced the preliminary agreement of the IMF with the programmed adjustment for 1983. The bankers were introduced to the four Brazilian projects:

1) New loans of $4.4 billion for 1983, a figure which represented about 7 percent increase in exposure for the average bank creditor;
2) Renewal during 1983 of maturing medium-term loans of approximately $4.7 billion for an additional eight-year period;
3) Keeping open existing trade related credits estimated in $8.8 billion;
4) Commitment to the interbank credit lines to Brazilian banks operating internationally at the figures outstanding on June 30, 1982—an estimated $10 billion.

The IMF hinted, as in the Mexican case, that bank approval of the four-point Brazilian request was a pre-condition for Fund approval of a $4.8 billion, three-year program for Brazil. The bankers balked at the Brazilian request for a prompt reply, i.e. before December 31, 1982, a few working days later.

As predicted, time was rather short for such complex negotiation. On December 30, anticipating bank approval of project 2 above, Brazil declared that it would not pay the amortizations due in 1983 as they came to maturity.

This was considered an implicit moratorium, a recognition that the four projects were in fact a renegotiation of Brazil's debt and—as declared by the director for the foreign area of the Brazilian Central Bank—a threat to the banks: either the agreement would be reached before the new deadline—established tentatively as March 1—or Brazil would be forced formally to declare a moratorium on its foreign debt.

An agreement was eventually reached, and the four projects were solemnly signed in New York on February 28. But this was hardly the end of the story. For, contrary to the overly optimistic assessment of the Brazilian authorities and their banker advisers, the "voluntary" lending levels contemplated in Projects Nos. 3 and 4 simply did not materialize. International banks in fact continued to withdraw their credit lines from both Brazilian trade and Brazilian banks abroad. Lacking international reserves to cover the corresponding cash shortfalls, Brazil's Central Bank started to fall increasingly into arrears in its foreign payments.

Moreover, in its first quarterly review of the Brazil loan, the IMF staff found that both the budget deficit and domestic credit creation were much higher than promised by the Brazilian government in its two initial Letters of Intent (dated January 6 and February 24). Hence, the decision was made to suspend disbursement of the second installment of the IMF loan to the country (which was due May 31), pending a revision of the adjustment program. In the terms of the agreement with the banks, this automatically also suspended further disbursements of the loans agreed under Project No. 1. There followed a six-month period of tense negotiations, while a compromise worked itself out. On the Brazilian side, politically the most spectacular consequence was the passage by Congress of a new wage law, by which average wages' growth was reduced from some 100 percent to about 86 percent of the inflation rate in the past six months. The banks eventually agreed to an additional $6.5 billion loan, to cover the financing needs of both 1983 and 1984, while postponing for the future all debts maturing in 1984, and strengthening the centralized controls over the compliance of more modestly planned levels for Projects Nos. 3 and 4. Lower private commitments under Project No. 3 were compensated for by additional export credits from the Eximbanks of the U.S. and other Western governments. A Paris Club meeting was also held, in which an understanding was reached for the rescheduling of Brazil's official debt. The whole package was conditioned on the Fund's approval of the Letter of Intent of September 15, as amended on November 14. This was finally obtained in late November. Then, a massive bookkeeping operation could take place, with new credits being disbursed and arrears cleared, for the satisfaction of the financial world, which now could turn its attention to the upcoming negotiations with Alfonsin's Argentina.

Meanwhile, the consequences of the austerity measures were being hard felt by Brazilians. GDP per capita fell by an additional 5.7 percent in 1983, total industrial output by 7.0 percent, and capital goods production by 20 percent. Droughts in the Northeast and floods in the South, a 30 percent maxidevaluation in February, the elimination of food and energy subsidies, and some "corrective inflation" in public utilities prices sent the wholesale price index skyrocketing to 211 percent per year in December 1983.

The good news was in the financial side and in the external accounts. In December, the Fund staff unsurprisingly found that the Brazilian government had met the ceilings for credit expansion and the budget deficit which were set one month before. Furthermore, the Fund board tried to minimize the possibility of future problems arising in its quarterly reviews of the program, by instructing its staff, first, to set the performance criteria on a pay-as-you-go-basis, and, second, to add to its own preferred measures of the budget deficit and domestic credit expansion—those favored by the Brazilian government.

On the external front, a spectacular 43 percent reduction of capital goods imports was the single most important factor behind a trade surplus even higher than the $6 billion initially forecast, in spite of a weaker than predicted export showing. The services account also did better than predicted. As a consequence, the country succeeded in impressively reducing its current account deficit from $14.8 billion in 1982 to $6.2 billion in 1983.

In March 1984 the Brazilian government disclosed the second volume of its Economic Program, with its plans for the remainder of the year.[7] Domestically, the major objectives were the attainment of a small operational surplus in the pubic sector accounts and the reduction of money supply expansion to 50 percent on a yearly basis. The main purpose was "to permit a drop of as much as one-half in the rate of inflation during 1984."[8] Externally, the objective was to obtain a trade surplus of $9.1 billion (as compared with $6.5 billion in 1983) and a current account deficit of $5.3 billion (which compares with a deficit of $6.2 billion in 1983). This should have been consistent, according to the government, with a positive change in gross international reserves of $4.4 billion during the year.

The government expected that these targets would also be consistent with no additional negative changes in the country's GDP. However, it should have been clear that even if it succeeded in halving the rate of inflation, which it was not, observance of the monetary target would have implied a drop of the money supply of as much as 27 percent during the year. It was unlikely that the economy would have been able to withstand such drop in real money without a further decline in real activity. Thus, what could be said in mid-1984 was that either the monetary targets were not to be met or else there were no end in sight to the longest and deepest recession in recorded (postwar) Brazilian economic history.

Moreover, under IMF conditionality, the country was also unlikely to resume a sustainable growth path over the medium run. For the adjustment program was designed to generate a trade balance surplus large enough to permit an early retirement of Brazil's foreign debt. More specifically, the country was expected not only to shrink its current account deficit but to generate an increasing current accounts surplus. The IMF belief that this was consistent with the maintenance of a meager 4 percent real GDP growth rate after 1984 was based on overly optimistic assumption about OECD economic growth and real international interest rates.

A more realistic assessment of the domestic and external economic circumstances would have indicated the need to dissipate the fiction that the debt could have been repaid in the near future, according to the original contractual obligations. In fact, an effective consolidation of the external debt seemed, then as now, to be imperative, as a step towards resumption of sustained GDP growth in Brazil. Proposals for such debt consolidation have been spelled out elsewhere by such authors as Bailey, Fishlow, Kenen, and Rohathyn, and, therefore, need not detain us here. They must underlie a renegotiation of the terms under which past debt was contracted and the program with the IMF was set up. In order to avoid Babel, certainly a more determined negotiating posture is necessary on the part of those responsible for economic policy making in Brazil in the second half of the 1980s.

This essay presented a general outline of the controversial processes through which the Brazilian economy arrived at its present perilous situation. We adopted this historical perspective not so much to provide another interpretation of an irrevocable past, but to draw lessons for the present and future construction of what we hope will be a democratic Brazil.

From the drama in three acts played between 1968 and 1982, it is possible to extract several lessons. From the first act, from 1968 to 1973, we learn that growth euphoria has its costs, the most serious of which is the illusion that the secular constraint imposed by the balance of payments on the Brazilian economy had become a thing of the past; and that Brazilians were destined to reproduce in the tropics the consumption patterns characteristic of industrial societies. It is true that the first oil shock was unforeseeable, but the fact that the pre-1974 growth and distribution process could not be maintained for the rest of the decade was noted by various observers from the beginning of the 1970s.

From the second act, 1974 to 1978, it is possible to extract the lesson that a decision to avoid recession, as an abrupt adjustment device to deal with new adverse relative prices and international conditions, was essentially correct. But it erred in not perceiving the need for much greater selectivity in investment choice and resource allocation. Adjustment through borrowing rendered the Brazilian economy excessively vulnerable to additional external shocks. Again, it was difficult to foresee the second oil crisis and the interest rate shock which marked the close of this act in 1979. But it was possible to observe the increased vulnerability of the Brazilian economy to international conditions, as a consequence of a debt contracted at floating rates, without which the achievement of the rates of investment of the 1974–76 period would have required a much more significant export promotion effort associated with a contraction of domestic consumption spending.

From the policy errors of the third act (1979 to 1982) the more specific lessons remain that short-term economic policy cannot be managed without a strategic vision and an adequate examination of international conditions. Between March and August 1979, one identifies "a prudent planner's" attempt at readjusting the economy through growth deceleration.[9] This was abandoned in favor of a contradictory expansionist policy, which led nowhere but to its own extinction at the end of 1980. There began the current contractionary phase, which initially led to the recession of 1981. It was a recession produced to attempt to restore the policy makers' credibility with the international financial community, and to avoid having to turn to the IMF. The recession was rendered useless by the disarticulation of the international private credit market in September 1982. Brazil was then obliged to turn to the Fund and to renegotiate its debts with the banks.

This renegotiation was marked by four mistakes of analysis. The first was the expectation that international credit markets would quickly return to normal, and resources would start again flowing to Brazil as soon as the accord was signed with the IMF. The second consisted in the undue haste to arrive at such an accord in the belief that it would not be enforced, and that there would be a political decision on the part of the IMF not to demand the fulfillment of the performance goals originally agreed upon. The third, related to the first two, consisted in underestimating the amount of new resources that Brazil would need in 1983. The fourth was that the Brazilian debt could be dealt with within the terms of a Gregorian calendar: Once the year of 1982 was "closed," then the financial necessities of 1983 could be discussed, and so on into the future.

The lesson of this period is that the arduous process of renegotiation requires, necessarily, taking into account a longer-term horizon, a more realistic evaluation of the need for new resources, a viable and medium-to-long-term agreement with the banks and with the governments which control international financial institutions. This agreement must condition the transfer of real resources to the rest of the world (via trade surpluses) to new capital inflows, to international interest rates and to the performance of Brazilian exports.

It is necessary, desirable, and possible to avoid a confrontation, but it is a mistake not to assess the economic and political costs of strict punctuality in the payment of contractual obligations associated with the external debt without new inflows on better terms. Any solution will have to be negotiated with the banks and with governments, especially with the U.S. administration, but the payment of the debt cannot overly compromise the future growth of the Brazilian economy.

This work is critical of an austerity policy that takes domestic recession as the only solution, until something positive happens in the international arena. It is true, in any case, that as long as the question of Brazil's cash flow in convertible currency is not taken satisfactorily into account through a more comprehensive renegotiation, the space for maneuver in economic policy—whoever is conducting it—will be extremely reduced.

This essay insists on the imperative necessity of resolving the problem of the debt overhang as the condition *sine qua non* for overcoming the present difficulties, while keeping in mind the importance of the debate on the distribution of the costs of the current "phase of adjustment."

The internal distribution of these burdens is an essentially political-economy process, which justifies serious discussion of an apparently technical problem, such as Brazil's foreign indebtedness, in a book dedicated to the broader issue of the prospects for democracy in a country with a long experience with both authoritarian and liberal traditions.

From the Miracle to the Fund: Part Two

Against even the most optimistic predictions, in 1984 and 1985 Brazil succeeded in generating trade surpluses large enough to keep current on the interest payments of its foreign debt, without the need of new money from the banks. This allowed Brazil to ignore the fiscal and monetary targets of the three-year IMF program signed in early 1983, which was eventually canceled in the beginning of 1985. After falling by 2.8 percent in 1983, Brazil's GDP started growing again, by 5.7 percent in 1984 and a whopping 8.4 percent in 1985.

When inflation, thanks to a general price freeze, was brought down to nearly zero in mid-1986, a new Brazilian miracle seemed to have happened. Speaking to a Washington audience in late June 1986, Paul Volcker admiringly sustained that Brazil was the living proof that the Baker Plan could work, as the largest debtor had adjusted its external accounts, started growing forcefully, and was finally dominating its entrenched inflation. Brazil's achievements were also praised in an influential 1986 book on Latin America's growth crisis, by Balassa, Bueno, Kuczynski, and Simonsen. Their conservative manifesto presented Brazil as an example to other Latin countries, saying that "its progress has profound importance for our overall topic—both in evaluating the policies applied and, perhaps, in offering indications for what can be done elsewhere."[10]

Only two years after this second coming, Brazil was back to the IMF, after suspending for nearly a year the interest payments of its long-term debt with commercial banks. In 1988, Brazil's inflation was the highest ever in the country's history and its GDP growth rate was brought down to zero. What's the story behind this second Brazilian traverse from a "miracle" to the Fund?

Brazil's trade surplus evaporated at the end of 1986, because of the expansion of domestic demand caused by the Cruzado Plan, a monetary reform and price-freeze program implanted in February 1986.[11] A measure of this extraordinary expansion is furnished by the National Household Survey (PNAD) of October 1986. Average personal income was then found to be from 25 to 40 percent higher than a year before, depending on the price index chosen for deflation. Part of these higher incomes led to another jump in GDP, by 8 percent in 1986. Part of them led to an exhaustion of Brazil's trade surplus. Another part, finally, led to a resumption of two-digit monthly inflation rates, as soon as the price freeze was revoked in 1986.

After the failure of the Cruzado Plan, in February 1987 the Brazilian government declared a moratorium on interest payments on its long-term debt to commercial banks, as its foreign reserves were then nearly exhausted. This move caused a commotion in international financial markets, but in the end it was of limited consequence, for the government was unprepared to follow through with a consistent set of proposals backing its new stance towards the country's creditors. However, the implications of the Brazilian action for speeding up other countries' debt negotiations and for changing perceptions about the challenge posed by the debt problem should not be overlooked.

In September 1987 the government resumed negotiations with commercial banks on the basis of a new set of proposals. Preliminary agreement was reached in June 1988. The Brazilian request for a standby arrangement with the Fund was submitted to the Executive Board in July 1988. Negotiations for re-establishing access to bilateral official credits immediately ensued.

Brazil is engaged in the process of "normalization" of its relations with its creditors, but this truce can only be understood as temporary. True enough, in 1988 the country is proving capable of resuming the generation of mega-trade surpluses, but this is being done at the cost of major disorganization of domestic economic relations. To witness, in 1980 Brazil invested 22 percent of its GDP. Financing came mostly from internal sources—20 percent—but 2 percent came from net transfers from abroad.[12] In 1987, investment had dropped to 17 percent of GDP. Internal savings was approximately constant at 21 percent of GDP, but—in spite of the moratorium on interest payments—net transfers from abroad turned negative, to the tune of −4 percent of GDP.[13] Thus, the counterpart of the mega-trade surpluses has been a major reduction in Brazil's capital formation.

Brazil's trade surplus is produced by the private sector. But the public and publicly guaranteed debt responds for over 80 percent of Brazil's medium- and long-term foreign debt. To be able to buy from the domestic private sector the foreign currency it needs to keep current on the interest payments abroad, the government—unable to raise additional taxes on an impoverished economy or to cut personnel expenditures—has been forced first to reduce infrastructure investment and second to expand its domestic debt. Were it not for the interest payments on domestic and foreign debt, Brazil's public sector would have its accounts under balance. Brazil is undergoing a major fiscal crisis, as a consequence of the burden on the government accounts of the adjustment of its external accounts to the debt crisis.

For the time being, the government is trying to cope with this fiscal crisis by maintaining a cooperative mood towards its foreign creditors. But the economy is stagnant and inflation threatens to become explosive. In November 1989, Brazilians are going to elect a new President, thus effectively ending the transition from authoritarianism. By then, either the creditor nations will have found the means to solve the international debt crisis or else Brazil's first democratically elected administration in thirty years will quite naturally develop its own initiatives to cope with the problem, given the growing pressures for social change and modernization, both of which require a reactivation of the development process.

NOTES

1. R. Cooper, "Problems and Prospects for the World Economy: Round-Table," in Rudiger Dornbusch and Jacob A. Frenkel (eds.), *International Economic Policy: Theory and Evidence* (Baltimore: Johns Hopkins Univ. Press, 1979), 325.

2. Examples of this line of argument are Roberto de Oliveira Campos and Mário Henrique Simonsen, *A Nova Economia Brasileira* (Rio de Janeiro: Livraria José Olympio, 1974) and Donald Eugene Syvrud, *Foundations of Brazilian Economic Growth* (Stanford: Hoover Institution Press, 1974).

3. See Régis Bonelli and Pedro Malan, "The Brazilian Economy in the Seventies: Old and New Developments," *World Development* 5:1/2 (1977); and Edmar Bacha, "Issues and Evidence on Recent Brazilian Economic Growth," *World Development* 5:1/2 (1977). A revised version of the latter also appears as "Selected Issues on Recent Brazilian Economic Growth," in Lance Taylor et al., *Models of Growth and Distribution for Brazil* (New York: Oxford Univ. Press, 1980).

4. Albert Hirshman, *The Strategy of Economic Development* (New Haven: Yale Univ. Press, 1958), 167.

5. P. Nogueira Batista, Jr., *Mito e Realidade na Dívida Externa Brasileira* (Rio de Janeiro: Paz e Terra, 1983).

6. J. W. Dizard, "International Banking: The End of the Let's Pretend," *Fortune*, Nov. 29, 1982, p. 6.

7. Central Bank of Brazil, *Brazil: Economic Program—Internal and External Adjustment*, Vol. 2 (Brasilia, D.F., March 1984), p. 4.

8. Ibid.

9. Carlos Díaz-Alejandro, "Some Aspects of the 1982–83 Brazilian Payment Crisis," *Brookings Papers on Economic Activity* 2 (1983): 31.

10. B. Balassa, G. Bueno, P. P. Kuczynski, and M. H. Simonsen, *Toward Renewed Economic Growth in Latin America* (Washington, DC: Institute for International Economics, 1986), 17.

11. For an analysis of the Cruzado Plan see D. D. Carneiro, *Brazil Country Study 11. Stabilization and Adjustment Policies and Programs* (Helsinki: WIDER, March 1987), and E. Modiano, "The two cruzados: The Brazilian stabilization programs of February 1986 and June 1987," *Journal of Economic and Monetary Affairs* (1988, forthcoming).

12. Net transfers from abroad are the difference between net capital inflows and net foreign capital income. It can also be measured by the difference between imports and exports of goods and non-factor services. Internal savings is the difference between GDP and consumption. All values are in constant 1980 prices, as reported in the Brazilian national accounts.

13. In 1985, previous to the interest moratorium and the Cruzado Plan, Brazil was investing 16 percent of GDP, with internal savings at 23 percent of GDP and net transfers from abroad at −7 percent of GDP.

PART III

Democratizing Pressures From Below

5

The "People's Church," the Vatican, and *Abertura*

RALPH DELLA CAVA

The definitive narrative of the Roman Catholic Church's role in the decade-long process of *abertura* (1974–85) which recently drew to a close will likely require another decade to be written. But its general lines are now beginning to emerge and are worth tentatively tracing here.*

But this tracing is not without purpose.[1] On the one hand, it seeks to underscore the highly "conjunctural" character of the Brazilian Church's effort to return Brazil to the "rule of law." As that effort got under way in the 1970s, the press as well as the Church's partisans and protagonists tended to overstate the case for Catholicism's participation (against that of other institutions), understate its internal divisions (which have once again come to the fore), and finally pay scant heed to the Church's enduring institutional interests as a factor in its recent "political" activity.

On the other hand, tracing the Church's role in the *abertura* without sentimentality will also give us some insight into the longer-run question of the permanence of the Church's grassroots communities, the world renowned *comunidades eclesiais de base* (CEBs). Through some 80,000 CEBs organized along the length and breadth of the country, the Brazilian hierarchy (350 bishops strong and the third largest in the Roman Catholic world) has emerged from this decade as perhaps the single most important voice for the nation's lower classes. Moreover, from the Church's point of view, the

*The original version of this essay was completed in 1985 and subsequently revised as part of a long-term research project, entitled "Catholicism and Society in Post-War Brazil," co-directed by the author and Dr. Paula Montero of the University of São Paulo. The project is sponsored by the Pontifical Catholic University of São Paulo under grants from the Ford Foundation and the Tinker Foundation, Inc.

CEBs have become as much an "alternative" form of cultic organization as they are "schools" for educating the exploited in their inalienable human rights. Finally, it is the CEBs (and several other ancillary, Church-related structures such as the *Comissão Pastoral da Terra*) which have articulated the most radical, people's critique of Brazilian capitalism and equally spirited defense of a new socialist order. Consequently, the question of the durability of the CEBs is not only an academic or ecclesiastical matter, but one of the utmost political importance.

Yet another long-range issue which this analysis seeks to illuminate is the relationship between the Brazilian episcopacy and the Vatican concerning the future of Catholicism in the modern world. The Brazilian hierarchy in the last three decades has emerged as one of the most pastorally innovative, organizationally complex, and theologically progressive episcopacies in the world. Moreover, its powerful vision of the Church's future is thought to stand in sharp contradistinction to the one increasingly embraced by the Holy See over the last decade.

Now, let us turn to the questions at hand: what does the role of the Catholic Church in authoritarian and democratizing Brazil directly reveal to us about the *abertura* as a historically closed process, and what may it indirectly indicate to us about the unfolding future of CEBs in Brazil and the two major competing visions of the Church as Catholicism embarks upon its Third Millennium?

From the military coup of 1964 to the inauguration of the New Republic in 1985, the Church's interaction with Brazilian society can be better understood by dividing the period into four distinct historical moments.

The first extends from the 1964 coup to the simultaneous 1968–69 ascent of military hardliners both to the high command of the armed forces and to the presidency of the Republic.

In this period, the ideological and political divisions which had rent the Catholic laity, clergy, and episcopacy into "progressives" and "conservatives" during the previous decade continued to prevail. Conservatives came to hold the upper hand with the 1965 election of their candidates to the Secretariat of the National Conference of Bishops of Brazil (whose Portuguese acronym is CNBB) and the strong support exhibited by the Holy See for their position.

This development greatly facilitated the State's purge of youthful militants, among them almost all the Catholics identified with the Church's para-ecclesiastical structures such as the Young Christian Students, Young Christian University Students, and Young Christian Workers (respectively, JEC, JUC, and JOC). A change in international Church policy further precipitated their demise. By 1968, the most innovative Catholic experiment since the French worker-priests had forcibly come to a halt. Henceforth, the episcopacy alone would speak for the Church; in matters of politics, both clergy and laity would have to hold their tongues.

But unlike the laity (who were now either indifferent to religion or exiled, imprisoned, or forced into hiding by the repression), the Brazilian clergy—priests and nuns—took to the streets against the arbitrary acts of the regime. Nor did they spare criticism of bishops who preferred silence in these matters or who in Church affairs were slow to favor the winds of change then blowing off the Tiber after the closing of Vatican Council II (1962–65). On a variety of issues (from celibacy to family planning, from a democratization of the sacraments to the election of bishops), the episcopacy proved intransigent.

Priests, historically in short supply, requested laicization en masse. Nuns who outnumbered males four to one followed suit, although at a slower rate. Vocations which had momentarily flourished from the ranks of specialized Catholic Action (JEC, JUC, and JOC) virtually withered, while replacements from Europe were harder to come by as the "priest-crisis" racked the whole of the Church Universal.

This decline in an "ordained" and celibate cadre—a fundamental structural impediment to Catholicism's ability to "reproduce" itself institutionally and as a religious enterprise keep pace with Brazil's demographic growth—could not have occurred at a more inopportune juncture. Within two years, a new industrial boom in São Paulo and the full-scale economic "conquest" of the vast untapped interior would draw anew migrants from the traditional Catholic hinterlands of the Northeast and Minas Gerais, just as São Paulo's great post-World War II industrialization had done after the late forties.

In both these instances, once-Catholic rural workers converted in large numbers to Pentecostalism and to various Afro-Brazilian and other cults on the outskirts of big cities and in the small boom towns of the great jungles.

Pentecostalism especially had been and remains until today the driving wedge against Catholicism's once impregnable religious monopoly. Abandoned by the late nineteenth-century elites for positivistic rationality and by the middle classes some half-century later for Marxist or consumerist talismans, the Church in this last half of the twentieth century has been "evacuated" on ever larger scales for Pentecostalism by ordinary workers of the cities and countryside.

As this first period drew to a close, a conservative hierarchy, solicitous of the military's promises to accede eventually to civilian demands for democratic restoration, confronted in Pentecostalism the challenge of an emergent "people's religion" with far fewer clergy, almost no laity and without effective means to recruit either one.

A " civil war" opens the second period in 1968–69. A campaign for civil rights, instituted in 1973 by various Christian denominations (except for the Pentecostals), sees it to a close.

The " civil war" initially pitted the armed forces against the urban guerrilla movement, a veritable "Children's Crusade" comprised of high school and university students and spearheaded by dissidents of Communist and Marxist parties. In time, the generals also marked the Church as an enemy. In May 1969 an outspoken priest from the Northeast, reviled for his association with a long-standing opponent of the military, Dom Helder Camara, Archbishop of Olinda and Recife, was assassinated by agents of the regime for his non-violent political activities among students, who are always suspect. In turn the Church made of him their first martyr of the repression.

That martyrdom spoke more eloquently than all the resolutions adopted at the Second Conference of Latin America's bishops held at Medellín (Colombia) the previous October (1968) in a monumental effort to "translate" the teachings of Vatican Council II into the realities of this Third World (whence three in every five Catholics would hail by the end of this century). The bishops' denunciations of "institutionalized violence," their commitment to be "one with the poor," and their *mea culpa* for four centuries of alliance with the ruling classes had been drafted by Church theologians and intellectuals, and were embraced enthusiastically by all the bishops and even endorsed by the then reigning pontiff, Paul VI.

For the moment, Medellín was the "saving grace" of the Brazilian Church and, for that matter, for Catholicism in every other Latin American nation where the scandal of military rule impiously took root over the coming years. Momentarily, the massive exodus of disenchanted clergy from the priesthood came to a halt, the preoccupation over Pentecostal victories was turned more fruitfully to the defense of the Church's "corporate integrity" against the assaults by the illegal regime, and the once bipolarized conference of bishops (CNBB) converged into a single centrist defense of civil liberties and human rights. But only by the end of this period.

At the beginning, the Church of the Brazilian Northeast, the most impoverished area of the country, literally stood alone. Its bishops had pioneered since the mid-fifties new forms of church organization, consciousness-raising campaigns for adult literacy, and the first, albeit ecclesiastically controlled and ideologically anti-Communist, mobilizations of rural workers.

Furthermore, the increased repression, symbolized by the martyrdom of a young priest and the growing impoverishment of Northeastern workers despite the military-induced "economic boom" of the early seventies, drove the regional Church to action. Document upon document gave the lie to the regime's propaganda, its arbitrary alterations of the legal system, its flagrant disrespect for civil liberties and human rights.

Towards the end of this second period, the Church in the industrialized South came and joined the struggle of their Northeastern confreres. Until then, the Southern bishops as a whole had vacillated towards the military rulers. Despite repeated cause for condemnation of military violations against workers (in the 1968 Osasco strike), against Church pastoral agents (in São Paulo in 1969 and 1971), the conservative hierarchs held out olive branches to the Church's enemies and the people's.[2]

Two dates mark the reversal of this policy. In 1970, in a story whose details are still unclear, the Vatican's Pontifical Commission for Justice and Peace and later Pope Paul VI himself denounced torture in Brazil. In the following year, Dom Paulo Evaristo Arns, the newly appointed Archbishop of São Paulo, as head of the largest Roman Catholic archdiocese in the world publically condemned the torture of Church workers in the prisons of São Paulo, more specifically in the prisons and torture chambers of the soon-to-become infamous Second Army Command headquartered in that metropolis.[3]

The die was cast.

As this second period drew to a close, the Brazilian Church as a whole became galvanized into leading a world-wide campaign against torture in Brazil. Not even repeated efforts of the regime to label churchmen "subversives" and encourage ultra-conservative Catholic groupings to recklessly attack priests and bishops as Communists could prevent the CNBB from forging a united front.

Moreover, prompted by the Northeastern critique of the "economic miracle" (high annual growth rates resulting from declining real wages and labor repression), the Church in the South, especially in São Paulo, soon followed suit and took up the cry.

By late 1973, on the occasion of the twenty-fifth anniversary of the United Nations Universal Declaration of Human Rights, the Christian Churches of Brazil (with the exception of the Pentecostals, who are also seen as "competitors" by such "mainline" denominations as Presbyterian and Methodists) launched a nation-wide campaign for human rights.

The first step towards denying legitimacy to the regime had been taken. Moreover, in the absence of viable voluntary associations and political parties, the Churches in general and the Catholic in particular had by now become the single largest opposition force to military rule. In the case of the Catholic, no other institution except for the military, enjoyed a nation-wide network of cadres, a system of communications (even if only door-to-door) that functioned despite censorship and, unlike the military, a world-wide organization on which it could draw for support and bank on for an international "hearing."

As the Church marshalled the opposition to the regime, the latter's military leaders (in the person of the President-designate of the top brass, General Ernesto Geisel) proposed a "gradual and slow" devolution of political power to civil society. Referred to at the outset as *distensão* (the reduction of tensions), the policy spelled out neither a timetable nor a program.

The slow unveiling of *distensão* in late 1973 and early 1974 opens the third period which for our purposes can be said to have drawn to a close in 1978, when auto workers, with the sympathy of the middle classes and even a few entrepreneurs, unleashed the first strikes in a decade and signaled both the depths of discontent to which society as a whole had been driven and the fragility of military rule.

It is now commonly agreed that the policy of *distensão* had originated within the military for the primary, if not exclusive, purpose of curbing a minority hardline faction within the armed forces. Initially composed of ideologically intransigent conservatives, the hardline by now came to include both officers who had actively combated the urban and rural guerrillas and those who came to control a rapidly and greatly expanded national intelligence agency (the SNI or Serviço Nacional de Informações).[4] At no time was this policy or its subsequent modifications actually intended to return the executive branch of government to the opposition as such. In a word, *distensão* was in effect a policy prompted by internal military factionalism and aimed at more artfully ensuring long-term military rule.[5]

Indeed it was the hardline's reaction to their evident loss of power that triggered the ensuing rash of outright acts of terrorism perpetrated, it now appears, with SNI's knowledge or outright collaboration against the Church, the National Bar Association (OAB), the journalists' associations, research centers (such as the São Paulo-based CEBRAP) and in desperate acts of public carnage (such as "RioCentro"). Against these and other acts of unprecedented violence, even against "elite" citizenry, the moderate military faction remained intent on preserving both military discipline and their own continuity in office. They tacitly made alliances and reluctant concessions to the middle- and upper-class sectors of Brazilian society. By the mid-1970s these sectors began referring to themselves as "civil society" and to their ongoing effort to acquire minimal civil liberties as "opening up spaces" (*abrir espaço*). Moreover, after the surprising electoral upsets of government forces in 1978 and 1980, the term *distensão* definitively yielded to *abertura*, or opening, as a civilian movement intent on returning to the rule of law gained irreversible momentum.

Against this background, the Church was called upon and played its most significant role of the *abertura*, especially in São Paulo. Certainly not the personality alone of Dom Paulo Evaristo Arns, a Cardinal since 1973, accounts for this, although his impeccable character, political acumen, and personal courage would inalterably mark the future. Rather, just as São Paulo most pronouncedly embodied the country's economic and political contradictions and rose up to advocate a better future for the nation, so too did the Church in that metropolis come to speak for the Church throughout Brazil.

The specific conditions that made that possible are reviewed below. But the symbolic emergence of that reality and its political repercussion can be mentioned now. The death under torture of Vladimir Herzog, the Paulista journalist, a Jew, and alleged Communist in October 1975, proved the "last straw" of regime brutality for civil society. An ecumenical service in the São Paulo Cathedral represented the closing of ranks—above party, class, religion, race, and region—of civility against barbarism.[6]

From the point of view of civil society, the Church was now morally empowered to serve as its surrogate. In the campaign against torture, the recently created Peace and Justice Commission of the Archdiocese of São Paulo (P&JC-SP) shared that task with the OAB. In doing so, it also became in effect a nation-wide force of civil society (even though the headquarters of the official National Peace and Justice Commission were in Rio de Janiero, but where the conservative Archbishop, Dom Eugenio Cardinal Sales, who "intervened" with generals as "one power to another," had effectively muzzled it). In its pursuit of the goals of untampered elections, an end to censorship, and amnesty for political prisoners, the Church in effect gave legitimacy to civil liberties and to liberal democracy. In doing so, the church's policies now coincided with the objective of those economically privileged, but politically disenchanted, classes who sought a return to the "rule of law."

It was among the less privileged classes that the Church had most to gain—and most to lose. In the shanty towns (sometimes called *favelas*) on the far-flung outskirts (or *periferia*) of Greater São Paulo populated by largely jobless migrants from the rest of the country and in the satellite towns of industrial workers (like São Bernardo), the Church's "presence" was being severely challenged by competing faiths and ideologies.

Among the jobless, almost all "historic" Catholics, Pentecostalism and a variety of Afro-Brazilian, syncretic religions, most significantly, umbanda, proselytized with astonishing success in the early seventies. The Church, encumbered by the immobilism of its medieval European parish structure, had neither the finances nor the manpower to keep pace (as its policy since the forties had dictated). Among industrial workers, secular trade unionism frankly made more sense. Indeed, the Church's successive strategies since the fifties to gain a foothold in the factories had made little or no headway (as the experience of the Círculos Operários, the JOC, the Frente Nacional do Trabalho, and the "imported" worker-priest experiment in Osasco of French Fathers Domingos Barbé and Jean Wauthier attest). In the long view, the unbridled economic growth and ensuing impoverishment of São Paulo (memorialized in the study commissioned by the P&JC-SP and, entitled *São Paulo: Growth and Poverty*) merely aggravated during the decade realities that had persisted for over a quarter-century.[7]

These concerns clearly lay behind the "trial and error" efforts to catechize between 1969 and 1975 on the heels of the vocation crisis and the demise of specialized Catholic action (JUC, JOC, and JEC)[8]. Elsewhere in the country and especially in the Northeast, however, the CEBs had just begun to take root.[9] Their potential as a new form of Church structure and methods of evangelization was gradually underscored by the two "National Encounters" convened in Vitória (ES) in 1975 and 1976. Finally in 1976 and 1978, the Archdiocese of São Paulo adopted respectively their First and Second "Two-Year Plans" in which the "pastorals" (plans for religious and social action) for promoting CEBs among the jobless on the periphery and among unionized workers in satellite towns would be given priority.[10]

But whatever the original intent of ecclesiastical architects, the CEBs would take on a life of their own. In retrospect the reasons now appear obvious. São Paulo's internationally recruited clergy, inspired by Medellín and frequently harassed by the regime, envisaged a return to the fraternity and equality of primitive Christianity. Their lay cadres—mostly Catholics, but not a few Marxists—came with previous often clandestine political experience and current political concerns in so far as other "social spaces" were under heavy regime surveillance. Liberation theologians (whose works first began to appear only in the early seventies), for their part, found it easier to bring their theologies "down to earth" in the social specificity of the CEBs, while the Church pedagogues, especially attuned to working-class idiom, developed the techniques of applying biblical exegisis to the surrounding social problems. And then, in a parallel development (still to be analyzed) Church intellectuals completely "reworked" the field of "folk religion" (sometimes referred to as popular religion or popular Catholicism); heretofore condemned as "superstitious," the beliefs and practices of the unlettered were now appreciated as potential wellsprings of personal and collective transformation.[11]

Clearly, the receptivity of ordinary and long-suffering believers to this "revolution within the Church" was itself extraordinary (and unfortunately almost without written record). But three factors that help explain the rapidity with which this "process" reunited, renewed, and relaunched an institution and its constituency on a largely overt political-religious trajectory are worth mentioning briefly before concluding this period.[12]

First and foremost, a new hegemony, i.e. new hegemonic group—let's hereinafter call it, for want of a better term, the People's Church (Igreja do Povo)—had through an ongoing struggle come to prevail within Brazilian Catholicism by the early seventies. Indeed, the very history of this entire postwar period with respect to religion and society in Brazil is the history of this group's struggle. Thus, until *this* story can be told—with names and documents—our accounts per force must take recourse to the institutional and public record and its (necessarily distorted) reading of our times.[13]

But for our purposes here the new hegemony of the People's Church is best understood as follows. Since the 1950s, several generations of Catholic activists—men and women, lay and clerical, hierarchs and the grassroots—have forged a common cause and common links. Like the Church itself, this cadre is trans-national in character and indeed has its existence as much in the trans-national Church as it does in the specific arrangements of classes, regions, and ethnicities of Brazilian society itself.

It is this trans-national cadre—anchored ideologically in the liberal European Catholic critique of authority and tradition within the Church and, in a Latin American "Third World" critique of capitalist underdevelopment and the excesses of State power—that is directly responsible for today's progressive stance of Catholicism.[14] Vatican Council II, Medellín, and, in part, "Puebla" (after the Mexican city in which the III Conference of Latin American bishops was held in early 1979) are their "work." So too, it can be said, is the People's Church in Brazil in the seventies.

The formation of this cadre (like their struggle against those who held power in the Church through the first half of this century) still goes untold. But it is clearly linked to institutionally identifiable "movements, currents, forces, and schools" (as Raymond Williams might say) and can be with patience historically reconstructed. In that latter task, prominence must one day be accorded the Catholic University of Louvain and the generation of priest-sociologists trained there, the JOC, JEC, and JUC movements in Brazil and its offshoots and extensions (such as Ação Popular and the Movimento de Educação de Base), the São Paulo-based Dominicans, and to a lesser extent the Rio de Janeiro-based Centro João XXIII of the Jesuits.[15]

Finally, the struggle itself takes place at all levels within the Church (nationally and trans-nationally, from the grassroots to the hierarchs) and has enveloped bishops, pastors and nuns; schoolteachers and pastoral agents; sociologists, economists, political scientists, and historians; journalists and theologians; pastoral agents and Church office workers—in a word, the "organic intellectuals" of Gramscian thought. In the early seventies, their struggle accorded the People's Church an extraordinary ascendance within the institution as a whole. To the outside world, the former's commitment to the poor and a more just social order was perceived (or misperceived?) to be that of the entire Church.

The second factor which brought institution and constituency together with such rapidity was the CNBB's unequivocal legitimization of the "process" that had begun to unfold.[16] As early as 1970, a more liberal slate took office as well as the day-to-day control of the Secretariat and permanent commissions. In August 1973, it seized the offensive with regard to *distensão*, and in March 1974 authorized Brazil's four Cardinals to attend General Geisel's inauguration. The CNBB answered attacks against individual clergymen (such as Dom Helder or Dom Adriano Hypólito, the bishop of Rio de Janeiro's industrial suburb of Nova-Iguaçu) and consequently the "corporate autonomy" of the Church with unanimous denunciations. It countered the arbitrary measures of the regime with the frank espousal of democratic principles. In record time, the CNBB and its annual assemblies of the nation's bishops became the bellweather of the struggle for democracy. Four consecutive documents between July 1974 and late 1979 so attest.[17]

But it is within the governing structures of the Church as a whole that the CNBB played an even more substantial and innovative role: it formally sanctioned new institutions through which critical social segments of the faithful could mobilize against the onslaught—for the most part boldly economic, and at times coercively violent—of the regime.

Three of these merit mention here: the Council for the Indigenous Mission (CIMI), established in 1972; the first Inter-Ecclesial National Encounter (in 1974), a deliberately "non-official" structure to convene the diverse CEBs for the purpose of "exchanging" experiences; and finally the Land Pastoral Commission (CPT), created in 1975 and which like CIMI and the National Encounter was at once autonomous of the CNBB, but—through its bishop-members and officers—an indissoluble entity of the CNBB itself.[18]

The three structures shared several features in common: for the most part, they held jurisdiction in the "frontier" areas of the country where capitalist intervention in the form of government corporations, private enterprise, and the multinationals advanced savagely and unbridled; they "spoke for" local communities of Indians, rural day-laborers, squatters, and small farmers whose lives were daily threatened by "economic progress" and where neither government (in the guise of the National Indian Service and rural unions) nor political parties took up their defense; finally, they were structures in which Church cadres (from bishops to laymen) genuinely shared the lives and destinies of their humblest constituents and exhibited in the face of daily danger extraordinary personal courage.[19]

Of the three, the CPT is the most controversial, perhaps because it is so consistently militant in both its attacks on the regime and its defense of rural workers. It is also controversial within the Church and as such stands as a metaphor for the "problematic" surrounding all grassroots undertakings. Are these new structures, for example, merely new "instruments" for the Church to manipulate its historical rural (and in the case of Brazil, indigenous) "wards"? Or do these new organizations reflect a genuine Church effort to keep pace with the dramatic changes in class relations now occurring and that in time will give way to more appropriate interest groups, like a workers' party or rural trade union?

These questions also lay at the heart of frequent conservative accusations that the Brazilian Church favors only one class, that of the "oppressed" to the detriment of its universal ("polyclass") mission. They also are at the root of an ongoing debate among Church intellectuals as to whether the "people" are or are not capable of shaping their own destiny without the "guidance" of a (usually middle-class) "vanguard" of pastoral workers. To this debate,[20] precipitated fully only after 1980, it will be necessary to return again. Nor is it irrelevant to the third and final factor bringing institution and constituency together.

That factor is the emergence of new secular social movements among the popular classes in the peripheries of São Paulo between 1973 and 1978. They took various forms—from mothers' clubs to youth groups, from daycare centers to cost of living associations. But regardless of their particular finality, in each of them "the role of the Church was central and direct."[21] For outsiders, the exact relationship has remained unclear. A study now in progress may shortly clarify that nexus.[22]

But these popular movements, like the CPT and the CEBs, posed questions about grassroots organizing similar to the "problematic" mentioned before. Above all, they posed a fundamental political question: could these movements which arose over local and specific campaigns (for paved roads, sewerage systems, new schools, etc.), and more often than not geographically delimited to one or another periphery neighborhood, resist the return of electoral politics or the need for truly national organizations like parties and trade unions? Moreover, could the participatory experience made necessary by the regime's excesses (against liberties) and inadequacies (towards popular welfare) ultimately transform elitist politics (whether military or civilian) into a truly democratic politics? Into a truly democratic praxis?

For the Church, this political question might prove itself to be nothing less than "subversive." If indeed the CEBs were the essential nucleus on which—according to many observers—the new "Popular Movement" (as these secular social movements came to be called collectively) was built, then what would be the consequences of a democratic social praxis upon the Church? A democratic praxis within the CEBs? Within the Church as a whole?[23] Precisely this vision of democracy within the Church was one of the central projections of Leonardo Boff, the Franciscan theologian.[24] During this period, Boff had come to be ranked among Latin America's leading exponents of Liberation Theology, that interpretation of Catholicism which argued that the Faith had to free men from their sins and social injustice simultaneously, an interpretation whose orthodoxy was continuously challenged since the early seventies.

In the ensuing and, for our purposes, last period, the preceding questions came forcibly to the fore. New ones also arose. But, from 1978 until military rule formally ended in March 1985, facile description of the political process finds no haven.

As for the Church the period was replete with paradox and frought with dilemma. With regard to the polity, it would seem, the very "opening to democracy" (or more modestly put, the transition to civilian rule) to which the Church had so much contributed now obliged it to "disengage" itself from politics and as an institution to disavow the partisan preferences of its rank and file. With regard to world Catholicism, forces within the very trans-national structure out of which the hegemony of the People's Church had emerged were now—with the Holy See's apparent approval—marshalling forces in Brazil and abroad to dismantle that same People's Church.

Obviously, both developments are only analytically separate. Historically and from the point of view of the abiding struggle over power and authority within the Church, they are of a single cloth, inextricably woven together. In the account which follows, however, it seemed advisable to unfold the paradoxes and dilemmas of the period around three nearly chronologically sequential questions: The papacy of John Paul II and the rise of a "Euro-Latin Alliance" within world Catholicism; Brazil's "new parties" law and the conflict between the Church's hierarchy and grassroots over party options; the debate over "the Two Churches" and what I shall call the "conservative restoration" within Brazilian Catholicism.

It is now unarguable that the 1978 election of Karol Wojtyla to the papacy responded to the deep-seated need of his fellow Cardinals to rally to a helmsman who would at least set the supposedly rudderless Montini-esque barque of Peter on a surer course. In Latin America, that sentiment was deeply shared by profoundly conservative hierarchs who slowly but surely came to the conclusion that their young intellectuals and theologians had deceived them in 1968 at Medellín. In fact since 1972, with the election of Alfonso López Trujillo, the ambitious and brilliant conservative bishop of Medellín, to the Secretary-Generalship of CELAM, a systematic region-wide purge of progressive cadre began in earnest, first within CELAM bureaucracies and then throughout the continent.[25] Supported theologically, morally, and financially by a conservative wing of the West German

Catholic hierarchy, the newly forged trans-national, (let us henceforth call it) "Euro-Latin Alliance" set out to "conquer" Rome. At the Fourth Synod of Bishops convened there literally on the eve of Wojtyla's election the die was cast to pull the People's Church out of "politics" and push it back into the sacristy.[26]

Puebla was to be the stage on which the Euro-Latin Alliance and the recently proclaimed Pontiff, John Paul II, were to join forces in the rollback of the People's Church. For more than two years, the CELAM Secretariat had systematically laid plans for Medellín's reversal.[27] Propagated in a preparatory document, the subtle text, labeled the "Green Book," had the opposite effect. Everywhere, the progressive currents reacted. They mobilized national hierarchies, especially the Brazilian and Peruvian, to discuss the proposal publically and, thus seizing the initiative threw the conservatives off guard. A compromise text was hammered out and served as the starting point for the meeting which—postponed by the death of John Paul I—at last commenced in January 1979.

The outcome fell short of victory for the Euro-Latin Alliance. First of all, the progressives, denied status as "experts" (*periti* in Latin) and entry to the assembly hall (a seminary chosen for its distance and incommunicability from the center of Puebla), simply set up shop at local hotels. A system of runners kept them in contact with the handful of progressive prelates "within the walls." As each topic came to the top of the agenda inside the seminary, the *periti* "outside the walls" retorted without delay. Moreover, it was the *periti*—rather than the secretive CELAM press office—who kept the world press abreast of the daily "score." Finally, the Holy Father himself apparently stole the conservatives' thunder. Scandalized by the misery of Mexican peasants (that made their Polish counterparts look like plutocrats), the Pope reportedly discarded a speech prepared for him by CELAM officials and unabashedly embraced the cause of the poor.

Indeed the repudiation of Medellín's teachings about the "poor" had been central to the attempted rollback at Puebla. But the progressives succeeded in approving a final resolution that put the Church on record as endorsing the now celebrated formulation, the "preferential option for the poor." But on other matters, Puebla was in my opinion a "draw," even though the progressives would subsequently portray the event in all the media accessible to them as a victory for the People's Church.[28]

That eventual strategy was well-founded for Brazil, where John Paul II was scheduled to make his second Latin American tour in July 1980. On the one hand, the Pope had indeed publically repeated in Mexico the conservatives' attack on Medellín. He brandished their doctrinal epithets both against the "politicized" view of Christ as a "Liberator" (instead of the theologically "timeless" Saviour of Souls), and against the "parallel magisterium" (alluding to the progressives' supposed "upstaging" of the doctrine of Papal infallibility).[29]

On the other, there were signs that the Euro-Latin Alliance was not without Brazilian sympathizers, or high-placed allies in Rome who could directly pressure the Brazilian Church. In late 1979, the press reported that Leonardo Boff's theology (already attacked seven years earlier by a prominent Brazilian cardinal) was about to be condemned by Rome.[30] In March 1980, rumor had it that the Cardinal Archbishop of São Paulo had knuckled under to "outside admonitions" and withdrawn his invitation to host the IV International Ecumenical Congress of Theology at the Pontifical Catholic University in the center of São Paulo. Months later, unidentified voices clamored for his removal.[31]

Not surprisingly, even the planning of the papal visit was rife with the divisions within the Brazilian hierarchy, while the visit itself appeared as a careful effort to balance one current against another.[32] As to the event itself, it gathered up tens of millions of the faithful to cheer on "John of God" as if he were some heroic Brazilian center-forward and left otherwise loquacious and skeptical social scientists speechless and then adulatory. When they came to their wits, almost all concurred in a single finding: that this great people's "feast" demonstrated how deeply "Christian discourse" underlay the daily language of the masses and that this language antedated and perhaps even resisted many subsequent discourses.[33]

A retrospective balance of the papal visit has yet to be undertaken. But at this juncture three rather daring propositions can be put forth. First, in Brazil and the rest of Latin America, religiosity or faith is almost as extensive as the masses of the poor. But by no means is the People's Church—nor for that matter, the CEBs—even minimally co-extensive with that religiosity. Second, in contradistinction to the multiplicity of representations of the Brazilian Church, the Pope was the single symbol of all religiosity. Moreover, in Brazil as in Mexico (and wherever he travels) the Pope appropriated that faith as "Catholic culture" and as such pronounced it anterior to both the Brazilian nation and all the current specificities and divisions within the actual Brazilian hierarchy. Rome, if you would, transcended class and nation and church and thus bid to command all three.

Finally, for its part the People's Church decided to "appropriate" John Paul II for both of the above circumstances. Just as in the aftermath of Puebla, now on the heels of the papal visit it proclaimed itself, if you will, "the Pope's party." In my opinion, this was a calculated political judgment: the new Euro-Latin Alliance was seen at this moment as the "party of the Roman Curia" in the throes of a comeback that the new Pope, still the champion of the workers of Kracow and Nova Huta, would perhaps in time put in its place. Perhaps.

It was certainly just a matter of time, however, before conservative voices of the Brazilian Church made the Pope's words their own. Selected words, that is. Words, as one wise and sober observer had predicted, that would be carefully chosen and clipped out of context; "words of caution and possible admonishments."[34] The occasion was the race to form new parties, draw up slates and platforms, and launch campaigns for the elections of November 1982. Until the Pope's visit had concluded, the issues posed to the Church by the "new parties" law (the *Lei Orgânica de Partidos* of 1979 which sought to divide the anti-regime opposition by granting the right to form new political parties) simply had to take a back seat.

Conservatives, of course, read every subsequent move as proof of the politicization of the Church. In that context, Dom Eugênio Sales, their spokesman, publically seized upon the "failure" of the CNBB to publish a papal letter of December 1980. This letter warned against temporal "distractions" to the Church's religious mission and against the "betrayal" of putting social welfare above religious salvation. Its most pointed paragraph appears in retrospect to have been "tailor-made" for the moment (as if it had been drafted in Brazil):[35]

> Most grave would be the loss of identity if, under the pretext of acting in society, the Church allowed herself to be dominated by political contingencies, if she became an instrument of certain groups or put her pastoral programs, her movements and her [grassroots] communities at the disposition or at the service of party organizations.

But even progressives and moderate bishops for that matter were hardly advocating a "Church party." In fact, the Cardinal Archbishop of São Paulo had earlier ruled out reviving the onetime confessional Christian Democratic Party whose pre-1964 electoral base in São Paulo was substantial. Moreover, CDP flag bearers were nestled comfortably in the *Partido do Movimento Democrático Brasileiro*, PMDB, the newly formed majority opposition party (that in November 1982 carried Franco Montoro, the leading ex-Christian Democrat, to the governorship of the state).

The problem was elsewhere: in the ecclesial base communities. There the "new parties" law had sown confusion (at best) and partisan militancy (at worst—in the minds of some). In April 1981 at the IV Inter-Ecclesial Encounter, held in Itaicí, which brought together CEB members, pastoral agents and their advisers (*assessores*), (including theologians, bishops, and Church intellectuals), the day devoted to the discussion of politics "was highly problematic."[36] In the end, the Encounter gave its blessing to politics as "a great arm" in the construction of justice and to political parties as possible programmatic and practical instruments of action in society. But the fundamental conclusion was for the CEBs to have no party at all: "We also believe that the ecclesial base community is not and cannot be a party cell (*núcleo partidário*), but rather a place where we must live out, deepen and celebrate our faith . . . so that we might see if our political action is in agreement with the Plan of God."[37]

As early as February 1981, some CEBs, however, had already opted for the Partido de Trabalhadores (PT), the entirely new political grouping of auto workers, intellectuals, and university youth, principally from São Paulo.[38] Led by Luís Inácio da Silva, nicknamed "Lula," the charismatic leader of São Bernardo auto workers, the PT—despite its initial hostility to the Church and to its competitive union cadres (the *oposições sindicais*)—was by now doubly identified as the "Church's party," or at least, the "party of the CEBs."

The decision of the São Paulo hierarchy to convert Church buildings into union halls during the strike of 1980 (after the government had closed them down by force) partly contributed to that picture. So too did the presence of Frei Betto (Carlos Alberto Libanio Christo), a Dominican friar, as Lula's constant companion and house guest during those strikes. Moreover, as pioneer in the early seventies of the CEB experience in Vitória (ES) and author of the principal tract of 1981 that boldly called upon the CEBs "not to be indifferent" or "omissive" to the efforts of workers and popular movement activists to embark on a new "political conduct," Frei Betto and his friendship with Lula gave credence to the "indissoluability" of the PT and the Church.[39]

The hierarchy of the Church certainly did all it could to dispel that perception. In July 1981 it summarily called back the primer, prepared by a commission of the São Paulo Archdiocese and entitled "Faith and Politics." A text accompanied by slides (and also available in cartoon editions), the document was intended to educate CEB members in their political responsibilities and orient their options before the 1982 elections. But the press corps attending the off-the-record briefing about the primer found one cartoon edition to be hardly "neutral." As a voter approached a crossroads in one instance, he came face to face with a traffic sign. Arrows pointed helter-skelter, except for one that pointed to the "right road." On *that* arrow, the initials "PT" were prominently marked. The following day, headlines viciously attacked the hierarchy of partisanship. Subsequent editions of the primer had all initials removed.[40]

Other parties seen as friendly to the "oppressed's" cause—such as the PMDB and the primarily Rio de Janeiro-based Democratic Workers' Party (PDT) of Leonel Brizola—were also endorsed. But throughout the remainder of the electoral campaign the impression of a CEB-PT alliance was indelible.[41]

The PT's poor showing at the ballot boxes in 1982 did more than stun political neophytes. For the elite political class, it assuaged them that the grassroots Church was less of a threat and required less of a hearing than they had previously imagined. For the CEBs and the intellectuals of the People's Church, it called for a major re-evaluation of their position before the new political order and before the Church as a whole.

That evaulation is now in course. Just prior to the elections, it unfolded in an atmosphere of "confusion . . . fear . . . withdrawal . . . and disorientation. . . ." Grassroots CEB members tried to make sense out of the constantly changing pressures exerted upon them by the parties, pastoral agents, and Church hierarchy.[42] Since the elections, it had proceeded with greater dispassion and deliberation. Clearly, this is not the place to set out all the issues under review. However, the preliminary insights of, in my opinion, the most balanced and lucid analyst, the Jesuit priest Cláudio Perani, deserve mention.

Perani, whose writings appear almost exclusively in *Cadernos do CEAS*, the bimonthly journal of his Order's Centro de Estudos e Ação Social in Salvador (BA), suggests in a recent article that after a decade of labors at the grassroots level the Church's work has fallen seriously short of the mark.[43]

First of all, there remains a greater lack of grassroots leaders than of their middle class (lay and clerical) advisers; second, that relationships between the CEBs and secular voluntary associations are as unclear today as a decade ago; third, that the purpose of the Church in "conscienticizing" CEB members about politics and especially partisan politics remains ambiguous; fourth, that there has been a pronounced tendency to use "faith" as a justification for political options, rather than as a source of questioning about such options; fifth, there is a latent impulse to convert the CEBs either into a new form of "Christendom," albeit popular and even "leftist" or some lower-class analogue to the specialized Catholic Action movements of the fifties in which the hierarchy was the first and ultimate locus of decisions.

Indeed, of all the militants of the last decade, Perani seems to be alone in his forthright eschewing of rhetoric about the *povo* (lower classes) bursting in upon the Church and "converting" it. To the contrary, his criticism centers precisely on how weighty the influence of the institutional Church continues to be with regard to the CEBs and their until now hegemonic cadres of the "People's Church."

In his most recent article, Perani reminds us that there is still "a long road ahead" before power is "redistributed" in the Church. "We must recognize," he adds, "the great capacity of ecclesial authority to maintain or recoup its power."[44] Bishops weigh heavily on the institution and their "election" still takes place in Rome. Nor has anything changed in regard to the nomination of vicars to local churches. Moreover, the ever larger number of bishops who have recently endorsed (*aderiram*, in Portuguese) the CEB movement, the CPT, and the CIMI is not necessarily a sign of their "conversion" to the popular cause.[45] Further, the continuing lack of clergy and vocations and, in my own opinion, the rapid numerical expansion of the episcopacy is not necessarily a sign of the "democratization" of the Church, but perhaps rather of the increasingly greater direct intervention of bishops into the day-to-day life of the faithful. Finally, Perani, in a phrase borrowed from a CEB text from Maranhão, seems to exhort CEB advisers to "leave Noah's Ark" now that the military deluge has subsided, for the people are greater than any "grand design" of the CEBs. It is time, Perani proposes, for renewal, new militance and a healthy respect for pluralism at the grassroots.

Perani's call for a fresh start will come as a surprise only to less astute observers of the Brazilian CEB movement. But it is also an accurate indicator of the entire Church's awareness that—with the restoration of political parties, the electoral process, and an elitist political system—Catholicism has come to a new crossroads in its centuries-long inherence in Brazil.

Equally true is that Brazilian Catholicism has come to a new crossroads within the Church Universal. Thought to be impossible only a decade or two ago, conservative forces within the Brazilian Church have come to stake their future on the new Euro-Latin Alliance and, not surprisingly—despite the unilateral appropriation by the People's Church—of the Holy Father himself.

This is not to say, as one analyst who insists it "can never come to pass," that the components of a trans-national church external to Brazil are about to crush in one fell swoop the country's 80,000 CEBs.[46] Nor is it to say, as the same analyst again denies as possible, that a papal visit or even the Papacy is able to "change the Church of Brazil" and its intimate ties to the lower classes.

This is to say, however, that a partial "conservative restoration" within the Brazilian Church (hierarchy and clergy for certain; among the laity it still remains to be seen) has been in progress. Moreover, this "conservative restoration" (as the Brazilian faction shall be called) has forged links to the Euro-Latin Alliance that with each passing day appears to have greater access and truck with the papacy. In addition, it has created its own publications (the locally produced *Boletim do Revista do Clero* and the internationally edited, *Communio*, both distributed nationally out of the Archdiocese of Rio de Janeiro), established working seminars for middle-class professionals, and guarded zealously the teaching posts in many seminaries around the country. Its principal spokesman is Dom Eugênio Cardinal Sales; while a half dozen other bishops can be counted as supporters.

Certainly, the "conservative restoration" is hardly about to "come to power." But its several sources of strength should not be discounted. For one, this faction has never repudiated "democracy"; at least, it was perfectly at ease with the return of civilian elitist politics under Tancredo Neves; it is decisively not socialist, as are many of the CEBs (but hardly the Brazilian electorate). For another, it espouses "that ole time religion": pomp and processions; pilgrimages, novenas, devotions to the saints; in a word, the stuff from which popular religiosity, the Faith of the people, draws its strength, as the Pope's visit to Brazil made clear. Moreover, that kind of religion is purportedly on the rise. Never before have Catholic publishing houses sold so many pamphlets about the lives of the saints.[47] Never before has Pentecostalism had more converts (so that even the Vatican has had to request Brazil's bishops to help discern precisely what religious needs are going unanswered).[48] Finally, even the CEBs have recoiled from their earlier "political road" and reverted into the biblical circles and discussion groups from which they originated one or more decades ago.[49]

But perhaps, the greatest strength of the "conservative restoration" lies in its ties to the trans-national Euro-Latin Alliance, to the latter's access to European Catholic money, and to the expertise of the new "movements" on the rise in an essentially conservative Europe (such as Opus Dei or the Italian-based Communione e Liberazione)[50] or the alternative "models" to poverty developed in India (which for the last several years has had the highest number of vocations in the entire Catholic world) as exemplified by Mother Teresa's "apolitical," "integral and charitable" approach to the dying.[51]

Increasingly, it is the Euro-Latin Alliance that is appropriating Pope John Paul II as the pilot of their project—and with his apparent, enthusiastic support. One of a series of conferences held around the world to "explicate" the Pope's thinking and theology was convened in Rio de Janeiro in October 1984. Theologians from several countries took part as did a select group of local Catholic clergy and laymen. Presided over by Cardinal Sales, the conference served to confer upon him and his archdiocese a place of preeminence in "interpreting" the Holy Father's thought and theology within Brazil.[52]

One last source of strength is worth mentioning: the "conservative restoration" has a base within the hierarchical structure of the Brazilian Church at the very moment, as Perani noted, that the episcopacy is in full expansion and repossession of those tasks once reserved for clergy and laity. Unlike previous conservative movements within Brazilian Catholicism (such as the Society for Tradition, Family and Property or the intellectual grouping headed by the later writer Gustavo Corção and journalist, Leonildo Tabosa Pessôa, who briefly edited *Hora Presente* during the height of repression),[53] the "conservative restoration" cannot be easily dismissed as can a primarily lay movement (as was indeed the fate of TFP and HP in the seventies or specialized Catholic Action in the sixties).

But if this "conservative restoration" is only at the beginning of a clearly uncertain future, its recent past alone would require us to examine it with care. Its historical place is in the Euro-Latin Alliance, more particularly among the cadres of Monsignor López Trujillo, today a cardinal. Its historical task has been to dismantle the hegemony of the People's Church. To that task, it has asiduously architected "the theory of the Two Churches," the one in communion with Rome and the other, the People's Church, a potential danger to orthodoxy, truth, and piety.

The leading exponent of the theory is the recently consecrated bishop, the Brazilian Franciscan, Friar Boaventura Kloppenburg. Former editor of the respected quarterly *Revista Eclesiástica Brasileira*, former apologist of Catholicism against spiritism, Afro-Brazilian sects and pentecostalism, former advocate of ecumenism after Vatican Council II, former official of CELAM under López Trujillo and editor of CELAM's monthly, *Medellín*, Friar Boaventura is also the former professor of his younger confrere, Friar Leonardo Boff.

Kloppenburg's attacks on the People's Church date back to 1971 and 1972, the most vicious to 1977 (entitled in Spanish, *Iglesia Popular*, in which he "confirmed" for the dictatorial regimes of the day that there are "Marxists" and "subversives" among the clergy).[54] His attacks on his disciple began in Brazil in 1982 shortly after his return from CELAM and his elevation to the episcopacy. They were expanded in 1983 with a Brazilian edition of his 1977 work, *Igreja Popular*, and continued through 1984, when, in September, the Sacred Congregation for the Doctrine of the Faith (the former Holy Office or Inquisition) summoned Friar Boff to Rome.[55]

Space does not permit a lengthy analysis of the theological issues in dispute.[56] In fact, what was utterly surprising about the initial charges of the Congregation contained in the so-called "Ratzinger Document" of August 1984 was the lack of reference to either the theologians or the titles of their texts that had given such offense. Only with the subsequent Vatican clarification (of March 1985) was the public to understand that neither Friar Boff's ecclesiology nor Liberation Theology stands condemned.[57]

Then what is this controversy all about?

Two highly speculative assumptions must here be interjected. Let us assume that the People's Church's "calculated judgment" at Puebla and after the papal visit to Brazil—that progressives could constitute the "party of the Pope"—has simply ceased to prevail. No longer is the Pope, at that time hardly in office a year, to be contemplated as a faction or party potentially at odds with the Curia. To the contrary, he must now be viewed as one with Curia, if we mean by "Curia" the far larger, new Euro-Latin Alliance that has been restructuring itself within world Catholicism since the closing of Vatican Council II in 1965. For the moment, let us leave the reasons, historical, philosophical and of political actuality, in abeyance.

But let us make the second assumption: that somewhere at the highest levels of Vatican government—as befits the world's oldest, most continuous trans-national society—in the Papacy itself, but not in isolation from world events and key Churchmen throughout the world—a fundamental policy decision was made about the shape of John Paul II's papacy.

Precisely what that decision or set of decisions encompasses lies beyond the detailed understanding of anyone not privy to them. But without a doubt, the reaffirmation of papal authority in all matters and on all levels was, in my opinion, central to the shape of things to come: papal centralism is now both an end and a means to all other policies. In contrast to the ambivalence of his next-to-last predecessor, Paul VI, this Pope does not vacillate. Nor does he see theology as a field of inquiry free to take on debates, invent new relationships and (least of all) submit them to popular approval. Nor, does he acquiesce as did Paul VI, in a kind of Catholic "polycentrism," a policy by which national churches (or regional ones, as is the case of Latin America or Africa) take on qualities and autonomy of their own. Finally, whatever the long-term goals of this new papacy are, they will clearly be achieved with new forces, new cadres, new directions.

To accept the above is to give an order to what otherwise appear to be haphazard and inexplicable incidents.

Thus the rebuke to Leonardo Boff stands in a succession of rebukes—to Hans Kung and Edward Schillebeeckx and many others. That rebuke declares that theology is no longer speculative, while it is the Pope alone who will mark the time of Liberation Theology.

To intervene in the naming of reactionary bishops to the progressive Dutch Church, to detain under Roman house arrest the "witch-doctor" bishop of Lusaka (Zambia), to pit the "conservative restoration" against the People's Church in Brazil (or in Nicaragua), to withdraw support from Solidarity in favor of a Vatican-sponsored agricultural bank in Poland are consecutive affirmations of the new centralism that the Papacy is eagerly restoring.

Finally, to admonish the Jesuits, historically the "Pope's Shock Troops" since the sixteenth century, to elevate Opus Dei to an unassailable "personal prelacy," to exalt Mother Teresa as the exemplary servant of the poor, and Italy's Communione and Liberazione as the face of world youth, and finally to invoke a mythic Christian Europe bound by Marian shrines from the Urals to Hadrian's Wall, from Lujan to Aparecida, from Guadelupe to Washington, are part and parcel of the new forces, cadres, and directions now emerging (with no preordained guarantee, of course, that they will meet their goals).

And speaking of goals, what are the Vatican's for Latin America, where at the end of this century three in every five Catholics in the world will reside? And for Brazil, where the People's Church—still a vital, living force of countless believing militants—has evidently and irreversibly come into conflict with this Papacy's "grand design"?

Here I recall that unnamed analyst who earlier in this text denied that a Papal visit could change the Church of Brazil or that trans-national conservative cadres could crush the intimate ties of the lower classes to the resourceful, even if besieged, hegemonic cadres of the People's Church.

Caution! Anything is possible. Especially with the major shift under way today within the papacy. In the case of Brazil, a key determinant in resisting this shift that would silence not only the People's Church but also their socialist ideal for Brazilian society as a whole is—as this essay suggests—the *comunidades eclesiais de base*. Only in probing their real strength can the chances for survival of the People's Church really be measured; but that will have to be the subject of an entirely separate analysis.

Even if the People's Church fails to carry the future, whether for world Catholicism or, on a lesser scale, for the Brazilian Church, there is absolutely no doubt that in their moment of conjunctural unity in the early and mid-1970s, the Vatican, the Brazilian episcopacy, the network of CNBB intermediate organizations such as CPT, and the priests, nuns, laity and poor of the "People's Church" played a crucial role in augmenting democratic pressures in Brazil. They did so by carrying out their prophetic mission of denouncing the abuses of the authoritarian State, while at the same time, protecting and expanding democratic spheres and practices within Brazilian society.

NOTES

1. One of the more creative theoretical works is Vanilda Paiva, "Teses sobre a Igreja Moderna no Brasil" (23–25 Nov. 1983, mimeographed, 16 pp.). The most empirically comprehensive study of the Church during the last decade is Scott Mainwaring's, *The Catholic Church and Politics in Brazil, 1916–1985*" (Stanford: Stanford Univ. Press, 1986), parts of which have appeared in the journals, *Sintese* and *Revista Eclesiastica Brasileira* (hereafter cited as *REB*). His bibliography is indispensable. A useful (and sympathetic) account of the Church's role in recent politics is Maria Helena Moreira Alves, *Estado e Oposição no Brasil* (*1964–1984*) (Petrópolis: Editora Vozes, 1984).

A favorable view of the "progressive" Church is Helena Salem (ed.), *A Igreja dos*

Oprimidos, Brasil Hoje No. 3 (São Paulo: Editora Brasil Debates, 1981). The most hostile view of the same "progressive" Church is Plínio Corrêa de Oliveira et al., *As CEBs . . . Das quais muito se fala, Pouco se Conhece—A TFP as descreve como são Comunidades Eclesiais de Base* (São Paulo: Editora Vera Cruz, n.d.); likely date of publication was 1982. Its bibliography, however, is impartial and extensive.

2. A conservative apologia for the military regime is TFP [Sociedade para a defesa da Tradição, Família e Propriedade], *Meio Século da Epopéia Anticomunista*, 2nd ed. (São Paulo: Editora Vera Cruz, 1980). The charge that one conservative Catholic group went so far as to sanction torture is found in Charles Antoine, *O Integrismo Brasileiro* (Rio de Janeiro: Editora Civilização Brasileira, 1980); the original French edition was published earlier.

3. For an account of this period see Thomas Bruneau, *The Political Transformation of the Brazilian Catholic Church* (New York: Cambridge Univ. Press, 1974).

4. This viewpoint is expressed in Alfred Stepan, *Rethinking Military Politics: Brazil and the Southern Cone* (Princeton: Princeton Univ. Press, 1988).

5. See, Ralph Della Cava, "The Military, the Opposition and the United States: Three Forces Shaping the Future of Brazil," *The Brasilians* (New York), April–May 1978; also published in a shorter version as "Democratic Stirrings," *Journal of Current Social Issues* (Summer 1978): 22–28.

6. Two English-language accounts of this episode are "Brazil: The Sealed Coffin," *New York Review of Books* 22, no. 19 (27 Nov. 1975): 15; and, Ralph Della Cava, "Brazil: The Struggle for Human Rights," *Commonweal* 52, no. 20 (19 Dec. 1975): 623–26.

7. Archdiocese of São Paulo, Peace and Justice Commission, *São Paulo: Growth and Poverty* (London: Bowerdan Press, 1978); originally published as *São Paulo 1975: Crescimento e Pobreza* (São Paulo: Edições Loyola, 1976).

8. There is no study of the demise of a specialized Catholic Action nor of the "trial and error" catechetics of the period as a whole. The *cursilho* movement was especially "successful" during the period as were local youth movements, especially in the archdiocese of Rio de Janeiro.

9. On the initial primacy of CEBs in the Northeast, see Affonso Gregory (ed.), *Comunidades Eclesiais de Base* (Petrópolis: Editora Vozes, 1973), 46–96, esp. p. 49. The best succinct overview of the CEBs is Cândido Procópio Ferreira do Camargo, B. Muniz, and A. F. de Oliveira Pierucci, "Comunidades Eclesiais de Base," in *São Paulo: O Povo em Movimento*, ed. Paul Singer and Vinicius Caldeira Brant (Petrópolis: Editora Vozes/CEBRAP, 1980), 59–81. The most comprehensive treatment is S. J. Marcello de Azevedo, *Basic Ecclesial Communities: The Challenge of a New Way of Being Church*, translated by John Drury (Washington, D.C.: Georgetown Univ. Press, 1987).

10. On the Two-Year Plans, see Thomas Bruneau, *The Church in Brazil—The Politics of Religion* (Austin: Univ. of Texas Press, 1982), 91; especially useful is chap. 5, "Current Church Responses and Strategies."

11. There is an extensive literature, but the critical Catholic "revision" is partly published in "Catolicismo Popular," the special issue devoted to the subject of *REB* 36, no. 141 (March 1976); an initial attempt to take stock of the question is Rubem César Fernandes, "'Religiões Populares,' Uma Visão Parcial da Literatura Recente" *BIB* 18 (2nd semester, 1984), pp. 3–26.

12. The notion of a "process" is found in Cândido Procópio Ferreira do Camargo, et al.'s article on the CEBs (cited above in note 9).

13. The phrase "People's Church" is borrowed directly from the title of Cândido

Procópio Ferreira do Camargo's "A Igreja do Povo," *Novos Estudos-CEBRAP* 1, no. 2 (April 1982): 49–53. The author may have purposely used this phrase in contradistinction to "Popular Church" (*igreja popular*), a term that since about 1982 is largely employed by critics of the progressive Church. Consequently, "People's Church" struck me as a politically neutral expression that nonetheless conserved the quintessential outlook of the progressives and so is the preferred usage here. On the emergent hegemony of the progressives, see the largely schematic work of Luís Gonzaga de Sousa Lima, *Evolução Politica dos Católicos e da Igreja no Brasil* (Petrópolis: Editora Vozes, 1979) and the recent (and first) overview by Scott Mainwaring, "The Catholic Left, 1958–1964," Chapter 5 of his *The Catholic Church and Politics in Brazil, 1916–1985* (Stanford: Stanford University Press, 1986).

Monographic sources include Emanuel de Kadt, *Catholic Radicals in Brazil* (New York: Oxford Univ. Press, 1970); Luís Alberto Gomes de Souza, *A JUC: Os Estudantes Católicos a Politica* (Rio de Janeiro: Editora Civilização Brasileira, 1984); and Scott Mainwaring, "A JUC e o surgimento da Igreja na base," *REB* 43, no. 169 (March 1983): 29–92.

14. The concept of a trans-national cadre remains to be elucidated and empirically anchored. Obviously, religious orders, congregations, and some lay institutes—terms for relatively autonomous, usually self-governing groupings within the Church—have almost always been "trans-national" and international. But the sense of "trans-national cadre" in this essay refers to no nominal grouping *per se*. Such cadre, as I employ the term, constitute an "informal" network of like-minded activists; this less restrictive sense is adumbrated to some extent in successive paragraphs of the text.

15. Other important groupings might have been included, such as the Commissão de Estudos de História da Igreja na America Latina (CEHILA), Instituto de Teologia, Recife (ITER), and the Instituto Brasileiro de Análise Socío-Econômica (IBASE).

16. On the CNBB, see, Paulo J. Krischke's *Igreja e Populismo: A Crise da Democracia no Brasil* (forthcoming), originally submitted as a doctoral dissertation in political science at York University (Canada), Sept. 1981.

17. See *Estudos da CNBB*, no. 2, "Igreja e política: subsídios teológicos," 4th ed., 1980 (originally published in 1974); ibid., no. 24, "Subsídios para uma política social," 4th ed., 1982 (originally published in 1979). Also see *Documentos da CNBB*, no. 8, "Comunicação pastoral ao povo de Deus," 3rd ed., 1977 (originally published in 1976); ibid., no. 10, "Exigências cristãs de uma ordem política," 9th ed., 1981 (originally published in 1977). Both series, *Estudos* and *Documentos*, are published by Edições Paulinas of São Paulo.

The CNBB assemblies are extensively reported in the daily press of São Paulo and Rio de Janeiro; official texts can be found in the monthly *SEDOC*, an information bulletin published by Editora Vozes.

18. A brief sketch of these three institutions can be found on João Batista Libânio, *O Que É Pastoral?* (São Paulo: Editora Brasiliense, 1982), esp. pp. 96–120. On CIMI, see, Fany Ricardo, "O Conselho Indigenista Missionário, 1965–1979," *Cadernos do ISER*, no. 10 (1980); on the National Encounters (*encontros*) of CEBs, consult the appropriate issues of *CEDOC*; on the CPT, see the official publication of the CNBB, *Pastoral da Terra* (São Paulo: Edições Paulinas, 1976) and a follow-up, "Pastoral da Terra: Posse e Conflito," *Estudos da CNBB*, no. 13 (São Paulo: Edições Paulinas, 1976).

19. The CIMI and CPT have recorded these events in their own publications, *Porantim* (Brasília) and *Boletim* (Goiânia), respectively; the CEBs have no central

office, but their several encounters are documented in *SEDOC*. A dated, but now classical account of struggles along the Indian frontiers—with reference to the role of several CIMI activists, such as the bishop of Araguáia, Dom Pedro Casadáliga—is, Shelton Davis, *Victims of the Miracle: Development and the Indians of Brazil* (New York: Cambridge Univ. Press, 1977).

20. Three starting points to this debate are: Luis Gonzaga de Souza Lima, *Evolução Política dos Católicos e da Igreja no Brasil*; Souza Lima's important article, "Comunidades Eclesiais de Base," *Revista de Cultura Vozes* 74, no. 5 (June–July 1980): 61–82; and Frei Betto, *O Que É Comunidade Eclesial de Base*, 2nd ed. (São Paulo: Editora Brasiliense, 1981).

21. This is the conclusion of Vinícius Caldeira Brant, "Da resistência aos movimentos sociais: a emergência das clases populares em São Paulo," in *São Paulo: O Povo em Movimento*, 9–28, esp. p. 14.

22. Letter from Paulo J. Krischke, São Paulo, 25 Feb. 1985: "inicia-se agora um estudo sobre a influência de setores ligados à igreja nos movimentos sociais urbanos de São Paulo na última dêcada;" the project is centered in CEDEC, São Paulo.

23. Souza Lima, "Comunidades Eclesiais de Base."

24. Leonardo Boff, *Igreja—Carisma e Poder* (Petrópolis: Editora Vozes, 1977); this contains a number of previously published articles.

25. The "swing to the right" within the Latin American Church has been no secret. It is documented in, among other sources, Penny Lernoux, *Cry of the People* (New York: Penguin, 1982); another side of the question is presented in José Comblin, "A América Latina e o presente debate teológico entre neo-conservadores e liberais," *REB* 41, no. 164 (Dec. 1981), 790–815.

26. Enrique Dussel, "A Igreja Latino-Américana na Atual Conjuntura (1972–1980)," in *A Igreja que surge da Base*, ed. Sérgio Torres, et al. (São Paulo: Edições Paulinas, 1982), 160.

27. My account rests on Penny Lernoux, "The Long Path to Puebla," 3–27, esp. p. 23, and Moisés Sandovál, "Report from the Conference," 28–43, both to be found in *Puebla and Beyond*, ed. John Eagleson and Philip Scharper (Maryknoll: Orbis Books, 1979). Another useful American source on Puebla prior to its originally scheduled dates, 12–28 Oct. 1978, is the entire issue of *Cross Currents* 28, no. 1 (Spring 1978).

Position papers of the conservative Churchmen are contained in a very early work, *Teología de la Liberación* (Burgos: Ed. Aldecoa, 1974); this includes the revealing analysis of Mon. López Trujillo, "Panorama de la Liberación en América Latina," 295–326, which argues first that liberation is "integral and universal" and, second, that it is a gratuitous gift from God; he then concludes: "Una liberación que, si bien tiene incidéncia en lo social-económico-político-cultural, no se agota en ellas, y adquiere plena significación humana y cristiana a partir de la liberación en Cristo" (p. 26). This has been the basic (rather reasonably formulated) position of the otherwise aggressive campaign against the Liberation theologians.

28. See, Frei Betto, *Diário de Puebla* (Rio de Janeiro: Editora Civilização Brasileira, 1979) and the popular account in cartoons, *Puebla para o Povo* (Petrópolis: Editora Vozes, 1979).

29. Sandovál, "Report from the Conference," 32–34.

30. "Scherer critica o livro de Boff," *Estado de São Paulo*, 29 Aug. 1972; "Núncio ignora punição de teólogo," *Estado de São Paulo*, 23 Dec. 1979; "O teólogo brasileiro se defende e avisa: se fôr punido . . . ," *Jornal da Tarde* (São Paulo), 24 Dec. 1979; "CNBB nega debate sobre Boff," *Estado de São Paulo*, 29 Dec. 1979; and, "No Vaticano, o 'dossiê Boff'," *Estado de São Paulo*, 20 Jan. 1980.

31. Interviews conducted in São Paulo, July–Aug. 1981.

32. Luís Alberto Gomes de Souza, "E agora, depois da visita?" in *O Povo e o Papa*, ed. A. L. Rocha and Luís Alberto Gomes de Souza (Rio de Janeiro: Editora Civilização Brasileira, 1980), 210–15; an earlier and slightly different version appeared as "Balanço da Visita do Papa," *REB* 40, no. 159 (Sept. 1980). Other accounts of the papal visit to Brazil are, Joviano Soares de Carvalho Neto, "O Papa no Brasil: Impacto e sentido de uma viagem," *CEAS* 69 (Sept.–Oct. 1980): 9–24; and Rubem César Fernandes (ed.), *O Papa no Brasil, Aspectos Sociológicos, Cadernos do ISER*, 11 (1980).

33. José Oscar Beozzo, "Religiosidade Popular," *REB* 42, no. 168 (Dec. 1982): 744–58.

34. Gomes de Souza, "E agora depois da visita?," 213.

35. See, "A Carta de João Paulo II aos bispos do Brasil," (10 Dec. 1980) in *REB* 41, no. 161 (March 1981), 152–57, esp. pp. 153–54.

36. Luís Alberto Gomes de Souza, "A política partidária nas CEBs," *REB* 41, no. 164 (Dec. 1981), 708–27; a substantially similar version also appeared in *CEAS*, 76 (Nov.–Dec. 1981): 36–49.

37. Gomes de Souza, "A Política partidária nas CEBs," 724.

38. Ibid., 724–27.

39. Frei Beto, *O Que É Comunidade Eclesial de Base*. An earlier effort to examine the relationship between the CEBs and politics is, Clodovis Boff, *Comunidade Eclesial, Comunidade Política: Ensaios de Eclesiologia Política* (Petrópolis: Editora Vozes, 1978), in which some essays date to the mid-seventies and thus constitute an important document for a more detailed historical examination of the evolving political perceptions of the CEBs' church architects.

40. The primer was published as, Arquidiocese de São Paulo, *Fé e Política: Povo de Deus e Participação Política* (São Paulo: Comissão Arquidiocesana de Pastoral dos Direitos Humanos e Marginalizados de São Paulo, 1981); the relevant slides/sketches are numbers 23 and 24 (p. 17).

The primer's publication caused an uproar that is recorded in the daily press; see, "CEBs de São Paulo ensinam ao povo como deve votar," *Jornal do Brasil*, 2 Aug. 1981; and, "Igreja lança curso de política," *Folha de São Paulo*, 2 Aug. 1981. Even earlier, criticism of the archdiocese's use of graphics to instruct the popular classes on current political issues was widespread; symptomatic was "Assim atúa a 'ala progressista' da Igreja," *Estado de São Paulo*, 27 Sept. 1981.

41. Joviniano de Carvalho Neto de Carvalho Neto, "Os partidos políticos no Brasil: de 1945 aos nosso dias," *CEAS* 73 (May–June 1981): 7–20. Pertinent is the author's observation: "O PT surgiu e cresce como uma perigosa aliança de operário e Igreja de Base, uma rebelião apoiada pela igreja" (p. 19). This is but one of the many affirmations of the alliance, frequent denials to the contrary notwithstanding.

42. Cláudio Perani, "Comunidade Eclesial de Base e Movimento Popular," *CEAS* 75 (Sept.–Oct. 1981): 25–33; also see, Roseli Elias, "CEBs: movimento de base da Igreja," *CEAS* 69 (Sept.–Oct. 1980).

43. Cláudio Perani, "Pastoral Popular: Poder ou Serviço?" *CEAS* 82 (Nov.–Dec. 1982), 7–19.

44. Cláudio Perani, "A Igreja no Nordeste—Breves Notas Histórico-Críticas," *CEAS* 94 (Nov.–Dec. 1984): 53–73.

45. See, "Testemunho dos bispos em apoio das CEBs," issued on the occasion of the Fifth Inter-Ecclesial Encounter held in Canindé, Ceará, in July 1983, in *REB* 43, no. 171 (Sept. 1983): 595–97.

46. Luís Gonzaga de Souza Lima, "A visita do Papa: a versão das elites e as clases populares," in *O Povo e o Papa*, 11–14.

47. "A moda do povo: católicos veneram santos como no passado," *Veja*, 13 March 1985, p. 75.

48. "Vatican Asks Bishops' Advice on Countering Sects," *New York Times*, 17 March 1985; and, "Atuação dos novos missionários preocupam à Igreja," *Folha de São Paulo*, 3 March 1985. The latter refers to a CNBB study of the situation which was recently dispatched to the Secretariat of Christian Unity.

49. Perani, "A Igreja no Nordest."

50. José Comblin, "Os 'Movimentos' e a Pastoral Latino-Américana," *REB* 43, no. 170 (June 1983).

51. Sometime between 1981 and October 1984, Mother Teresa had been invited to the Archdiocese of Rio de Janeiro. In some quarters the prospect of her establishing a house there was perceived as a challenge—in practice and conception—to the CEBs and their approach to the "oppressed."

52. This was held in Rio de Janeiro in October of 1984.

53. On *Hora Presente*, see, Antoine, *O Integrismo Brasileiro*, and Samyra B. S. Vieira, "Integrismo Católico X Conservadorismo Político: Os intelectuais da revista *Hora Presente*" (relatório de pesquisa em pos-graduação, Universidade de São Paulo, June 1983, mimeographed). On the TFP, see Thomas Niehaus and Brady Tyson, "The Catholic Right in Contemporary Brazil: The Case of the Society for the Defense of Tradition, Family and Property (TFP)," in *Religion in Latin American Life and Literature*, L. C. Brown and W. F. Cooper, eds. (Waco, Texas: Markham Press Fund, 1980).

54. These are the charges which Leonardo Boff brought against his former professor in "Igreja: Carisma e Poder—uma justificação contra falsas leturas," *REB* 42, no. 166 (June 1982): 227–60.

55. One of the important polemics arose after the publication of Leonardo Boff's *Igreja: Carisma e Poder* (Petrópolis: Editora Vozes, 1981). See Boaventura Kloppenburg, "A Eclesiologia militante de Leonardo Boff," *Jornal do Brasil*, 27 June 1982, and Boff's reply, "Segurança eclesial e a eclesiologia militante," *Jornal do Brasil*, 25 July 1982.

56. I made a partial attempt in an interview entitled "Liberation Theology on Trial," *Abigarrada* (Queens College) 6, no. 2 (Nov.-Dec. 1984); early Brazilian efforts can be found in *Comunicações do ISER* 3, no. 11 (Nov. 1984).

57. "Vatican and Friar: Not a Judgment on New Theology," *New York Times*, 22 March 1985. The "Ratzinger Document" was issued by the Sacred Congregation for the Doctrine of the Faith under the title, "Instructions on Certain Aspects of Liberation Theology."

6

Grassroots Popular Movements and the Struggle for Democracy: Nova Iguaçu

SCOTT MAINWARING

This essay analyzes the role of grassroots popular movements in the struggle for democracy.* The subject is important for several reasons. First, although I agree with the other analyses that the *abertura* was largely an elite process, I also argue that grassroots movements did have an impact on the political situation, especially after 1978. Second, grassroots movements awakened great hopes and expectations in Brazil, particularly between 1978 and 1982, when many people saw them as the answer for a true democratizing process. For years the military regime had successfully contained the opposition; the amalgam of movements that burgeoned in the late 1970s suggested not only an erosion of the military regime, but also the birth of a more autonomous civil society. These expectations were somewhat exaggerated, but grassroots movements are central to any analysis of the prospects for strengthening civil society in Brazil. Third, popular movements are an important mechanism of representation of the poor majority of the Brazilian population. Their efforts to win basic services (some of which are considered "rights" in many societies) assume particular significance in light of the egregious inequalities in Brazilian society. They are also an important channel of popular participation, a "space" in which poor people learn to assert themselves and develop new conceptions of authority, legitimacy, and justice. Finally, the quality of democracy will be closely connected to the dynamism of grassroots movements and their relationship to the State.

*The author wishes to thank Elizabeth Allen, Ruth Cardoso, Peter Flynn, Margaret Keck, Paulo Krischke, Alfred Stepan, and various colleagues at CEDEC for their helpful suggestions.

Whether Brazil's democracy will allow for more popular participation than the political system did in the past, and whether it will address the gross inequalities, depend partly upon the strength of popular movements.

In this essay, I analyze the contributions and limits of urban popular movements in the struggle for democracy in Brazil. I argue that these movements played a meaningful role in the process of democratization, but that this very process created new dilemmas and internal conflicts for the movements. Movement leaders expected political liberalization and the transition to democracy to strengthen social movements, but this hope proved ill founded. These movements face major barriers in working towards a democracy more responsive to the vast contingent of the urban poor. An analysis of urban popular movements can detect some propitious changes in the popular process and in the Brazilian polity, but on balance, the weight of conservative political actors appears likely to remain preponderant.

Even more so than is the case for most other political actors, the analysis of urban popular movements presents difficult methodological problems. This is because of the highly fragmented character of urban popular movements, which in Brazil have no national level organization and rarely even have state-wide organizations. There is, therefore, no way of analyzing one particular organization and avering that it is typical of most (or even a large number of) urban popular movements. In this sense, there is a contrast to parties, the Church, the military, the industrial bourgeoisie, and even the labor movement. It is similarly impossible to construct a "general" model of grassroots movements in contemporary Brazil. When grassroots movements emerged, how they developed, when they faced periods of demobilization, what linkages they constructed to political parties and the State, how the State responded—all of these issues vary considerably from case to case. Moreover, grassroots movements deal principally with the State at its most decentralized level—specific organs of a given municipality—which means that the "State" which one particular movement faces may differ substantially from the "State" which another one faces.

What, then, can be done to avoid having the study of urban popular movements become excessively idiosyncratic on the one hand, or excessively general on the other? No matter what methodology one chooses, it is essential to have an awareness both of general proclivities and of particular differences. Within this guideline, carefully done case studies can be a fruitful means of analyzing urban popular movements. Case studies provide a sense of the dramas lived out by specific actors in concrete situations. They illustrate the kinds of dilemmas popular movements generally face, but they show how a given movement has attempted to deal with these dilemmas. Finally, case studies suggest some of the ways in which popular movements contributed to the struggle for democracy, as well as the immense difficulties they have encountered in working for governments more committed to popular causes.

For these reasons, I chose to focus on a case study, that of the Movement of the Friends of the Neighborhood (MAB) of Nova Iguaçu, which I have followed intermittently since December 1980.[1] By most standards—ability to mobilize the local population; local, state, and even national projection; recognition by the public authorities—MAB is an exceptionally well-organized movement, probably the most successful in the state of Rio de Janeiro, which in turn has the strongest state-wide neighborhood federation (FAMERJ) in the country. Focusing on an unusually successful movement allows us to assess the political impact of urban popular movements in a limit case. Thus it presents an interesting case for analysis, although not a "typical" one. While focusing on the case of Nova Iguaçu, the essay reflects upon general trends and dilemmas, as well as upon the general impact of grassroots movements.[2]

The Socio-Economic and Political Context in Nova Iguaçu

Nova Iguaçu is a poor city located some twenty miles to the north of Rio de Janeiro. With a population of approximately 1.5 million, it has become the seventh largest city in Brazil, having grown at a rapid rate since 1950. Although today Nova Iguaçu is a large city, in the recent past it was known primarily for its agricultural production. Located in the Baixada Fluminense, a large lowlands which have a hot climate, Nova Iguaçu became one of the most important orange-producing regions in the country around the turn of this century. The orange cycle entered a decline in 1926 when diseases started to kill the trees in parts of the Baixada. The municipality's population grew from 33,396 in 1920 to 105,809 in 1940, but the population was still predominantly rural. By the end of World War II, orange production had dropped off dramatically.[3]

After 1945, Nova Iguaçu began a new phase characterized by being a distant periphery of Greater Rio. As Rio grew, real estate prices pushed the popular classes into favelas or the outlying periphery areas like Nova Iguaçu. From 145,649 inhabitants in 1950, the population increased to 359,364 in 1960 and 727,140 in 1970, making Nova Iguaçu the fastest growing major city in the country. In 1950 46.60 percent of the municipality's population still resided in rural areas, but by 1980 this figure had dropped to 0.29 percent. This growth slowed during the 1970s, but the population still increased to 1,094,805. It is predominantly a working-class (generally unskilled labor) city with a high percentage of migrants.

By any standards, local services are grossly inadequate. In 1980, only 37.7 percent of the municipality's population had running water, and 27.2 percent were without electricity. Only 30.3 percent had sewers; sewage is disposed through open canals and rivers, which is severely damaging to the local ecology and helps account for the bad health conditions. The city had only 265 doctors, 27 dentists, and 961 hospital beds, in all cases approximately one-eighth of Rio's per capita level.[4] Partially because of a shortage of schools, in 1978, according to the mayor's estimate, 150,000 school-age children were not enrolled,[5] and most schools were in poor condition and

Table 1

Year	*Population*	*% of Population in Rural Areas*
1920	33,396	
1940	105,809	
1950	145,649	46.60
1960	359,364	28.34
1970	727,140	0.39
1980	1,094,805	0.29

seriously deficient in supplies. The illiteracy rate for people over 10 years old was 17 percent in 1980. As of 1978, only about 15 percent of the municipality's garbage was collected, leaving some 500 tons of garbage per day in open sewers and unoccupied land. Inadequate police facilities have led to one of the highest crime rates in the country, and transportation facilities are equally deficient.[6]

Popular movements have historically been weaker in Brazil than in many Latin American countries. Within this general context, Nova Iguaçu has been one of the Brazilian cities with a stronger tradition of popular mobilization. As early as 1945, there were isolated attempts to organize the population to obtain better urban services. In 1950, the first neighborhood associations were formed. As the national and local climate of the late populist years (1958–64) stimulated a rich political debate throughout the society, the neighborhood movement expanded. In 1960, the leaders organized the First Congress of the Commissions for Urban Improvements of the Neighborhoods of Nova Iguaçu. The Congress mobilized many neighborhood associations and obtained some concessions from the city administration. The pre-coup years saw other experiences of popular mobilization in the Baixada Fluminense, including an important labor movement and movements of peasants and rural workers. The post-'74 neighborhood movement would draw upon this history of popular mobilization; several leaders in the post-'74 movement had participated in these earlier struggles.

The coup wiped out the most important popular movements. Key leaders of the neighborhood movement were imprisoned, and the repression prevented efforts to coordinate the movement between different neighborhoods. The associations and commissions that survived articulated their demands individually, and there was little public sensitivity to them. The repression and the weakness of local opposition forces made any popular organizing outside the Church almost impossible.

In the years following the coup, the city continued to grow at a rapid pace, creating new social tensions. The limited local resources were not always well used, and the city administration was incapable of providing urban services to keep pace with the population increase. Politically, too, things were difficult. In addition to the official repressive apparatus, the infamous Death Squad was very active in the Baixada. By 1979, the Death

Squad had executed some 2,000 people in Nova Iguaçu, and another paramilitary organization executed 764 in the first semester of 1980 alone.[7] The progressive local leaders of the official opposition party, the MDB (Brazilian Democratic Movement), were imprisoned, and by 1970, the party entered a crisis. Statewide, the MDB fell into the hands of a conservative group closely linked to the military regime and became noted for being corrupt.[8] The local government party, ARENA, was conservative even in comparison with its counterparts in other major cities. It was notorious for corruption and was largely uninterested in resolving the problems of the population. Despite the local MDB's problems, ARENA was defeated in 1974 and subsequent elections.[9]

Meanwhile, the Catholic Church was undergoing the changes that would make it the bulwark of popular movements. The diocese of Nova Iguaçu was created in 1960, and until 1966 it followed a relatively conservative orientation. That year, D. Adriano Hypólito was named bishop and began to encourage the changes that led the Church to become closely identified with the popular sectors. In 1968, the diocese voted to establish base communities as one of its principal priorities. During the most repressive years, the base communities, which started to flourish during the early 1970s, were virtually the only popular organizations that promoted critical political perspectives. Although they were involved only in rudimentary political actions such as signing petitions for urban services, their prior existence would facilitate more extensive organization and mobilization when the repression relaxed. Many leaders and participants in the neighborhood movement were motivated by their experiences in the CEBs.

Political liberalization followed the same general contours in Nova Iguaçu as nationally, with a gradual easing of repression, especially after 1978. The *abertura* helped make possible the re-emergence of popular movements by reducing the fear of participating, enabling the left to engage in popular organizing, and enabling the movements to construct linkages to other institutions, like the Church, the opposition party, the press, and human rights groups.

Nevertheless, some caution is necessary in discussing the relationship between MAB and the *abertura*. Although the *abertura* began in 1974, repression against popular movements did not ease up significantly until 1978. In Nova Iguaçu, the far right continued to engage in terrorist practices. The most spectacular incidents involved the kidnapping and torturing of D. Adriano Hypólito in 1976 and the bombing of the cathedral in 1979. Even after 1983, paramilitary groups continued to operate with impunity; the Baixada Fluminense won a reputation as Brazil's most violent urban region.

More generally, the liberalization process did not noticeably affect politics in Nova Iguaçu until the late 1970s, given the weakness of the local opposition, the authoritarian nature of the city administration, and the presence of the Death Squad. The city administration and the local PDS were particularly discredited. The administration remained unresponsive to popular demands, and the MDB's (and later PMDB's) problems continued through Tancredo Neves's election in 1985. In contrast to other cities, where some MDB politicians supported the popular movements, in Nova Iguaçu the neighborhood movement remained relatively isolated, with the Church as its most significant ally. Furthermore, it would be erroneous to posit a linear relationship between liberalization or democratization and the expansion of social movements. In Nova Iguaçu, as elsewhere, democratization created new dilemmas and problems for popular movements, even as it allowed new spaces for popular organization.

The Neighborhood Movement, 1974–1985

The Nova Iguaçu neighborhood movement began in 1974, when two young doctors committed to working with the poor started to work in one of Nova Iguaçu's outlying neighborhoods.[10] Initially, they treated the population without charge and offered health courses. Medical treatment, however, had only palliative effects in a region with widespread malnutrition, open sewers, no garbage collection, and other health problems, so they began to think about organizing the population to help change those living conditions.[11]

In 1975, the diocese's branch of Cáritas, an international organ of the Catholic Church for serving the poor, hired these doctors and two others to start a health program. These four doctors played a leading role in transforming the previously isolated neighborhood efforts into a coherent popular movement. In November 1975 the diocese began to hold health discussions led by the four doctors. The group's orientation was expressed in a March 1976 document:

> The solution of health problems depends more on the population's unity and action than on the presence of a doctor. Having a health post is important, but it does not resolve health problems. Therefore, all the forms the population has of uniting to reflect on its problems and develop its consciousness and unity are important. Actions which are purely palliative, which are not concerned with the population's conscientization, discourage true learning and do not resolve health problems.[12]

In this early phase, the majority of people attending the courses worked at health posts. The doctors were satisfied with these courses, but they were also interested in reaching the poor themselves. In 1976 they held health courses in six neighborhoods throughout the municipality, and at this point more people from a working-class background started to come. The discussions started to include all problems faced by the population rather than just health issues. Simultaneously, the population began to organize neighborhood associations to address these needs.

In May 1977 the movement began to call itself *Amigos do Bairro* (Friends of the Neighborhood). The movement continued to expand, involving a growing number of neighborhoods, throughout 1976 and 1977. This expansion dictated the need for more formal leadership structures, and at the thirteenth meeting, in March 1978, the movement created a Coordinating Commission. This leadership structure was an important step in expanding beyond fighting for isolated material needs to developing a mass movement with broad political horizons. Other important steps in the same direction occurred around the same time, including turning the health newspaper into a newspaper for the movement. MAB was starting a period of consolidation and rapid expansion.

By May 1978, the bimonthly meetings involved people from eighteen neighborhoods of Nova Iguaçu. At that time, the movement adopted its definitive name, *Movimento de Amigos do Bairro*, MAB (Friends of the Neighborhood Movement). The local associations continued to be the primary instrument for organizing the neighborhood, and MAB became the means of coordinating the efforts of different associations and turning these efforts into a cohesive project, capable of pressuring the State into becoming more responsive to local needs.

In May 1978, one neighborhood association took a petition with 1,500 signatures to the city administration, but the mayor refused to receive it, stating that he would accept petitions only from people who had paid their property tax. The residents wrote to several council people (*vereadores*) protesting this policy, and the movement got the local press involved in the issue. The pressures forced the mayor to partially retract his initial statement. On July 25, he stated his willingness to receive all petitions, but still declared that in allocating public resources, he would give priority to people whose taxes were paid. This was MAB's first major victory in pressuring the city administration to re-evaluate its policies towards the popular sectors. Equally important, it was the first time MAB received considerable press attention and won allies among local politicians.

To protest the mayor's initial decision to refuse to receive petitions and his failure to listen to the demands of the local population, MAB decided to hold an assembly. Held on October 14, 1978, with 700 participants representing 38 neighborhoods, the assembly began a new period in MAB's development, marked by stronger linkages to local politicians and the press, and with more extensive participation. The development of local allies would give the movement greater impact than it previously had.

During its early years, even though MAB had moved towards becoming a mass movement, it was still almost exclusively concerned with the immediate material needs of the local population. This began to change in late 1978 and early 1979 as the leadership opened out more towards national politics. MAB participated in a solidarity movement with the 1979 auto workers' strike in Greater São Paulo and with a teachers' strike in Rio and started to send representatives to local demonstrations. The leaders began to support issues related to the democratization of the society, such as party reform, political amnesty, and reform of local government.

MAB's dynamism created a new problem for a city administration accustomed to ignoring popular demands. MAB used the administration's unresponsiveness to further delegitimize the city government. The movement publicized the government's repeated failures to meet promises, the disrespect it had shown for MAB participants, the financial scandals which surrounded the administration, and its failures to attend to the needs of the local population. In response to the problems, MAB held a second major assembly on July 15, 1979, with 3,000 participants representing sixty neighborhoods of Nova Iguaçu. The importance the movement had acquired was seen in the publicity the assembly received and in the presence of important political figures, including a federal senator. The mayor agreed to weekly meetings with representatives from different neighborhoods of Nova Iguaçu. MAB, by now the most important popular movement in Nova Iguaçu, was in a new, more mature phase.

The party reform initiated in 1979 was one of the most important steps in the *abertura*. It deeply affected the subsequent political struggle, including the popular movements. MAB's leaders had always experienced some internal divisions, but these divisions were accentuated with the party reform. Some MAB leaders opted for the PMDB, some joined the PT, a minority decided upon the PDT, and still others chose to avoid making a partisan commitment. Among the members of the original Coordinating Council, eleven opted for the former and eight for the latter (support for the PDT grew after 1982).

The party issue would have been less significant if it had not paralleled basic differences in philosophy about what should be done at various political conjunctures and how popular movements should be led. Some leaders (mostly PT) were more concerned about grassroots discussions and about making sure the common people led the process, while others (mostly PMDB) emphasized the importance of creating a broad front that would participate in the redemocratization process. In the state of Rio de Janeiro, this schism was further accentuated by the PDT's penetration in popular channels, which increased after 1982. Ironically, then, political liberalization, which facilitated the growth of the movement, also created conditions for internal competition and division.

Until December 1981, although there were some tensions between PT, PMDB, and PDT leaders, the existence of competing conceptions about how to run the movement helped MAB to articulate a careful balance between grassroots work and broader political issues. By late 1981, about ninety neighborhood associations were participating.

In December 1981 the movement began a period of greater internal conflict at the leadership level and some demobilization of the grass roots. A key problem was the accentuation of internal tensions, principally stemming from partisan disputes. In December 1981, MAB held the Second Congress of Neighborhood Associations of Nova Iguaçu (the first had been held in 1960), became a federation, and held elections for a new Coordinating Council. The elections for the Coordinating Council led to rancorous disputes that followed complex intra- and inter-party lines. There were tensions between the new leaders and some of those who left, and charges of manipulation were made on both sides. Never before had MAB experienced such deep internal disputes.

The 1982 elections for governor, federal and state Congress, and local government caused further serious conflict within the movement. Officially, MAB adopted a position of autonomy vis-à-vis political parties: as a movement, MAB did not opt for any particular party and was open to all individuals, regardless of party affiliation. At the same time, many MAB leaders recognized the importance of electing individuals more sympathetic to the movement, so over a dozen MAB leaders ran for office, generally in the PMDB or the PT.

The election results proved a major disappointment to the movement's leaders, most of whom had worked for the PMDB or PT. None of the popular candidates from the PT or PMDB of Nova Iguaçu were elected. Only one individual (a PDT Councilman) had participated in MAB won, and his participation in MAB was more limited than that of other popular candidates. Leonel Brizola won by a large plurality in Nova Iguaçu, and the PDT easily won the municipal elections. In the dispute for mayor, the election yielded the following results:[13]

PDT	129,789
PDS	67,484
PMDB	66,252
PTB	20,084
PT	7,262

The dismal electoral performance of the PT speaks for itself; otherwise, the results for MAB were even worse than the numbers suggest. Both in Nova Iguaçu and in the state of Rio, the progressive factions of the PMDB were vanquished. In Nova Iguaçu, a conservative faction of the PDT came to power. Although less repressive and more open than past local governments, it also faced problems of corruption and lack of responsiveness.

State-wide, Brizola implemented populist practices aimed at developing popular support. Faced with the severe economic crisis, the federal government's strategy of reducing its allocation of resources to opposition governors, and a PDT minority in the state parliament, Brizola had difficulties in effecting major changes. The fact that the opposition could not introduce substantial improvements in popular living conditions created new problems for MAB.

In December 1983 the movement held new elections for the Coordinating Council, and the different factions improved their relations, leading to another phase of growth. The new Coordinating Council had 11 representatives from the PMDB, 7 from the PT, and 1 from the PDT. In early 1984 the movement participated in the campaign for direct elections. In November 1984, along with the other two neighborhood federations of the Baixada

Fluminense, MAB held one of its largest demonstrations ever, with about 4,000 people, around issues of public health. But the improved relations among the movement's different political factions proved ephemeral, as did the period of greater mobilization. In December 1985, new elections for the Coordinating Council once again produced bitter divisions leading to the decision of some people to withdraw from the movement.

By January 1985, when Tancredo Neves was elected President, MAB represented 120 neighborhood associations in the Baixada Fluminense. Considerable optimism initially surrounded the inauguration of a new democratic government, but among poor people this optimism dissipated over time. Among MAB leaders, there is a consensus that the movement is living difficult days. In the words of one prominent popular organizer, "The popular movements are experiencing a crisis. MAB used to mobilize 3000, 4000, 5000 people. It can no longer do so. The people are uninterested, disenchanted with politics. We thought that mobilization would increase with the *abertura*, but that hasn't happened. It's not that the popular movement has declined, but it has not managed to grow as we hoped."[14]

MAB and the State, 1974-1985

Consistent with the widespread opposition emphasis on strengthening civil society and criticizing Brazil's statist tradition, the grassroots movements that emerged in the post-'74 period generally espoused an ideology of autonomy vis-à-vis the State. Nevertheless, MAB and other movements constantly interacted with State organs to obtain urban services. MAB's leaders may have been interested in broader political issues such as the restoration of democracy, but most grassroots participants were motivated by the prospect of material benefits for their neighborhoods. Winning these material benefits required not only public mobilization but also demands targeted at administrative organs of the municipality of Nova Iguaçu and the state government of Rio de Janeiro. Futhermore, even though the innovative character of some social movements stemmed in part from their questioning the traditional State-oriented politics, the capacity of social movements to promote change depends partially on their ability to influence the State.[15]

It is important to distinguish between the State's response to popular movements and its orientation towards the popular sectors. Neighborhood movements generally present themselves as movements which represent the entire population of a given geographic area, but even in well-organized movements, participation is limited relative to the overall population. MAB represented 164 neighborhood associations by October 1986. No membership figures are available, so we can make only rough estimates about how many people participate. I would estimate that an average of 50 people participate per association, though the core group of active participants is considerably smaller. Perhaps as many as five times that number have been involved at one time or another. This would mean that the number of participants is about 8,000 (an estimate that seems reasonable in light of the fact that some MAB demonstrations have mobilized 5,000 people) and perhaps 40,000 have participated at some point. While these numbers are impressive, the fact remains that only a small percentage (at most 3%) of the population has ever participated in the movement. A much larger number of people have never heard of MAB or of the neighborhood association that operates where they live. As a result of this limited participation, the State can respond to some demands raised by the popular movements while at the same time essentially ignoring or even being hostile to popular movements, which can be seen as a threat because of their autonomy. This has in fact happened in Nova Iguaçu.

The other general consideration that needs to be highlighted before embarking on the specific analysis of Nova Iguaçu is the need to disaggregate the State in analyzing social movements. Social movements have some choice about which State organs they focus on. As a result, they can take advantage of political divisions and struggles within the State itself, especially in a period of relatively open politics, when politicians compete for popular sympathies. Whether it is for ideological reasons (a commitment to improving popular living conditions) or purely self-serving (winning popular support to enhance one's political career), the leaders of some State agencies are far more disposed than others to responding to popular needs.[16]

This internal differentiation of the State was not significant in Nova Iguaçu until the late 1970s, given the uniformally uninterested response of State organs towards popular movements and living conditions. But beginning around 1978, the state government started to become more concerned with winning popular support, while the local government was noticeably less responsive. From that time on, the cleavages between the city and state governments are important in analyzing MAB's relationship to the State.

Table 2 summarizes changes in State orientations towards MAB and popular needs in general. The following pages examine these changes in greater depth.

Table 2.

	Local Government Orientations		
	1974–78	*1978–82*	*1982–85*
Toward MAB	Repressive	Unresponsive	Unresponsive
Toward Popular Needs	Neglect	Neglect	Minimally responsive
	State Government Orientations		
	1974–78	*1978–82*	*1982–85*
Toward MAB	Repressive	Unresponsive	Moderately responsive
Toward Popular Needs	Neglect	Conservative Clientelism (Minimally Responsive)	Progressive Populism (Responsive)

During the first four years of MAB's existence (1974–78), the State was consistently unresponsive to MAB. The leaders of Nova Iguaçu saw no reason to respond to the demands formulated by the nascent popular movement. The city administration perceived the popular movement as an enemy to which it did not need to make concessions. At best, the attitude towards MAB was one of neglect, and on occasion the movement suffered repression. Several MAB leaders were threatened, and at least two were assaulted and warned to curtail their participation. MAB leaders described the administration's personal treatment of movement people as one of "flagrant rudeness" during this period. MAB maintained little direct contact with the state government during these nascent years.

The period between June and October 1978 marked the beginning of a second phase in the relationship between MAB and the city administration. The city government began to change from a blatantly authoritarian approach to one which included some elements of conservative clientelism. This period, which lasted until 1982, was characterized by occasional minor concessions to the movement. Intermittently, the administration was less hostile to the movement, even though it provided a few concrete improvements in urban services. In October 1978, the mayor finally agreed to attend a MAB meeting, indicating greater concern with his public image among the popular sectors. In front of 700 people, the mayor recognized the fairness of MAB's demands and agreed to begin holding weekly meetings with the movement to discuss the population's most pressing demands. In early 1979 the administration announced a new community development program, the Community Operation of Social Integration of Nova Iguaçu, which would supposedly involve popular participation. In practice, the project was barely implemented except for a program of polio vaccinations, but its existence indicated concern with responding to popular demands. Despite the minor improvements in the city's response to popular demands, the movement still faced a series of frustrations in this regard.

Partially in response to the frustration of years of popular organizing which had little impact on local government, in 1980, MAB leaders decided to attempt to negotiate directly with the state government. A demonstration in Rio on June 13, 1980, with 700 participants, was the first time MAB went to the state government to demand urban improvements. This move towards the state government marked an important step in MAB's visibility and capacity to negotiate. It initiated a strategy of forcing the government party, the PDS, and the now defunct Popular Party (PP) to compete in providing services. While Nova Iguaçu was governed by the PDS, the state governor of Rio, Chagas Freitas of the PP, was the only opposition governor in the country. Chagas Freitas was a conservative figure within the most conservative opposition party and had close links to the federal government, but his strategy for dealing with the popular movements was less confrontational and repressive than that of the PDS of Nova Iguaçu. Known for his clientelistic practices with the popular classes, Chagas Freitas was more tuned into the exigencies of the process of political liberalization than the PDS. By 1978, his dominant response to the popular sectors was conservative clientelism. By clientelism, I mean the practice of establishing a patron-clientele relationship with the popular sectors. Chagas Freitas's political machine used the State to provide services and favors on a piecemeal basis in attempts to win popular sympathies. The political machine was run in a personalistic fashion: politicians who were part of Chagas's coterie competed for access to State money to obtain services and favors for their own clientele, which consisted of certain neighborhoods, cities, and regions, or of interest groups like samba schools. This orientation contrasts with an entitlement approach that makes the provision of urban services less dependent on the personal mediation of a machine politician.

Typical of Chagas Freitas's policies towards popular movements was his successful attempt to create a more conservative federation of favela associations to compete with a more militant federation. In doing so, Chagas divided the favela movement, took the steam out of the more autonomous movement which already existed, and implemented clientelistic policies in a number of favelas throughout the city. The repressive removal policies of the 1964–73 period gave rise to a policy of "urbanizing" (providing basic urban services for) the favelas.[17]

MAB hoped to encourage competition between the PP and the PDS at a time when the parties were scrambling to win popular support for the 1982 elections. Movement leaders feel, however, that this endeavor met with limited success. Although Chagas Freitas used some state resources for urban improvements in Nova Iguaçu, he did so to a limited extent. MAB rarely won access to state government. Rather than responding to an autonomous popular movement, Chagas attempted—unsuccessfully—to reproduce, in Nova Iguaçu, the strategy of creating neighborhood associations from above.

The decision of twelve movement leaders to run for office in 1982 must be understood within this context of eight years of frustration with the State's response to popular demands. MAB leaders felt that the only way to effect significant changes in local government was by electing someone to represent popular demands within the administration. The defeat of the popular candidates and the PDT victory added one more chapter to the history of frustrations besetting MAB. At the time, none of the nineteen members of the Coordinating Council belonged to the PDT. While a large number of grassroots participants had voted for the PDT, the PDT victory meant a continuation of the lack of any privileged channel of access to state officials and resources. As noted above, one person who had participated in MAB was elected Councilman on the PDT ticket, but this individual was critical of MAB's leadership.

The period beginning January 31, 1983, when Paulo Leone (PDT) became mayor of Nova Iguaçu, and March 15, 1983, when Leonel Brizola became governor, marked the opening of a third phase in the relationship between the State and MAB. During this period, MAB directed most of its demands towards the state government, which was more concerned about popular needs and had more resources than the municipal government. Brizola's predominant approach to popular movements was one of progressive populism. In his discourse, Brizola emphasized participation and social justice. Despite the problems created by the severe economic crisis that affected his first two years in office, Brizola's government evinced a concern for ameliorating the most pressing popular needs, and its most publicized material achievement was in the area of schools in poor neighborhoods. Election results in 1985 and 1986 indicated that Brizola and the PDT enjoyed broad support in poor areas of the major cities (especially Rio and Nova Iguaçu). Furthermore, through his efforts, Brizola attracted the support of a large number of leaders of popular movements, including some in Nova Iguaçu. Nevertheless, Brizola was generally cool towards autonomous popular movements.

Typical of the relationship between MAB and Brizola was an incident in late 1984. In October 1984, MAB and the two other federations of neighborhood associations of the Baixada Fluminense held an assembly and encouraged Brizola to attend. The governor did not even send a representative, so the three federations decided to hold a demonstration in front of the Government Palace in Rio on November 18, 1984. Brizola took the demonstration as a personal affront, accusing MAB of attempting to undermine his popularity. At the same time, Brizola initiated programs that responded to the demands MAB had made in the areas of health, education, and transportation.

Equally interesting in this sense was the PDT's attempt to create new neighborhood associations in Nova Iguaçu, thereby reproducing the PP's earlier attempts to do so from above. Again, however, the endeavor met with little success; most of the associations that survived eventually joined MAB. Brizola did, however, win the support of many popular leaders in Nova Iguaçu, including some MAB leaders. Brizola's cooptation of a significant part of the favela movement in Rio de Janeiro indicates that other incursions into the popular sectors were quite successful.

The relationship between the city administration and MAB was more conflictual, as MAB leaders ultimately came to feel that the new administration was almost as bad as the previous one. In Nova Iguaçu, the PDT won the 1982 elections largely on the basis of Brizola's charisma and electoral legislation that mandated straight party voting. The local PDT turned out to be incompetent and relatively unattuned to popular needs. When he assumed office, Mayor Paulo Leone announced his intention to work closely with the popular sectors, and in June 1983 held a large assembly, with 4,000 people present, to discuss the city's needs and projects. In October 1984 the city administration sponsored the First Congress for the Development of the Municipalities of the Baixada Fluminense, again welcoming popular participation. The administration's discourse was favorable to popular participation and to MAB. For example, a publication on the administration's first two years mentioned the "support the administration gives the neighborhood associations of Nova Iguaçu, represented by the Federation of MAB. In previous years, access to the Executive Power was difficult to obtain. Today, thanks to the philosophy implemented by Mayor Paulo Leone, everyone is listened to. Demands and suggestions are noted and directed to the pertinent authorities."[18]

From the viewpoint of MAB leaders, however, the relationship was not good. MAB leaders criticized the gap between the administration's discourse and its practice. They complained that the Leone administration was almost as corrupt and inefficient as its predecessors. Brizola, normally very accommodating, had a falling-out with Leone and tried to have the mayor expelled from the PDT. Leone eventually joined the center-right PFL in 1985. From the time of his inauguration, the local PDT was plagued by problems similar to those ARENA and the PDS had faced earlier: considerable in-fighting, charges of corruption, and limited resources with which to administer a highly populated municipality with pressing needs.

In material terms, Nova Iguaçu did not experience any progress after Brizola and Paulo Leone took office, as much of the local population suffered from what was perhaps the worst economic crisis in Brazil's modern history (see the essays by Fishlow and by Bacha and Malan). This fact points to the difficulties popular movements face in securing the services that would appreciably improve the living conditions of the poor people who reside in Brazil's urban areas. Yet even if MAB's direct impact in obtaining material demands has been limited, the movement helped legitimate popular demands, a fact that is manifested most clearly in the attitudes politicians have evinced towards the movement. During the early years of MAB's existence, not only did the mayor continually show disrespect for the movement, but local PDS politicians, including some town council representatives, denounced the movement as a nexus of Communist infiltration. By contrast, the Leone administration recognized MAB as an important and legitimate expression of popular demands in Nova Iguaçu. For the first time, the Leone administration created institutionalized means of receiving popular demands in the form of a permanent planning commission which included MAB representatives.

Dilemmas and Problems

When grassroots popular movements proliferated in major urban areas in the second half of the 1970s, the first analyses about their political impact were optimistic.[19] Many early works assumed or posited a secular increase in mobilization and efficacy of these movements. In fact, the movements have been subject to ebbs and tides, confirming the expectations of some theoretical works on the subject.[20] Without pretending to be exhaustive, the following pages analyze some of the outstanding dilemmas MAB has faced.

1) Popular consciousness and popular participation. The success of a movement like MAB depends largely on the leadership's ability to encourage popular participation. In the absence of such participation, the movement cannot claim the representativeness it needs to effectively bargain with the State, nor does the population accumulate political experience. Yet encouraging popular participation is difficult given the local living conditions and popular consciousness.[21]

The vast majority of the population of the Baixada Fluminense have difficult lives. When they can get jobs, men work long hours for low pay and frequently travel as many as four hours per day to and from work. Women take care of the home and the children, and generally work at least part time as well. The exhausting nature of daily life by itself represents an obstacle to popular participation; people do not find the time or energy to add one more commitment to their already difficult lives. In addition, going to meetings usually entails round-trip bus fares, which further burdens the already tight family budget.

More important than these problems is the question of popular consciousness. Even though popular consciousness reflects an awareness of poverty, in most cases it is not a consciousness that radically opposes the extant social order. People may think that the wealthy rip them off, but they also believe that the system offers enough opportunities that enterprising individuals can get ahead. Most people seek to improve their material situation not through collective movements, but rather through individual enterprise.

As a rule, the popular sectors are somewhat skeptical about the possibility of effecting political change.[22] Politics is seen as an elite struggle, and the State is perceived as a realm beyond the popular sectors. Politicians seek votes and make promises during electoral campaigns, but once the campaigns are over, life goes on as usual—which means that the politicians forget the poor. In the limited spare time they have, poor people typically focus on a range of issues that have little to do with politics: family life, sports, relaxing. It is tempting for the social scientist to brand this "alienation," but given the realistic assessment that it is difficult to effect political change through collective action, this evaluation is at best facile. Popular attitudes nevertheless limit the possibility of collective action, for only at the point when people believe in the legitimacy and efficacy of collective action is it possible to organize a social movement. Even then, as analysts from Olson to Dahl have noted, some people (the "free riders") want others to carry the ball, and others are not interested enough to find the time to participate.[23]

This generalized disbelief in the possibility of effecting political change is exacerbated by extremely limited political awareness on the part of most poor Brazilians, a fact made apparent in a number of surveys. In a 1985 survey conducted during the weeks immediately preceding the elections, IDESP researchers asked people if they remembered what parties existed. Only two of the parties had an identification ratio of over 35 percent. Only 52 of 690 people were able to identify within reason what "right" means in political terms. There was a generalized disbelief in the ability of people to vote. Only 17.2 percent responded "yes" when asked whether people know how to choose candidates; another 20.4 percent responded "more or less," but the dominant response (60.0%) was "no" (another 2.3% did not respond). Lack of political awareness cut across the entire social structure, but was especially pronounced among the poor.[24]

Further compounding this difficulty is the fact that even among the poor there are conflicting material interests. A group of people may all be poor but still have different material priorities. For some, land titles are the salient issue; for others, water, sewers, or electricity are. The State, especially at the local level, has limited resources with which to respond to a wide panoply of demands. This means that a minor victory for one poor neighborhood may diminish the possibility that the State will provide services to another poor neighborhood. Divisions and cleavages not based on material differences also impede the formation of social movements whose existence is predicted upon working for a common good. Differences in religion, race, and status permeate poor neighborhoods, undermining the potential for collective action.[25]

In light of this situation, it is not surprising that MAB's leaders perceive the difficulty of getting people to believe in the efficacy of popular organizations as one of the movement's greatest obstacles. The lengthy history of popular exclusion from the most important political decisions, partially excluding the 1955–64 period, has played a major role in shaping popular consciousness. Popular movements have begun to affect Brazil's elitist political culture, yet the lengthy heritage of elitism still weighs significantly.[26]

2) Frustration through continuous defeats. Even though MAB stands out as an unusually successful movement, its concrete victories have always been limited and partial, especially in relation to the tremendous effort spent in mobilizing the local population. The number of defeats and the amount of energy spent in winning minor material benefits have led many movement participants to drop out or reduce their participation. Many associations have had cyclical histories, with periods of growth followed by others of demobilization. Since 1974, at least thirty associations in Nova Iguaçu have collapsed or dropped out of MAB, underscoring the fragility of these movements.

In response to this problem, MAB always paid some attention to cultural questions. Many associations saw their work as community building, which emphasized personal relations, as well as presenting material demands. In 1984, MAB began a popular theater, which performed for local neighborhood associations, attracting as many as 300 people. The individuals who created the theater group saw their efforts as a means of extending the movement in a cultural direction, thereby hoping to avoid the cyclical effects of movements that focus exclusively on material demands.

3) Frustration because of the difficulty of capitalizing on victories. Continuous defeats are almost certain to have a demobilizing effect, but even when a movement wins some victories, it can be hard to capitalize on them.

Only under extraordinary conditions can social movements maintain high levels of mobilization for an extended period of time, and such conditions did not obtain after 1982. The optimism of the late 1970s created a climate favorable to mobilization; the climate after 1982, with the ephemeral exception of the first four months of 1984, was unpropitious. This change in social climate and expectations reflects, in part, what Hirschman has called "shifting involvements": after a period of public involvement, people refocused on private issues.[27] Social movements won some victories, but in the context of unresolved indigency and generalized lack of involvement in politics, these victories were not sufficient to motivate renewed involvement.

State strategies also helped defuse social movements.[28] Government officials treat new services as a State concession and achievement even when they respond directly to demands posed by social movements. This stance by government leaders, which is entirely understandable, can make it difficult for movement leaders to claim their victories. Since 1978, this has been a major problem for MAB.

This difficulty was evident, for example, in MAB's efforts in 1981–82 to get an already completed public hospital open for public use. The State never responded to the movement. Finally, on November 12, 1982, three days before the elections, President Figueiredo, hoping to swing enough votes so that the PDS would win the state of Rio, inaugurated the hospital. MAB had worked for over a year to get the hospital opened, yet the local and federal government were ultimately in the best position to capitalize.

This difficulty was compounded when Brizola and Paulo Leone took office. Both politicians had a discourse favorable to popular demands, and both implemented some programs favorable to the popular sectors. Yet neither was directly responsive to popular movements, preferring to present their programs as State initiatives. When the government was openly antipopular, it was easier for the movement to project an image which combated this view.[29]

Perhaps the most frustrating "victory" came in the political rather than the material sphere. Leaders in MAB and many other social movements worked hard to effect political change at the municipal, state, and national levels. Generally speaking, they were disappointed with the results of these efforts. By 1985, when he joined the PFL, Mayor Leone had no supporters among the leaders of Nova Iguaçu's popular movements. The disappointment with the changes in the federal government was also acute, especially at the grassroots. The conservative nature of the transition to democracy and of the new democratic government generated frustration within MAB, especially at the grassroots level. Only Brizola's government was partially spared this negative evaluation among movement leaders—and then only among a select group.

The disenchantment with the results of the opposition governments was not unique to Rio de Janeiro. In São Paulo, many movements experienced demobilization as a result of disappointment with the Montoro government and the economic crisis.[30] As organized social movements declined and the economic crisis accelerated, a number of spontaneous violent movements (supermarket invasions, etc.) occurred. While such movements were far from unprecedented, their magnitude was a new and disturbing phenomenon. The tenuous alliance that had existed among opposition forces of different stripes under ARENA and PDS governors broke down. The most progressive bishop in Greater São Paulo, D. Angélico Sândalo, stated in early 1984 that, "It is time for Governor Montoro to listen to the population and cease fearing it; to quit making excuses for a repressive government, which treated popular demands with clubs and arms; and to follow through on his campaign promise to promote revolutionary changes in the countryside. . . . Today, in its rhythm, its discourse, and its behavior, the Montoro government is like all the rest."[31] Meanwhile, Montoro, like Brizola, attributed his difficulties to the economic crisis and the federal government's attempts to weaken the opposition governments by cutting off resources. And both complained about the unrealistic expectations of social movements and the progressive Church. Senador Fernando Henrique Cardoso of São Paulo stated, "At times, the Church demands moral solutions for a structural crisis, leading to a certain lack of communication."[32]

4) Internal tensions. Internal tensions and conflicts inhere in virtually any movement or organization that attempts to follow democratic principles.[33] They need not adversely affect the movement or organization, but there is always the possibility that they can do so. In popular movements, this kind of enervation occurs when the leaders are so embroiled in internal disputes that the primary objective becomes defeating opposing positions within the movement rather than mobilizing the population for urban services.

Ever since MAB was created there have been conflicts among movement leaders, but these conflcts were relatively minor until 1980, when the change in party legislation led to the re-emergence of a multiparty system. Until that point, MAB leaders were united around the banner of opposition to the dictatorship. Among those who supported a political party, the option was unanimously for the MDB. This consensus broke down with the party reorganization of 1979–80. Party organization, which for years had been a banner of opposition leaders, accentuated divisions among movement leaders. MAB leaders remained united in their opposition to the military government, but especially as the 1982 elections appeared on the political horizon, they also competed to further the cause of their own party. Even when partisan disputes did not become prominent, the issue for the leadership changed from how to oppose the dictatorship to how to construct a democratic regime responsive to popular needs. Conflicts at the leadership level have revolved around both substantive issues (how to lead the movement, what position to assume vis-à-vis the State), questions of power (how to enhance the situation of a particular party), and a combination of the two (since party differences are reflected in substantive questions and vice versa).

From 1979 until May 1985, when the Congress introduced new party legislation, the leadership of MAB could be divided into three main groups, each of which is subdivided as follows:

1) a Catholic left, which is subdivided into PT x PMDB;
2) an independent left, which was subdivided into PT x PMDB x PDT;
3) an "organized" left, with linkages to clandestine (illegal) political organizations. Here, in addition to the PT x PMDB divisions, there were additional cleavages. The MR-8, Brazilian Communist Party (PCB), Communist Party of Brazil (PC do B), Libelu, and Socialist Convergence were represented by at least one leader. Some of the clandestine groups (MR-8, PCB, and most of the PC do B) supported the PMDB, while others (Socialist Convergence, Libelu, and a splinter of the PC do B) supported the PT.

After 1985, with the changes in party legislation under the democratic government, the PCB and the PC do B became legal political parties; the other groups continued to be clandestine political organizations.

The cleavages among these various groups and parties were complex, cutting across many different issues. On some issues, there were divisions between the secular and the Catholic left. On others, differences arose between the PT and PMDB leaders, or between MR-8 and PC do B positions. And on other questions, the cleavages did not correspond to any clear party alignments.

When Brizola took office, few MAB leaders supported the PDT. Over a period of time, however, Brizola won the sympathies of a number of popular leaders in the Baixada Fluminense. Brizola's willingness to offer public employment to popular leaders partially explains his support among this group. More important, many popular leaders felt that Brizola was preferable to the other alternatives in view of the PT's weakness in the state and of the PMDB's uncharacteristically conservative profile. Brizola's sympathizers point to the fact that he did a lot for the poor population while governing under difficult circumstances. They argue that Brizola opened in unprecedented fashion the public administration to the needs of the poor. On the other hand, his detractors sharply criticized Brizola's austerity program, his centralized decision-making style, his neglect of the state in order to promote himself for President, and his willingness to construct alliances with the most conservative political forces. Thus one of the biggest schisms in MAB's leadership pitted pro- and anti-Brizola forces. The PMDB, the PCB, and the PC do B were viscerally anti-Brizola; the PT, while maintaining its autonomy, sympathized far more with Brizola than it did with the PMDB of Rio.

During two periods the popular movement clearly suffered from internal disputes. The first occurred between December 1981 and mid-1983. The bitter disputes that accompanied the elections for MAB's Coordinating Council in December 1981 were originated by the parallel divisions over the 1982 elections. Power considerations (enhancing the position of one of the clandestine groups) rather than substantive issues were primarily at stake in MAB's election. The disputes were so bitter that some leaders considered creating a parallel movement. Both sides later overcame the animosity, recognizing that the movement had suffered as a consequence. The defeat of all the popular candidates in the 1982 elections probably contributed to the awareness of the fragility of the popular movement and the importance of uniting forces. But this defeat was not sufficient to overcome internal divisions for long. In the December 1985 election for MAB's Coordinating Council, vituperative conflicts arose and once again led to the enervation of the movement.

5) *Cooptation.* Most discussions of the cooptation of social movements have been somewhat simplistic; they have almost universally treated cooptation as an evil that social movements must avoid, while its opposite (autonomy) is a goal they must strive for. As is frequently the case, reality proves more complex than one would gather by examining the discourse of many movement sympathizers or of many analyses of the movements. Many movements and analyses have confused autonomy with non-involvement in partisan issues or conversely have confused cooptation with involvement in parties or State agencies. Although it is important for social movements to avoid becoming a servile instrument of parties or politicians, it is also important that they influence political parties and the State. What is sometimes portrayed as cooptation may be good judgment on the part of movement leaders in accepting a party or State position that will enable them to move effectively realize the same political goals they pursued in the movement. The decision of a movement leader to head a state or municipal agency may be a form of "cooptation," but it might also ensure greater receptivity to the movement's demands within the State bureaucracy. A discussion of autonomy/cooptation should therefore focus on trade-offs rather than simply positing the importance of "autonomy" in some undefined sense. The question is in what senses a movement is autonomous or coopted, and what opportunities it uses or loses in its relationship with politicians, parties, and the State.

The possibility of cooptation implies an exchange between the State and a popular movement or leader. The State can coopt only if it provides some resources in return. In this sense, we again see that political liberalization created new dilemmas for social movements: only when the State was willing to make some concessions to the movements did cooptation emerge as an issue. Moreover, the issue got more complex over time; the victories of opposition governors in 1983 and then the transition to democracy in 1985 made the problem of cooptation more salient.

In Nova Iguaçu, until 1979 there were limited possibilities of cooptation since municipal and state government leaders generally ignored popular movements. Beginning in the late 1970s, Chagas Freitas and his political machine attempted to coopt popular movements in Nova Iguaçu and elsewhere, including the ill-fated attempts to create neighborhood associations linked to the Popular Party in the Baixada Fluminense. However, in Nova Iguaçu these efforts failed, largely because MAB was already a well-structured movement whose leaders opposed Chagas's political machine. The relationship between MAB and the State became more complex when the

PP and PMDB merged in December 1981, leading ex-foes to join forces. Because of the weight of Chagas and his political machine, the PMDB in Rio opposed the merger but was left with no alternative when the national party voted for it. However, by the time the merger was solidified in Rio, the 1982 elections were almost at hand, and the long established antipathies between Chagas and MAB were left unresolved.

The issue of cooptation surfaced with greater force with the municipal and state-wide PDT victories in the 1982 elections. After some early attempts at cooperating, the municipal administration and MAB developed an essentially antagonistic relationship that precluded the possibility of cooptation. But with the state government, the issue of cooptation and autonomy appeared in all its complexity. Brizola attracted the support of many popular leaders in Nova Iguaçu and provided public jobs to some of them. Anti-Brizola leaders perceived this as a blatant case of pork barrel politics aimed at coopting popular leaders. The individuals who supported Brizola, in contrast, argued that no other governor had done more for the poor population, and that in their positions they could get access to State resources that would help their neighborhoods. While Brizola's popularity among middle-class voters declined precipitously,[34] he remained popular among the poor people of the Baixada Fluminense. In no other part of the state did the PDT fare as well in the 1986 elections.

Compared with many popular movements in the state of Rio, MAB was unusual in the extent to which it maintained autonomy with respect to Brizola and the PDT. MAB's consolidation as a well-organized popular movement whose leaders were generally sympathetic to other parties undoubtedly contributed to avoiding cooptation. Nevertheless, the question of autonomy/cooptation was a complex one that often divided movement leaders in how aggressive they should be in facing the Brizola administration.

Change and Continuity in the Popular Process

One of the most important questions regarding popular movements in the post-'74 period is their relative novelty. What has changed and what has remained the same? The issue is of central importance in assessing the prospects for overcoming Brazil's elitist political heritage.

As noted earlier, neighborhood associations are not new in Nova Iguaçu, and other studies have shown that this is also true elsewhere.[35] Furthermore, the changes in neighborhood associations are not as dramatic as some people suggested. Nevertheless, the character of the post-'74 movement in Nova Iguaçu has changed in relation to the pre-'64 movement in several important ways.

1) *Relationship between the movement and political parties.* The post-'74 movement had greater autonomy vis-à-vis the parties than the 1950–64 movement. In the earlier period, the Brazilian Labor Party,

the most progressive of the three major parties, played a major role in supporting neighborhood associations in Nova Iguaçu. The outstanding leader of the movement, for example, was the head of Nova Iguaçu's chapter of the PTB. One neighborhood leader of the period stated that PTB helped finance cultural events, organized the local association and obtained legal recognition. He also indicated that the party encouraged politicians to be responsive to popular organizations. All of the pre-'64 leaders who were interviewed noted that the PTB helped organize the 1960 neighborhood Congress. As was characteristic of the late populist period, the linkages between the PTB and the popular movements of Nova Iguaçu were informal. Leaders of popular movements asked PTB politicians to take care of specific requests. The Communist Party also helped support the neighborhood movement. Although its role was not as pronounced as the PTB's, the Communist Party was officially illegal but was still active in supporting many popular movements of the period.

In the post-'74 period, most of the key leaders first participated in MAB and later joined political parties. Neither the official opposition party, the MDB, which was dominated by conservative leaders without any linkages to the popular movement, nor the clandestine leftist parties, which were decimated, were important actors in the beginning of MAB. Some leaders of the movement later joined clandestine parties, and by 1982 the party dispute was sharp. Yet despite the complexities and difficulties of the relationship, MAB has retained greater autonomy vis-à-vis parties than the pre-'64 movement. Indeed, one of the greatest challenges the movement has faced has been reconciling diverse party positions within the movement.

2) *Relationship between the movement and the State.* MAB has had greater autonomy vis-à-vis the State than the pre-'64 movement. Especially between 1961 and 1964, the State played a major role in creating neighborhood and favela associations in Rio de Janeiro. Although the State did not create associations in Nova Iguaçu, the associations were closely linked to the State. For example, the State helped sponsor the Congress of Neighborhood Associations in May 1960. Leaders of the period emphasize that the relationship between the movement and the State was essentially harmonious. To some extent, the harmony between the popular movement and the State reflected the presence of a progressive governor, Roberto Silveira (1958–62), and a relatively progressive mayor, Aloísio de Barros, both PTB politicians.

By contrast, MAB emerged as a movement opposed to, and with no linkages to the authoritarian State, and over time it has remained relatively autonomous with respect to the State. MAB has also generally avoided the radical anti-statist positions which have paralyzed some grassroots movements. The movement has insisted on preserving its autonomy, but it has also seen negotiation with the State as essential to its development. Even though the movement has opposed many policies of the local and state governments, it has sought dialogue with the authorities. This autonomy vis-à-vis the State, without falling into a naive anti-State attitude, is one of the characteristics which has distinguished MAB as an unusually successful neighborhood movement. However, as noted previously, the issues of movement autonomy and cooptation are very complex and by no means have been permanently resolved.[36]

3) *Role of the left.* In Nova Iguaçu, the post-'74 neighborhood movement involved a Catholic left for the first time. The changes the Catholic Church underwent, both locally and nationally, were crucial in the formation of a Catholic left (see Della Cava's essay in this book). The Church in Nova Iguaçu supported MAB in a range of ways, including encouraging people to participate, providing limited yet essential resources, and legitimating the movement.[37]

In addition, parts of the Marxist left underwent a transformation and also were influential in the neighborhood movement. The broad contours of this transformation are indicated in Weffort's essay in this volume. After the dismal results of the guerrilla experience (1968–73), parts of the left began to rethink their politics. Among the significant changes was a new attitude towards liberal democracy, previously dismissed as a bourgeois façade. Having experienced the consequences of the absence of traditional democratic freedoms, much of the left began to criticize its past conceptions and to participate actively in the effort to construct a democratic regime.[38] A parallel and equally important change was the attempt to construct stronger linkages to the popular sectors. During the 1968–73 period, a "vanguardist" conception of politics permeated the left, and there were almost no linkages between the clandestine organizations and the masses. By 1973, the naiveté and tragic consequences of this approach were all too apparent, and the left began to seek new linkages to the masses.

Finally, during the 1960s, the left gave primacy to the conflict between labor and capital, and neighborhood associations were seen as secondary. Since 1974, neighborhood associations have received greater attention. In the state of Rio de Janeiro this change was reinforced by the relative decline of the labor movement. Before 1964, the state had some of the most powerful unions in the country. In the Baixada Fluminense, the labor movement played a visible role in local political struggles, while the neighborhood movement was secondary. In the post-'74 period, the region's labor movement never recouped the vitality it had before the coup,[39] while the neighborhood movement acquired unprecedented importance.

Some of MAB's outstanding leaders followed this trajectory from clandestine revolutionary organizations to working with the neighborhood movement. Although limited in numbers, the left's involvement in MAB dates back to the origins of the movement and has been critical in its entire development. The four doctors who began the movement had previously participated in the revolutionary left. As the movement expanded, other individuals with linkages (past or present) to leftist organizations also participated.

The impact of the Marxian left has been considerable, largely because of the dedication, political experience, and broader political vision of these people. People from the Marxist left participate in perhaps 20 of the 120 associations represented by MAB, and around five members of the Coordinating Council have been educated people from the Marxist left.

The Marxist left helped the movement transcend immediate material perspectives. Individuals from the left raised broader political issues and also worked to coordinate efforts between neighborhoods. The step from a movement solely concerned with the population's immediate needs to one where the leaders attempted to relate these needs to broader political issues was important. Popular movements can create pressures which cause authoritarian regimes to open up, but to do so, they must work beyond immediate material benefits towards issues related to democratization. Excessive focus on the broader issues easily leads to gaps between the leaders—who in a movement like MAB are politically sophisticated—and the rank and file—which is generally not very aware of the linkages between broader political issues and immediate material needs. Yet exclusive concern with immediate material needs prevents a movement from contributing to broader social change and makes the movement susceptible to internal crisis once it has obtained the benefits it initially sought or, conversely, once it becomes frustrated from repeated failure.

The efforts to coordinate work between neighborhoods also gave a new character to the movement. From a relatively early time, the movement was concerned about articulation among the participating neighborhoods. This was a marked contrast to previous neighborhood movements in Nova Iguaçu since only during the brief period before and after the 1960 Congress had there been serious efforts to coordinate work between neighborhoods. It is also one of the characteristics which made MAB an unusually well-articulated movement. Coordinating work between neighborhoods created the possibility of a mass movement with greater chances of pressuring the State.

The role the Marxist left played in helping organize the neighborhood movement is common in Brazil. The popular classes have always had some forms of resisting domination, but without the input of leadership generally drawn from outside circles, they have only exceptionally created effective political movements. Even the post-'74 popular movements, which have been more autonomous with respect to political parties and intellectuals, have generally relied on outside support, especially in the early phases.

Many people who participate in the neighborhood movement are critical of the Marxist left. One individual who has participated in the popular struggles of the Baixada Fluminense since the late 1950s stated, "The ideological proposals are generating some problems. The lack of credibility among politicians, and the constant attacks by the movement against the politicians, have impeded the growth of the popular movement. The popular movement has radicalized a lot. This radicalization has prevented its progress."[40]

Recognition of the Marxian left's role in MAB does not imply that the left's involvement in a popular movement is a panacea. The Federation of Favela Associations of Rio de Janeiro (FAFERJ) provides an example in which involvement by the Marxian left has, by the most favorable evaluation, produced mixed results. A frequent complaint in such cases is that the Marxian left manipulates the movement for its own ends.

4) *Grassroots participation.* Most popular movements in the pre-'64 period were hierarchically organized, and only in exceptional periods did the grass roots participate.[41] The available evidence suggests that at least the latter part of this generalization was true of Nova Iguaçu's neighborhood movement of the pre-'64 period. Only at one point, the 1960 Congress of Neighborhood Associations, did the movement mobilize a large number of people. After the Congress, participation once again declined. One participant of the period stated that the associations were formed on the basis of a small group of friends getting together and using their personal contacts to attempt to have problems resolved. While this method was successful in obtaining some short-term benefits, the movement did not mobilize large numbers of people or create solid associations.

In the post-'74 movement, the gap between rank and file and leaders in terms of political consciousness and involvement persists. Nevertheless, the efforts to promote grassroots participation and democracy have led to encouraging results in many associations. Some associations have as many as 500 members, and as many as 150 people participate in the weekly meeting. Moreover, the movement has mobilized over 1,000 people many times over the years. An individual who actively participated in the region's popular struggles in both the pre-'64 and the post-'74 periods commented, "The big difference between the popular movements today and the movements of the pre-'64 period is that today more people enter the struggle. Before 1964, a small group of people got together to try to resolve things. Today we try to increase everyone's awareness about the importance of participating. The number of people who participate increased a lot."[42]

Limits of change. The neighborhood movement which surfaced after 1974 had greater popular participation than ever before, and it was also more autonomous vis-à-vis the State and political parties than the pre-'64 movement. These changes suggest that the popular classes have emerged as a more conscious, active political force than in the past. Especially in light of the traditional relative exclusion of the popular classes, these changes assume considerable political importance. If popular movements continue to expand and are able to establish effective linkages to major political parties and to the State, they will contribute significantly to democratizing both the State and social relations.

Nevertheless, the changes which have occurred are generally subtle and fragile. In this sense, many analyses of grassroots movements have erred on the side of exaggerating the novelty, strength, and autonomy of grassroots popular movements.

While there has been some change in the level of popular participation, the contrast to the pre-'64 movement is not as sharp as some analysts have suggested. On the one hand, during some moments, there was significant participation in the pre-'64 movements. This is especially true of the labor and peasant movements, but was also occasionally true of some neighborhood movements. On the other hand, as the discussion of MAB's dilemmas indicated, it remains difficult to mobilize the local population. Popular participation in political life in general remains restricted. Furthermore, the gap between the political consciousness of the leaders and that of grassroots participants is profound. On the average, poor Brazilians are remarkably uninformed and apathetic about politics.

In Nova Iguaçu and elsewhere, the neighborhood movement was far from autonomous in relation to political forces outside the popular classes. In particular, the Catholic Church and the Marxian left played a fundamental role in the movement. Without support from the Church and the Marxian left, neighborhood movements have experienced considerable difficulty in developing, except in the face of a concrete threat from the outside, as in the case of favela removals or land expulsions. While the popular sectors may organize on their own, this organization still tends to be ephemeral, to focus exclusively on immediate material demands, and to fail to construct linkages to other neighborhoods or local institutions.

Despite the strengthening of popular movements, the State remains the dominant sphere of Brazilian politics. Political parties are still more effective than social movements in promoting political change, not because the parties themselves are strong, but rather because they are the primary instrument for gaining access to power in a democratic system. The relative fragility of popular movements and the need to effect change through the State were seen clearly in Nova Iguaçu in 1982, when over a dozen candidates from MAB decided to run for office because of the limits of the social movements. Despite the fact that grassroots movements of the post-'74 period were heralded by some as the answer to strengthening civil society, by 1982 many leaders of one of the country's most successful grassroots movements felt that change would be most effectively realized through the State. Equally noteworthy is the fact that all of the main political currents represented within MAB had candidates for office. There can be no doubt that grassroots movements have strengthened civil society—but it is also clear that Brazilian civil society remains subordinate to the State.

The 1982 election results further underscored the weakness of popular movements, both in Nova Iguaçu and generally. In Nova Iguaçu, with one exception, political candidates linked to popular movements were defeated. MAB considerably overestimated its electoral strength. Significantly, the only MAB candidate who won election was more conservative than all the others. It was a clientelistic approach which proved most successful in electoral terms. The unique party situation in the state of Rio de Janeiro contributed to the defeat of popular candidates, but it seems likely that under even the most propitious conditions, MAB's electoral impact will be limited.[43]

Despite the important changes Brazilian society underwent during the two decades of military rule, the State and political parties remain as central as ever in any attempts to democratize the political order.[44] The tradition of a strong State and weak civil society has undergone some alterations,[45] but the experience of MAB and other social movements suggest that this transformation has been limited. The case of São Paulo, where popular movements have elected a meaningful number of candidates, both in the PT and the PMDB, indicates that placing representatives within the State does not ensure greater public responsiveness to popular demands.

Despite the intentions of movements like MAB to challenge them, populism and clientelism are alive and well in Brazilian society.[46] The emergence of more autonomous, stronger popular movements in the second half of the 1970s did contribute to forcing political elites to change their discourse. The technocratic, authoritarian discourse of the most repressive period gave rise to a more participatory discourse. This change in itself is significant, but it should not camouflage the existence of populist practices. Nowhere was this so clear as in the state of Rio de Janeiro, with Governor Leonel Brizola.

Finally, despite the strengthening of popular movements, Brazilian politics remains fundamentally a struggle among different elite groups. This was clear in the period after April 1984, when the amendment which proposed establishing direct elections for President was defeated. Vast segments of the society, including many organized social movements, mobilized on behalf of direct elections. However, it was only through traditional elite negotiations behind closed doors that the transition to democracy was secured. Appositely, the master of elite negotiation and conciliation, Tancredo Neves, was the winner of the negotiations. Tancredo's victory represented the triumph of traditional Brazilian ways of doing politics—ways which largely excluded the popular sectors.

None of this is to deny the impact grassroots movements had in the struggle for democracy in Brazil, the role they may have in the future, or the changes Brazilian politics have undergone as a result of the grassroots movements. But any assessment of these movements must take into consideration the unusual resiliency of relatively traditional approaches to politics,[47] even though, if the argument here is correct, the traditional style has had to change to accommodate grassroots movements.

Grassroots Movements and Democratization, 1974–1985

At this point, it is time to leave a reflection centered on Nova Iguaçu and analyze, more generally, the contributions of grassroots movements in the struggle for democracy. While the Nova Iguaçu case illustrates many of the most important contributions, limits, and dilemmas faced by grassroots movements, it must also be analyzed within the context of an amalgam of heterogeneous grassroots movements.

As Bolivar Lamounier argues in his essay, the Brazilian regime reached the point where the most important decision arenas were open for dispute only in 1982. In this sense, the regime was successful in controlling the broadest contours of the *abertura* over a protracted period of time. Nevertheless, from 1974 on, the opposition forced the regime to redefine significant issues, even if these issues did not imply relinquishing the most important positions in the decision arena. Many changes were not foreseen by the originators of the *abertura*; they rather reflected an ongoing process of opposition initiatives, followed by subsequent regime responses and initiatives, with occasional negotiating between the two sides. Indeed, it was in part because the regime responded with relative sagacity to a wide array of opposition demands that it was able to move in a more liberal direction while still controlling the most important decision arenas. Without having responded to civil society, the regime would not have been able to compete in elections as well as it did. It was precisely its ability to compete in elections (aided, admittedly, by frequent manipulation of electoral laws, as well as by occasional intimidation and repression of the opposition) which enabled it to liberalize without being marginalized from the political process. In this sense, the Brazilian regime stands out with the Spanish one (1975–77) as one of the few authoritarian regimes that was able to promote liberalization while remaining a competitive political force.

It is within this overall context of the dynamic between the regime and the opposition that the impact of the grassroots movements is best comprehended. It would be misleading to attribute significant weight to popular movements at the beginning of the *abertura*. Indeed, elsewhere I have argued that the weakness of popular movements, rather than their strength, was an important factor in creating confidence within the regime that it could liberalize without adverse effects.[48] And especially in rural areas, the regime's approach to popular movements remained repressive until its demise in 1985.

Over a period of time, however, political liberalization allowed more space for popular movements, and these movements used this space to put new items on the political agenda. These items ranged from the right to strike and better work conditions to the right to land and urban services.

While the regime continued to resort to repression against popular movements, the relationship between popular movements and the State reproduced many aspects of the relationship between the opposition as a whole and the State. The regime resisted change and attempted to control it, but for its own survival it was forced to make some concessions to the popular sectors. Otherwise, in a period of increasing importance of the electoral process, it would not have been able to compete for the popular vote. While refusing to deal with popular movements, the regime attempted to meet some of the demands formulated by these movements. In doing so, it hoped to strengthen its own forces and weaken those of the popular movements.

Even when they appear to be inefficient political actors, social movements can play an important role by sensitizing other forces, especially political parties and the State, to the need to redefine the political arena. The movements themselves may die out, but they can promote lasting change by placing new questions on the agenda—questions which are ultimately adopted by political parties and acted upon by the State. In Nova Iguaçu, this indirect political role was seen through the way MAB helped encourage a transformation of political discourse, through the State's increased responsiveness to material needs of the population, and through the institutionalization of mechanisms of dialogue with the popular movements. In the country as a whole, the indirect role of social movements was manifested in the authoritarian regime's decision to formulate, for the first time, a strategy for dealing with popular demands in the late 1970s. After years of neglecting popular demands and repressing popular movements, the regime initiated some programs aimed at winning popular support. The various housing programs for the poor, the reformulation of wage policy to favor the poorest workers (1979), and Figueiredo's ill-fated attempts to cultivate a populist style and discourse (1979–81) were among the most important measures in this regard. Repression against popular movements continued, but the change in policies towards the popular sectors was clear. While it is impossible to "measure" the role of popular movements in promoting these changes, the government's concern about these movements is a strong indication that they had some impact.

If this argument is correct, it calls attention to the importance of studying social movements in relation to political parties, the State, and other institutions.[49] It also suggests that by themselves, social movements did not and will not have a great direct political impact. However, they are likely to continue acting as the "conscience" of the society, placing on the agenda issues of socio-economic justice, rights for the popular classes and minority groups, and popular participation.

Furthermore, even if the movements are ineffective in their political action, they can help redefine political culture. Particularly important in this regard is their role in legitimating popular demands and participation, and their role as a mechanism of popular political socialization. In this latter sense, it is relevant to ask not only how social movements affect the State, but also how they affect the participants. Through their engagement in social movements, poor people can work out a new understanding of authority and legitimacy; rather than remaining generally passive political actors, they become more active ones. Even if the material conquests of the movements are limited, they can be an important means of constructing a new popular identity that implies a fuller citizenship.[50] Whether or not the State changes in the short term, participation in popular movements can change the symbolic bases of politics. For this reason, it would be a mistake to limit the analysis to the interaction between social movements and the State.[51]

Toward the Future: Prospects for Grassroots Popular Movements

During the eleven years between the beginning of the *abertura* and the transition to democracy, many grassroots movements engaged in the efforts to redemocratize the society. According to the preceding argument, they were partially successful in these efforts. Paradoxically, however, every step along the way brought new and unexpected dilemmas. In MAB's case, for example, as the repression began to decline, the party dispute became more significant. The very conditions that made possible an expansion of the movement also increased possibilities for internal dispute and division. When the democratically elected mayor and governor took office in 1983, the movement faced new challenges stemming from the difficulties of mobilizing the population in a period of popular apathy.

History repeated itself with the transition to democracy at the federal level. Social movements participated in the 1984 campaign for direct elections, which helped make the transition possible. Yet after actively contributing to the campaign which resulted in Tancredo Neves's election, social movements once again found themselves marginalized from the centers of power.

Although Tancredo's election marked the demise of the military regime, the opposition effected this victory only with the active support of the large segment of regime defectors that formed the Liberal Front. The style and content of the new government clearly recognized this "compromise" character. The new government included, in the upper echelon of the decision-making sphere, several civilian leaders from the military regime. As Stepan has noted, there were six military ministers—a number unmatched by any other democracy in the world.[52]

After having struggled for more than a decade to effect political change, movement leaders and participants were bitterly disappointed by the performance of the Sarney government. Egregious corruption, a visible private appropriation of the State apparatus, extraordinary governmental incompetence, a serious economic crisis, and a decomposition of the democratic alliance were notable features of Sarney's first three years in office. The deterioration of the Sarney government, coupled with the absence of a viable opposition, left many movement leaders skeptical about the possibilities of a deeper democratization. In addition to the other disheartening developments described well in Souza's chapter, there was a tendency towards a disaggregation of networks in civil society that had formed during the years of resistance to authoritarian rule. A minimum condition for grassroots popular movements to re-achieve their vitality of earlier years is the existence of hope that things can be changed; for such hope to exist, a viable opposition in the party system must emerge.

However difficult the prospects for the grassroots movements may be, two points seem clear. First, despite adversity, these movements contributed to the struggle for democracy between 1974 and 1985. There is no a priori reason to suppose that they will necessarily decline under a democratic government.[53] This has occurred so far in the New Republic because the peculiar combination of a massively unpopular government and the absence of a viable party opposition has had a dampening effect on political participation. If public responsiveness to popular needs increases, or if popular movements find receptive allies in the party system so as to recreate a minimal sense of hope for political change, the present period of demobilization could be reversed. Second, although the consolidation of democracy does not depend on its responsiveness to social movements, the quality of democracy almost certainly will. For the social movements continue to raise the banners of socio-economic justice, rights for the popular classes, and popular participation. Given the country's lengthy history of elitism, as well as the profound socio-economic inequalities, no one can doubt the importance of these banners.

NOTES

1. Chapter 8 of my book, *The Catholic Church and Politics in Brazil 1916–1985* (Stanford: Stanford Univ. Press, 1986), relates other aspects of my work in Nova Iguaçu, focusing on the relationship between the neighborhood movement and the Catholic Church. Parts of the first two sections of this article are taken from my book and are reproduced here with the permission of Stanford University Press. Copyright © 1986 by the Board of Trustees of the Leland Stanford Junior University.

2. I use the terms "grassroots popular movement" and "urban popular movement" rather than the more commonly used "urban social movements. Urban social movements encompass a wide range of popular and middle-class movements. While there is some possibility that popular and middle-class movements will work together in confronting an authoritarian state, it is equally likely that their interests will be contradictory and competing. Therefore, to discuss urban social movements as a whole suggests an illusory unity.

3. For more detailed information on the history of Nova Iguaçu, see Leda Lúcia Queiroz, "Movimentos Sociais Urbanos: O Movimento Amigos de Bairros de Nova Iguaçu" (M.A. thesis, COPPE, 1981), chap. 2; M.T. de Segadas Soares, "Nova

Iguaçu: Absorção de uma Célula Urbana pelo Grande Rio de Janeiro" (Thesis of Livre Docência, Universidade do Brasil, 1960); and Júlia Adão Bernardes, *Espaço e Movimentos Reivindicatórios: O Caso de Nova Iguaçu* (Rio de Janeiro, 1983). For a socio-economic profile of the Baixada Fluminense's population, see Cristina Saliby et al., "A Política de Habitação Popular: Suas Consequências sobre a População do Grande Rio," unpublished manuscript, Rio de Janeiro, 1977.

4. Movimento dos Amigos do Bairro, "Primeiro Ciclo de Debates Populares do MAB," mimeo, Nov. 1980.

5. Ibid.

6. On the transportation situation, see Pastoral Operária de Nova Iguaçu, "A Condução do Trabalhador," in Carlos Rodrigues Brandão, (ed.), *A Pesquisa Participante* (São Paulo: Brasiliense, 1982), 63–85.

7. Maria Helena Moreira Alves, "The Formation of the National Security State: The State and Opposition in Military Brazil" (Ph.D. dissertation, Massachusetts Institute of Technology, 1982), 500. On the Death Squad's activities in Nova Iguaçu, see the *Jornal do Brasil*, April 13, 1983.

8. On the MDB in the state of Rio de Janeiro, see Eli Diniz, *Voto e Máquina Política: Patronagem e Clientelismo no Rio de Janeiro* (Rio de Janeiro: Paz e Terra, 1982).

9. In 1974, the top MDB candidate for federal deputy had 47,929 votes compared with 22,862 for the top ARENA candidate. The top MDB candidate for state deputy had 19,917 votes, compared with 9,974 for the top ARENA candidate; and the MDB candidate for federal Senator outpolled the ARENA candidate 99,628 to 43,352. Election coverage and data are found in *Correio da Lavoura* 2299 (Nov. 16–17, 1974). In 1978 the MDB won 118,774 votes for federal Senator while ARENA got 72,942. *Jornal do Brasil*, May 16, 1982.

10. Information on MAB's history comes from extensive interviews with movement, political, and Church leaders; participation in popular assemblies; and from MAB's publications. For detailed secondary sources, see Queiroz, "Movimentos Sociais Urbanos," and Adão Bernardes, *Espaço e Movimentos Reivindicatórios.*

11. On the connection between this health work and the early development of the neighborhood movement, see Estrella Bohadana, "Experiências de Participação Popular em Ações de Saúde," in IBASE, *Saúde e Trabalho no Brasil* (Petrópolis: Vozes, 1982), 107–28.

12. *Encontro* 2 (March 1976).

13. *Correio da Lavoura*, Dec. 24, 1982.

14. Interview, Oct. 26, 1986.

15. This point is forcefully argued by Renato Raul Boschi in his interesting paper, "Movimentos Sociais e a Institutionalizaçao de uma Ordem" (Rio de Janeiro: IUPERJ, 1983). See also the excellent paper by Ruth Cardoso, "Movimentos Sociais Urbanos: Balanço Crítico," in Bernardo Sorj and Maria Hermínia Tavares de Almeida, *Sociedade e Política no Brasil pós-64* (São Paulo: Brasiliense, 1983), 215–39.

16. For an excellent discussion of this issue, focused on the Mexican experience, see Jonathan Fox, "The Political Dynamics of Reform: The Case of the Mexican Food System, 1980–1982" (Ph.D. dissertation, Massachusetts Institute of Technology, 1986).

17. Lícia Valladares discusses the changes in housing policy in "A Propósito da Urbanização de Favelas," *Espaço e Debates* 2 (May 1981): 5–18.

18. Prefeitura de Nova Iguaçu, "Nova Iguaçu: 152 Anos."

19. For many years, the most influential formulations were by three Europeans. See Manuel Castells, *Movimientos sociales urbanos* (Mexico City: Siglo XXI, 1974); ibid., *Cidade, Democracia e Socialismo* (Rio de Janeiro: Paz e Terra, 1980); Jordi Borja, *Movimientos sociales urbanos* (Buenos Aires: Siap, 1975); Jean Lojkine, *O Estado Capitalista e a Questão Urbana* (São Paulo: Martins Fontes, 1981). A seminal work in Brazil was José Alvaro Moisés, "Classes Populares e Protesto Urbano" (Ph.D. dissertation, University of São Paulo, 1978). Moisés's work was also excessively optimistic, but it was less "economistic" than the European works just cited.

20. See, for example, Charles Tilley, *From Mobilization to Revolution* (Reading, Mass.: Addison-Wesley, 1978); Renato Raul Boschi and Lícia do Prado Valladares, "Movimentos Associativos de Camadas Populares Urbanas: Análise Comparativa de Seis Casos," in Boschi, (ed.), *Movimentos Coletivos no Brasil Urbano* (Rio de Janeiro: Zahar, 1983), 103–43.

21. The exposition that follows is more fully developed in my "Urban Popular Movements, Identity, and Democratization in Brazil," *Comparative Political Studies* 20 (July 1987), 131–59.

22. On popular consciousness, see Renato Ortiz, *A Consciência Fragmentada* (Rio de Janeiro: Paz e Terra, 1980); Teresa Pires do Rio Caldeira, *A Política dos Outros* (São Paulo: Brasiliense, 1984); Teresa Pires do Rio Caldeira, "Para que Serve o Voto? As Eleições e o Cotidiano na Periferia de São Paulo," in Bolivar Lamounier (ed.), *Voto de Desconfiança: Eleições e Mudança Política no Brasil, 1970–1979* (Petrópolis: Vozes/CEBRAP, 1980), 81–116; Youssef Cohen, "The Benevolent Leviathan: Political Consciousness among Urban Workers under State Corporatism," *American Political Science Review* 76 (March 1982): 46–59; Alba Zaluar, *A Máquina e a Revolta* (São Paulo: Brasiliense, 1985).

23. Mancur Olson, *The Logic of Collective Action* (Cambridge: Harvard Univ. Press, 1965); Robert Dahl, *After the Revolution?* (New Haven: Yale Univ. Press, 1970).

24. IDESP generously shared the unpublished results of this survey with me.

25. For an interesting discussion of this point, see Carlos Nelson Ferreira dos Santos, *Movimentos Urbanos no Rio de Janeiro* (Rio de Janeiro: Zahar, 1981).

26. Among the best works on the elitist nature of Brazilian politics are Roberto Da Matta, *Carnavais, Malandros e Heróis: Para uma Sociologia do Dilema Brasileiro* (Rio de Janeiro: Zahar, 1979); and Guillermo O'Donnell, "¿Y a mí, qué me importa?: Notas sobre sociabilidad y política en Argentina y Brasil," Kellogg Institute Working Paper 9 (Jan. 1984).

27. Albert Hirschman, *Shifting Involvements: Private Interests and Public Action* (Princeton: Princeton Univ. Press, 1982).

28. For an interesting broader discussion of how States defuse social movements, see Frances Fox Piven and Richard Cloward, *Poor People's Movements: Why They Succeed, How They Fail* (New York: Vintage, 1977).

29. For a brief critical assessment of the Brizola government, see César Guimarães and Marcelo Cerqueira, "O Governo Brizola à Procura da Identidade," *Novos Estudos* 10 (October 1984): 13–17.

30. See Pedro Jacobi, "Movimentos Sociais Urbanos e Política: Participando do Debate," VIII ANPOCS, 1984; and Maria da Glória Marcondes Gohn, "Movimentos Populares Urbanos e Democracia no Brasil," VIII ANPOCS, 1984.

31. "Igreja caminha para a decepção," *Folha de São Paulo*, March 14, 1984. For a critique of the Montoro government, written from the perspective of a PT intellectual committed to social movements, see Marco Aurélio Garcia, "Dezoito Meses de Governo Montoro," *Novos Estudos* 10 (Oct. 1984): 2–7.

32. "Igreja e Montoro se reaproximam mas ainda existem divergências," *O Globo*, July 29, 1984.

33. A conflict-free democratic organization or movement requires an extraordinary degree of common interest. See Jane Mansbridge, *Beyond Adversary Democracy* (New York: Basic Books, 1980).

34. According to public-opinion surveys, Brizola suffered a unilinear decline in popularity from the time he assumed office until the implementation of the Sarney government. By March 1985, he was the least popular opposition governor. Only 9.7% of the people surveyed said that his government was better than they expected, while 33.3% said it was worse and 11.2% said they had expected nothing. See "Richa e Garcia Sobem, Montoro Mantém e Brizola Cai," *Folha de São Paulo*, March 3, 1985.

35. See José Alvaro Moisés, "Experiências de Mobilização Popular em São Paulo," *Contaponto* III, No. 3, (1978): 69–86; Paul Singer, "Movimentos de Bairro," in Singer and Vinicius Caldeira Brant (eds.), *São Paulo: O Povo em Movimento* (Petrópolis: Vozes/CEBRAP, 1981), 83–108; Ana Luísa Salles Souto Ferreira, "Movimentos Populares Urbanos e suas Formas de Organização Ligadas à Igreja," *Ciências Sociais Hoje* 2 (Brasília: ANPOCS, 1983): 63–95; Eli Diniz, "Favela: Associativismo e Participaçao Social," in Renato Raul Boschi (ed.), *Movimentos Coletivos no Brasil Urbano* (Rio de Janeiro: Zahar, 1983), 27–74.

36. For an interesting study on the question of movement autonomy vis-à-vis parties and the State, see Eva Alterman Blay, "Movimentos Sociais: Autonomia e Estado," VII ANPOCS, 1983.

37. Daniel Levine and I discuss religion's role in motivating people to participate in popular movements, in "Religion and Popular Protest in Latin America: Contrasting Experiences," in Susan Eckstein, ed., *Protest and Resistance: Latin American Experiences* (Berkeley: Univ. of California Press, forthcoming). Interesting analyses of this issue in the Brazilian context are Ana Maria Doimo, *Movimento Social Urbano, Igreja e Participaçao Popular* (Petrópolis: Vozes, 1984); and Paulo Krischke, "As CEBs na 'Abertura': Mediações Entre a Reforma da Igreja e as Transformações da Sociedade," in Paulo Krischke and Scott Mainwaring (eds.), *A Igreja nas Bases em Tempo de Transição* (Porto Alegre: L&PM/CEDEC, 1986), 185–207.

38. On the renewed attention given to democracy, also see Bolivar Lamounier, "Representação Política: A Importância de Certos Formalismos," in Lamounier, Francisco Weffort, and Maria Victória Benevides (eds.), *Direito, Cidadania e Participação* (São Paulo: T. A. Queiroz 1981), 230–57; Robert Packenham, "The Changing Political Discourse in Brazil, 1964–1985," in Wayne Selcher (ed.), *Political Liberalization in Brazil: Dynamics, Dilemmas, and Future Prospects* (Westview: Boulder, 1986), 135–73; Carlos Nelson Coutinho, *A Democracia como Valor Universal* (São Paulo: Ciências Humanas, 1980).

39. On the decline of the labor movement in Rio de Janeiro, see Ingrid Sarti and Rubem Barbosa Filho, "Rio de Janeiro: O Sindicato vai à Luta. E Agora?," in CEDEC, *Sindicatos em uma Época de Crise* (Petrópolis: Vozes/CEDEC, 1984), 35–53.

40. Interview, March 27, 1985.

41. Most of the literature has emphasized the strength of corporatist mechanisms and the relative weakness of the popular classes as political actors. See, for example, Heloisa Helena Teixeira de Souza Martins, *O Estado e a Burocratização do Sindicato no Brasil* (São Paulo: Hucitec, 1979). Although they do not radically disagree with this perspective, other authors have emphasized the occasional capacity of the popular classes to intervene effectively in political life. See Francisco Weffort, "Sindicatos e Política" (Ph.D. dissertation, University of São Paulo, 1970); Moisés, "Classes Populares e Protesto Urbano"; Luiz Alberto Gómez de Souza, *Classes Populares e Igreja nos Caminhos da História* (Petrópolis: Vozes, 1982).

42. Interview, June 2, 1981.

43. Other authors have also suggested that the electoral impact of social movements was limited. See Teresa Pires do Rio Caldeira, "A Luta pelo Voto em um Bairro de Periferia," VII ANPOCS, 1983; Nísia Verônica Trindade Lima, "As Eleições de 1982 em Favelas do Rio de Janeiro," VII ANPOCS, 1983.

44. For a similar point, see the important contributions of Fernando Henrique Cardoso, "Regime Político e Mundança Social," *Revista de Cultura e Política* 3 (Nov. 1980/Jan. 1981): 7–25; Fernando Henrique Cardoso, "A Democracia na América Latina," *Novos Estudos* 10 (Oct. 1984), esp. pp. 50–56; and Paul Singer, "Movimentos Sociais em São Paulo: Traços Comuns e Perspectivas," in Singer and Brant (eds.), *São Paulo*, pp. 207–30

45. Much of the classical literature on Brazil has emphasized this combination of a strong state and a weak civil society. See Philippe Schmitter, *Interest Conflict and Political Change in Brazil* (Stanford: Stanford Univ. Press, 1971); Raimundo Faoro, *Os Donos do Poder* (Porto Alegre: Globo, 1958); Simon Schwartzman, *Basès do Autoritarismo no Brasil* (Rio de Janeiro: Campus, 1982); Maria do Carmo Campello Souza, *Estado e Partidos Políticos no Brasil* (São Paulo: Alfa-Omega, 1983); Hélgio Trindade, "Bases da Democracia Brasileira: Lógica Liberal e Práxis Autóritária (1822–1945)," in Alain Rouquie et al. (eds.), *Como Renascem as Democracias* (São Paulo: Brasiliense, 1985), 46–72.

46. See Paul Cammack, "Clientelism and Military Government in Brazil," in Christopher Clapham (ed.), *Private Patronage and Public Power* (New York: St. Martin's Press, 1982), 53–75.

47. For an interesting discussion of this point, see Frances Hagopian, "The Politics of Oligarchy: The Persistence of Traditional Elites in Contemporary Brazil" (Ph.D. dissertation, Massachusetts Institute of Technology, 1986).

48. See (with Donald Share) "Transitions through Transaction: Democratization in Brazil and Spain," in Selcher (ed.), *Political Liberalization in Brazil*, 175–215.

49. For a similar point, see Lúcio Kowarick, "Os Caminhos do Encontro: As Lutas Sociais em São Paulo," *Presença* 2 (Feb. 1984): 65–78; and Luiz Antônio Machado da Silva, "Associações de Moradores: Mapeamento Preliminar do Debate," VII ANPOCS (Oct. 1983).

50. I explore this issue in greater detail in "Urban Popular Movements." For related discussions, see Eunice Durham, "Movimentos Sociais, A Construcao da Cidadania," *Novos Estudos* II, 10 (Oct. 1984): 24–30; Tilman Evers, "A Face Oculta dos Novos Movimentos Sociais," *Novos Estudos* II, 4 (April 1984): 11–23.

51. For a similar argument, see Manuel Castells, *The City and the Grassroots* (Berkeley: University of California Press, 1983), especially pp. 291–300.

52. See Alfred Stepan, *Rethinking Military Politics: Brazil and the Southern Cone* (Princeton: Princeton University Press, 1988), 104.

53. In this sense I disagree with Guillermo O'Donnell and Philippe Schmitter, who argue that social movements almost inevitably decline in the period following transitions to democracy. See their *Tentative Conclusions about Uncertain Democracies*, Part 4 of O'Donnell et al., (eds.), *Transitions from Authoritarian Rule* (Baltimore: Johns Hopkins Univ. Press, 1986). O'Donnell and Schmitter seem to believe that institutionalized political channels (parties, the State, interest associations) become so dominant in democratic politics that less institutionalized channels (such as movements) lose their importance. I think this issue is more open-ended than they suggest; in some countries, most social movements decline after the transition, but in others, this is not the case.

7

Politicizing Gender and Engendering Democracy

SONIA E. ALVAREZ

Women emerged as key political actors in the Brazilian transition to democracy.* Starting in the mid-1970s, both university-educated, middle-class women and poor, uneducated women organized movements to press their gender-specific political claims on the Brazilian political system. Moreover, women also made up the backbone of many of the organizations of civil society and opposition political parties which successfully challenged authoritarian rule during the 1970s and '80s. Political liberalization or *abertura* in Brazil generated increased political oppportunity space for female political participation and for the articulation of gender-specific political demands. The gradual process of democratization reinforced and was in turn strengthened by an equally gradual process which I have labeled the *politicization of gender*—a process whereby issues previously considered private or personal are raised as *political* issues, thus to be addressed by political parties and the State.

*Author's note: The data on women and the Brazilian *abertura* process presented here were gathered during field research conducted in São Paulo and other major Brazilian cities during November-December 1981, October 1982 to October 1983, and July-August 1985. This fieldwork was made possible by grants from Fulbright-Hays, the Inter-American Foundation, the Social Science Research Council, and the Yale University Council on Latin American Studies. Research on post-1983 policy developments was made possible by a grant from the University of California at Santa Cruz's Faculty Research Committee. The data presented for '83–85 was compiled in collaboration with Miriam Bottassi and other members of the São Paulo-based *Centro Informação Mulher* and I thank them for their ongoing help with my research—their collaboration has defied cross-cultural and cross-continental barriers over the past three years. I would also like to thank Susan Szabo of the UCSC Latin American Studies Program for her invaluable assistance in the preparation of this paper and Janice Robinson for helpful editorial suggestions. An earlier version of this paper was presented at the XII International Convention of the Latin American Studies Association Albuquerque, New Mexico, April 18–20, 1985.

As these two processes unfolded in Brazil, demands for increased political representation and political clout for women as a group, for free, community-based and -administered day care, and for safe, accessible, and non-coercive family planning, were introduced into institutional arenas at all levels of Brazilian politics by women's movement organizations. These and other gender-specific demands were increasingly endorsed by most opposition political parties and by several of the post-1982 opposition state governments. In the recently inaugurated, post-authoritarian *Nôva República*, one might reasonably expect that women's political claims will receive at least verbal endorsement given the importance of women's participation in the opposition since the 1970s. The new regime owes a considerable amount of its popular bases of support to organized female constituencies. Autonomous women's movement organizations and women's branches of the major opposition parties unquestionably played a critical role in the nation-wide mobilizations for direct elections in 1984 and in the subsequent mobilization of women's support for the indirect candidacy of Tancredo Neves and the Democratic Alliance.[1]

But just how gender-specific political claims will be incorporated into the new regime's political institutions and public policies remains an open question. The political clout wielded by organized female constituencies during *abertura*, especially during the 1982 and 1984 electoral conjunctures, may well recede as Brazil returns to politics as usual under the post-authoritarian regime.

Comparative data on the politics of gender in other Latin American regimes in the past suggest a rather discouraging prognosis for the future of women within the Brazilian New Republic. Historical data reveal that the incorporation of women and women's issues into Latin American politics has most often reinforced existing patterns of gender-based inequality by relegating women and their gender-related political issues to a subordinate or secondary position within both male-dominant political institutions and political discourses. Women's political claims and women's movement organizations have most frequently been coopted, instrumentalized, or manipulated by political elites and the political apparatuses of the State in ways which serve the needs of the prevailing pact of domination—even when women have achieved limited gains through their increased participation in politics. Populist, democratic, authoritarian, and even socialist regimes have proven quite resistant to gender-based political claims,[2] albeit to different degrees.

While many analysts of contemporary Brazilian politics have noted the massive presence of women within the new social movements, few have considered its implications for *real* democratization (which would necessarily imply an end to institutionalized sexism) or pondered its potential for ameliorating gender-based inequality in Brazil. It is the central theoretical contention of the present analysis that gender-based inequality is inscribed in the very structure of State power in Brazil, as elsewhere, and that the incorporation of newly mobilized women and women's issues therefore represents one of the biggest challenges for the new democratic regime.[3]

The intractability of the political arena to women in particular and to gender-based politics in general is attributable to the fact that the modern State (whether capitalist, dependent capitalist, or socialist) is *not neutral* on gender issues. Feminist theoretical insights suggest that as the modern State represents the quintessential institutional separation of the public or political from the private or personal domains of human activity, it also institutionalizes gender power relations by circumscribing the female gender to the latter domain, politically reinforcing the boundaries which have confined women socially and historically. The political, then, becomes the domain of men and male issues, and issues which directly affect the lives of women, like reproduction, contraception, child care, rape and sexual abuse and battery, and so on, are pre-defined as outside the "proper" realm of politics.[4]

Contingent upon shifting social relations of production and reproduction, this public/private split[5] must be constantly ideologically redefined and those new definitions coercively enforced by the State. As has been amply demonstrated by recent feminist scholarship, the State must in fact regulate and delimit personal power relations in order to guarantee the continued functioning of the public sphere—hence, marriage, divorce, and inheritance laws, rape and pornography laws, State population control policies, and so on.[6]

Though "genderic,"[7] as well as economic and racial, power relations find their expression and articulation within the pact of domination represented within the State, the State does not monolithically represent male interests.[8] To coin a phrase, the State is *not* the executive committee of men, but rather is relatively autonomous of male interests, and therefore may act in the interest of women at particular historical conjunctures.

The State is relatively autonomous of patriarchal or male interests not because it is independent of those interests but because its legitimacy is partially derived from its ability to conceal the genderic, racial, and class interests represented within the pact of domination by granting some concessions to subordinate groups and classes which increasingly press their political claims upon it.[9] This feminist perspective on the relative autonomy of the State implies that class-based, racially based, *and* gender-based political struggles, led by social movements, can and *must* take place both within and without the political apparatuses of the State. Drawing from Carnoy's interpreation of Marxian class struggle theories of the State as represented in the work of Wolfe, Castells, and the later Poulantzas, one can suggest that the "white, male, capitalist" State can be moved against dominant interests by the

> development of movements inside and outside the State to force it to move against its fundamental role as reproducer of [gender, race, and] class relations . . . The capitalist [patriarchal and racist] State will not reform in a progressive direction without such movements pressing it. In other words, the capitalist [patriarchal and racist] State is inherently class-based [gender-based and racially based] and will act in that way unless pressured by mass organizations. The correct political strategy is to organize at the base, both outside and inside the State, bringing those organizations to bear on society's dominant institutions to reform them.[10]

The "gender struggle" view of the State proposed here also suggests that different political regimes, which represent different schemes of class, genderic, and racial domination and different policies for structuring the relationship between State and society, may also represent different *opportunity spaces* for social movements to act both within and without the political apparatus of the State and thus impact State policies. If the State does not monolithically represent male or patriarchal interests, then one can comparatively examine which regime characteristics and which political conjunctures appear most favorable for promoting changes in the status of women through public policy.

In spite of the structural-historical tendency toward State cooptation of women's movement organizations and their gender-specific issues, women are not the passive objects of State policy. Instead, women are, and have always been, active subjects in politics, even if their political participation has been largely confined to non-institutional political arenas. Thus, the analysis which follows also examines the effect of State policy outputs on women's movements' political strategies and political discourses. My inquiry into the relationship between women's movements and the State therefore conceptualizes that relationship as a dynamic and dialectical one rather than a linear one involving simply movement input→policy process→State output. As several critical studies of social movements have shown, it is crucial to examine how State policies shape, and sometimes determine the strategies and dynamics of social movement organizations.[11]

In light of this reconceptualization of gender politics, we cannot assume that merely because the post-authoritarian regime in Brazil is a would-be liberal, democratic one, it will necessarily restructure gender power relations in Brazilian society as it restructures State-civil society relations in general. Democratization is unquestionably important for women as a group, as it is for other social groups who were excluded or marginalized from politics under authoritarianism. But given the gendered bases of State power, the politics of gender in post-authoritarian Brazil are especially problematic, and the institutions created by the new regime to channel women's political participation are critical for mediating the structural-historical tendency toward State cooptation of women's movement organizations and their political demands.

This essay examines the relationship between women's political mobilization and shifts in gender-specific government policy during the final stages of the *abertura* process. Assuming that women were significant political actors in that process, how did the politicization of gender by women's movement organizations affect redomocratization? And assuming a dialectical relationship between the State and civil society, I address a related set of questions. What impact did the political liberalization process have upon the politicization of gender and the emergence and development of women's movement organizations in the 1970s and '80s? And what impact, if any, did organized, gender-conscious political pressure "from below" have upon the authoritarian regime's gender-specific policy outputs? And finally, what did democracy hold in store for women? Though liberal democracy clearly matters to women as citizens, as members of other social groups and classes, how would it matter to *women as women*, as gendered citizens whose lives and issues have historically been precluded from politics as usual?

In order to grapple with these questions while the democratization process was still under way, I undertook a within-nation comparison of two "micro-regimes" in the state of São Paulo. To elucidate the relationship between women's movements, political parties, and the State under authoritarian rule (in its transitional phase), I examined the relationship between women's movements' political inputs and the policy outputs of a pro-regime state government, the Maluf/de Barros PDS regime in São Paulo, 1979-83. And in order to explore the question of what liberal democracy held in store for women as women, I then turned to an analysis of how that relationship shifted under an opposition government, that of Montoro/Covas, 1983 to 1987.

I focused my analysis on day-care and family-planning policies as these issues have been especially prominent in women's movement politics *and* government policy in recent years. I also conceive these to be key issues in the analysis of the State's role in the preservation of the means of reproduction and therefore in the preservation of women's subordinate status.

These two issues are also of particular interest due to their class, as well as gendered, content. Both family planning and day care are relevant to *all* women as the primary reproducers of Brazilian society, but poor and working-class women do most of the "reproductive work" for their upper- and middle-class "sisters." Thus while many (if not most) middle-class women have domestic servants to alleviate their socially ascribed confinement to reproductive labor, day care is an issue of particular relevance to women who carry the double burden of paid labor (for bourgeois industrialists *or* bourgeois and petit bourgeois women) and unpaid (domestic) labor in their own homes. Similarly, though access to safe, non-coercive contraception is limited for all Brazilian women, women of the popular classes lack both the information and the capital with which to acquire the means of contraception more readily available to women of the middle and upper classes.[12]

Indeed, the impact of authoritarian regime policies in general varied dramatically according to women's class status. As the "wives, mothers, and nurturers" of family and community, working-class women were among the most significantly affected by regressive wage policies, rises in the cost of living, cuts in social welfare and educational expenditures, and so on. It was women of the popular classes who first clamored for their "right" to feed their families, school their children, and provide them with a decent life. And it was also motherhood, as a social institution, not a natural instinct, which prompted women to demand to know the whereabouts of their missing children, thus spearheading the human rights movement in Brazil as elsewhere in Latin America.

In short, the authoritarian development model, premised upon the political and economic exclusion of the popular classes, engendered significant changes in the domestic political economy of the lower classes. One result of these changes was the "politicization of motherhood," a factor frequently overlooked in the analysis of women's massive participation in popular movement organizations. Poor and working-class women mobilized as *women* to defend their rights as wives and mothers, rights which dominant authoritarian ideology assured them in theory, but which dominant political and economic institutions denied them in practice.[13]

Changes in the domestic political economy of the lower classes and women's resistance to those changes, then, must be seen as partially responsible for the emergence of women's movement organizations among women of the popular classes. But the creation of an organizational infrastructure for female political mobilization was also a critical factor.

As the Catholic Church turned towards the poor and against the military regime in the 1960s, it promoted organizations among the people of God at the community level, especially among those who had been progressively excluded and marginalized by the post-1964 regime.[14] And women were actively encouraged to participate as equals in these new community organizations. But the sexual division of political labor was not necessarily challenged by the people's Church, and separate women's associations, usually termed " mother's clubs," were often created by the newly militant clergy.[15]

Though mother's clubs and similar neighborhood women's associations have not necessarily raised women's consciousness of their class, race, or gender status in Brazilian society, they have provided the organizational context for *networking* among women of the popular classes.[16] Apolitical women's associations have thus provided the organizational base for political mobilization along class-specific and gender-specific lines.[17] The creation of this extensive mobilizational infrastructure, combined with the regime's political opening, and perhaps its hesitancy to be openly repressive to the "wives and mothers" of Brazil, increased the political opportunity space available for the expression of new political claims arising from perceived threats to the domestic political economy of the lower classes in which women are both the primary producers and reproducers.

Community mothers' clubs provided the organizational base for several political movements which expanded into city-wide, and even nation-wide, political campaigns. Militant motherhood provided the mobilizational referrent for the *Movimento Feminino pela Anistia*, the *Movimento Custo da Vida*, and the *Movimento de Luta por Creches* in the 1970s and '80s.[18]

These "feminine" movement organizations are usually devoid of specific ideological content—their demands are gender-related but do not necessarily challenge existing gender power arrangements.[19] Their political action has centered on direct revindication or lobbying vis-à-vis the municipal and state governments for concrete demands such as community day-care centers and better health care services for women and children; demands which, up until the 1982 electoral conjuncture, were rarely channeled through political party structures. The politicization of these women's groups around gender-specific issues is often influenced by more ideologically oriented, extra-community political actors such as party activists and middle-class feminists.

Feminist organizations have also proliferated throughout Brazil since the mid-1970s and there are presently more than four hundred feminist groups concentrated in the major Brazilian urban centers. Politicization and mobilization in feminist groups also center on women's socially prescribed roles, but these roles are ideologically defined as restrictive and oppressive to the full realization of women as "people" and "citizens." Thus gender provides the basis for feminist political mobilization as well, but the mobilizational referrent in this case derives from a direct, ideological challenge to prevailing gender power arrangements.[20]

The authoritarian development model had a radically different impact upon the lives of the middle-class women who predominate in the feminist movement. In the early years of the regime, the expansion of State sector employment and technical and professional university education actually resulted in some improvement in the status of white, middle-class women in Brazil. As Boschi points out, "the female economically active population (EAP) went from 18.5 percent in 1970 to 26.9 percent in 1980, a proportion which accounts for 41.2 percent of the increase in the total EAP over the decade." More importantly, the occupational structure of women's employment also changed, "the share of female EAP increasing in administrative occupations (from 8.2 percent in 1960 to 15.4 percent in 1980) and in professions of higher prestige (engineers, architects, doctors, dentists, economists, university professors and lawyers which went up from 19,000 in 1970 to 95,800 in 1980)." In 1980, the number of women enrolled in Brazilian universities practically equaled the number of men (689,000 men and 663,000 women).[21]

So networking[22] among would-be Brazilian feminists occurred in university and professional settings—settings which until the '60s had been largely male-dominated. Middle-class women's insertion into these previously male-dominant realms seems to have led some of them to question their own status as "lesser men" within those realms, to view gender-based inequality as a *political* problem.[23]

Middle-class women's increased involvement in higher education and the professions also increased the likelihood that they would come to be involved in the student movements, clandestine organizations, and politicized professional associations which challenged authoritarian rule in the 1960s and '70s. Participation in these opposition groups also contributed to the development of networks among middle-class women which would later be mobilized around gender-specific political issues. And as many of the feminists I interviewed suggested, the blatant sexism inherent in the theory and practice of many of these opposition groups also contributed to the development of feminist consciousness among some female militants.[24]

Political liberalization provided the political space within which that nascent consciousness could give rise to a full-scale social movement. In 1975, Geisel decided to pay lip service to the U.N.'s call for concerted government action toward eradicating gender-based inequality and allowed Brazilian women to organize meetings, conferences, and demonstrations in commemoration of International Women's Day. Those commemorations, held in Rio and São Paulo in March 1975, sparked the creation of feminist organizations and spurred feminist activism throughout urban Brazil in the ensuing years.

Feminists challenged the archaic civil code which made married women the vassals of their husbands; they demonstrated against the judicial sanctioning of violence against women and "crimes of passion" which inscribed women's sexuality as male property within the law; they clamored for equal pay and for the redistribution of domestic labor. And they aligned their new political causes with those of other oppressed and exploited groups in Brazilian society.[25]

The political action of feminist groups centered on protest actions—petitions, protest marches, mass media denunciation of sexist government policies, etc.—and on work with lower-class women's groups in peripheral neighborhoods. Many feminists have always been active in political parties—engaging in what the movement refers to as *dupla militância* (double militancy) in feminist movement organizations and "legal" or "illegal" political parties—and that double militancy was particularly accentuated during the 1982 electoral campaign.[26]

Indeed, by October of 1982, most of the existing women's movement organizations in São Paulo and elsewhere were deeply enmeshed, or, better, entangled in the political partisan struggle for institutional power which characterized the 1982 electoral conjuncture.[27] And gender-specific political demands had forcefully entered the mainstream of Brazilian politics.

From 1975 to 1981, women's movement organizations had mostly restricted their political activities to protest actions, mobilization, and consciousness-raising among women from various social sectors, and in some instances, direct revindication vis-à-vis the state and municipal government. The authoritarian State apparatus was perceived as unresponsive to gender-specific political demands (or any other demands, for that matter). And, though individual women's movement activists joined the only existing opposition party, the MDB, the movement as a whole viewed the legal opposition as lacking in effective power to carry through women's demands in the legislative arenas. Only the *creche* (day care) movement and other neighborhood-based women's groups had engaged in direct revindication vis-à-vis the municipal government (especially during the Reynaldo de Barros's administration)—but these groups rarely directed their demands through legal opposition channels until the reconstitution of the party structure in 1979-80. However, the importance of the 1982 electoral conjuncture in the Brazilian *abertura* process pushed women's movement organizations to rethink their relationship to contestatory partisan politics.

In 1981-82, many feminist and feminine movement organizations (like other organized sectors of civil society) had become demobilized and divided on partisan issues during the trajectory of the campaign. Women's *"frentes"* had been created for the avowed purpose of mobilizing women for greater participation in the campaign process. Many women had left movement militancy to engage solely in partisan militancy—political society had temporarily swallowed up sectors of civil society.[28] Women from neighborhood women's organizations and feminist groups had become candidates (for the *Partido dos Trabalhadores* (PT) and *Partido do Movimento Democrático Brasileiro* (PMDB) in São Paulo) for municipal, state, and federal level offices, and thus had become the self-proclaimed candidates for the women's movement. And issues previously considered private such as violence against women, day care, contraception and sexuality, and many other revindications raised by organized women over the previous decade, were prominently included in the platforms and programs of many individual candidates and national political parties. For the first time since the Brazilian suffrage movement in the 1920s and '30s, gender had become the basis for social mobilization, and gender inequality the object of generalized political debate.[29]

As a consequence, in part, of this unprecedented level of female political participation, gender-based political issues forcefully entered the Brazilian political arena in the post-electoral period. After March 15, 1983, family planning and the so-called Brazilian "demographic explosion" re-emerged as prominent issues in state and federal-level policy arenas. And during 1983 and 1984, the São Paulo municipal government once again was engaged in a struggle over the allocation of public funds for day care.

The following two subsections examine day-care and family-planning policies in the state of São Paulo prior to the new PMDB regime's installation of the State Council on the Status of Women *(Conselho da Condição Feminina)* in September of 1983. A later section reviews developments in these two policy areas after the Council's first two years in office in order to determine the extent to which the Council has intervened in the regime's policy process in ways which advance the status of Paulista women.

A gender struggle perspective on women in contemporary Brazilian politics draws our attention to the changing dynamics of movement-State relations over the course of political liberalization and democratization. Gender-related public policies are shaped by the State's need to maintain the public/private split as a means of ensuring the reproduction of the labor force while accommodating the changing and often conflicting demands of capitalism (for cheap female labor) and patriarchy (for women's continued confinement to the domestic sphere). However, if the State is understood to be relatively autonomous of gender interests, then it can also be viewed as an arena of gender struggle (and not just of gender-based social control). Gender-conscious political action, organized both within and without the State, may potentially influence or even alter gender-specific policy outputs.

The ability of women's movement organizations to affect State policy will be conditioned both by the political dynamics of the movement and by systemic shifts in the relationship between State and civil society—that is, by political regime change. Thus, micro- and macro-political variables (not just "functional" or economic ones) such as the social bases of support sought by a particular regime, the competing ideologies represented within that regime, the regime's ties to particular national and international interests and the presence or absence of organized gender-conscious political pressure from social movements, will play a key role in shaping the politics of gender.

The significance of organized, gender-conscious political pressure in moving the State against dominant interests is dramatically evident in recent developments in Brazilian population or family-planning policy. Whereas safe, accessible, non-coercive family planning has been a demand of women's movement organizations since their emergence in the mid-'70s, the Brazilian State had made few concessions in this regard until recent times.[30]

In fact, the post-1964 Brazilian regime has been overwhelmingly pronatalist, or, at best, ambivalent in terms of population politics. Purely economic considerations were clearly overridden by ideological and political factors in the shaping of Brazilian population politics prior to 1983.[31]

Sub-imperialist, expansionist ideology—the Miracle's vision of *O Brasil Grande*—led the regime to resist the implementation of a nation-wide family planning program in the '60s and '70s when most other Catholic Latin American nations were doing so. In the '60s and early '70s, the regime argued that the vastness of unpopulated Brazilian territory and the richness of its untapped national resources could accommodate an infinite level of population growth.

Coupled with this expansionist component of the regime's ideology in its early stages was a firm ideological commitment to the family. Indeed, bourgeois civilian supporters of the 1964 coup had mobilized thousands (primarily women) against Goulart, who allegedly threatened the very moral fabric of the Brazilian family through his "Communist" social welfare policies. The political right capitalized on the previous neglect of gender-specific issues by all Brazilian political parties during the democratic interlude and, during early 1964, organized marches throughout Brazil in the name of "Family, God, and Liberty" (the now infamous *Marchas da Família, com Deus, pela Liberdade*). The FAMILY, writ-large and abstractly, thus became one of the backbones of the new authoritarian regime in Brazil, just as it has often functioned as the bulwark of conservatism elsewhere in Latin America.[32] This gender-based component in the Brazilian regime's ideological support of the family must also be considered a significant factor shaping the pro-natalist policies of the regime in the '60s and early '70s.[33]

Until the late 1970s, the military regime adhered to a pro-natalist position, refusing to endorse the neo-Malthusian tide sweeping international politics during the '60s and '70s.[34] Though it allowed private family-planning agencies such as BEMFAM (*Sociedade Civil de Bem-Estar Familiar no Brasil*) to operate freely in several Brazilian states—primarily in the economically depressed Northeast—the State did not directly attempt to control population growth.[35] The government's position at the 1974 International Population Year Conference summarized the regime's posture toward family planning:

> Brazilian demographic policy is the sovereign domain of the government of Brazil. The government will not accept external interference, whether official or private in nature, in its demographic politics. Fertility control is a decision of the family nucleus which, in this regard, should not suffer governmental interference.[36]

In 1978, President Ernesto Geisel became the first Brazilian head of state to acknowledge publicly the State's responsibility for the provision of birth control methods, arguing that family planning was a means of elevating the quality of life in Brazil as it promotes "a necessary conciliation between demographic growth and satisfactory provision of employment, education, health care, housing and other social opportunities which are fundamental to a worthy life for all citizens."[37] In 1980, a federal health program entitled *Prev-Saúde* included an extensive section on family planning under the rubric of maternal-infant care, aimed exclusively at the distribution of birth control pills to fertile women (ages 15–49). However, this and other family-planning programs proposed in the late '70s were never effectively implemented on a national scale.

While BEMFAM had five clinics functioning in the state of São Paulo by 1978, and Dr. Milton Nakamura had established several "maternal-infant care" clinics in the capital which distributed birth control pills to "indigent" women, it was not until 1980 that the state government began to promote government-sponsored family planning. Under the administration of Paulo Salim Maluf, the Mobilization of Community Resources for Family Planning Program (*Programa de Mobilização de Recursos Comunitários para o Planejamento Familiar*) was instituted, followed in 1981 by Maluf's controversial Pro-Family program (*Pró-Família*).

Both these state programs, like the federal ones which preceded them, were aimed at low-income populations, arguing that family planning was a "human right" denied to poor populations because they lacked the economic means to purchase contraception and the public health facilities at which to purchase or obtain birth control methods. Funded by Japanese and American private family planning organizations and entrusted to the governor's wife, Dona Sílvia Maluf, the program involved the training of community volunteers to distribute birth control pills to women in peripheral neighborhoods in the capital and in rural areas elsewhere in the state, with limited or no medical supervision. The Governor's *Grupo de Assessoria e Participação* (known as GAP), a special *Malufista* organ created by the PDS in São Paulo to advise the executive branch on community issues, further suggested that people of color should be the primary targets of state population policy—or else the black population would come to predominate electorally and otherwise in the state of São Paulo.[38]

The openly neo-Malthusian targeting of low-income and black populations for birth control by the State elicited an immediate response from both the Paulista women's movement and the opposition-controlled state legislature. Labeling the Malufista family-planning programs *controlista*, the São Paulo legislature passed Projeto Lei no. 244 in 1980, sponsored by MDB State Assembly member Antônio Resk,[39] which prohibited the "implementation of any family planning program which seeks to, directly or indirectly, control population size, without the previous approval of the State Assembly." The legislative project was promptly vetoed by Maluf.

The Paulista women's movement also orchestrated a firm response to State population control initiatives. The first state-wide meeting of Paulista feminist groups, held in Valinhos in June of 1980, appointed a special commission to study the government's past and present population-control family-planning policies and to propose a feminist alternative "which would express the real interests and needs of Brazilian women." Proclaiming that "women's right to control their own bodies has long been one of the great banners of feminism" and that "both natalist and anti-natalist politics have utilized sexuality, the body of woman, as a social patrimony, denying her rights and her individuality," the movement documents distributed throughout the state of São Paulo in 1981 vehemently opposed the "ambiguous official proposal for intervention in the 'regulation of fertility' of women" and proposed the "right to have the necessary conditions to opt freely for maternity."[40]

Feminists, along with other progressive sectors of the opposition, pointed to the politico-economic rather than demographic origins of the Brazilian crisis:

> it is not the demographic explosion which is causing hunger, misery, and the *aggravation* of our historical situation of oppression but rather such aggravation is caused by the unjust distribution of national wealth and the lack of democratic freedoms which are there to preserve the privileges of a minority, to the detriment of the overwhelming misery of the majority.[41]

However, the feminist position differed markedly from that of the male-led opposition parties. The women's movement argued that the State *did* have the responsibility to provide women with safe, accessible, non-coercive methods of birth control:

> Today we struggle to have the conditions with which to exercise the right to opt freely to have or not have children, how many to have, and the spacing between one pregnancy and another. This is for us a legitimate and democratic revindication because it contains a series of aspects which are essential to the advance of the liberation of women such as: the strict respect for the free exercise of our sexuality; the demand that motherhood and domestic work be assumed as social functions; the battle against any and all forms of utilization of our bodies as a social patrimony, above our individual right to choose.[42]

Feminists also denounced the government's proposed programs for isolating women's reproductive function from the general conditions of women's health. By introducing the notion of reproductive choice as an essential precondition for women's liberation, the women's movement introduced a new, gender-specific element in the Brazilian pro-natalist versus anti-natalist debate—proclaiming that the heretofore "personal" control of fertility *was* an issue for public, political debate and State action, not to be dismissed solely as an "imperialist plot" to "kill *guerrilheiros* in the womb," as the political left historically had contended.

The women's movement and various opposition sectors sponsored a public debate to denounce *Pró-Família* in late 1981. The debate was to generate a massive campaign against Maluf's family-planning program. But the emergence of partisan rivalries within the women's movement and other organizations of civil society in 1982 prevented such a campaign from being effectively carried out. And it was not until March of 1983 that family planning re-emerged as a central issue in the Paulista women's movement—again in response to State policy initiatives.

Population control policies in the Third World have often gone hand-in-hand with strict monetarist policies imposed on national governments by the international aid community.[43] Facing the 1982–83 debt crisis and consequent renewed negotiations with the IMF and other international lenders,[44] the regime suddenly made population control one of its political priorities as of 1983. In his address to the newly elected National Congress on March 1, 1983, President Figueiredo argued that:

> In Brazil, during the last 40 years, demographic growth has surpassed 50 million inhabitants. This human growth, in explosive terms, devours, as has been observed, economic growth. The agent of instability, population growth causes social, economic, cultural and political disequilibriums which call for profound meditation . . . A wide debate on this subject, especially by the National Congress, will contribute to the fixing of objective, fundamental directions in this regard . . .[45]

The National Congress followed suit, immediately creating a special commission to study family planning and population problems, the Senate Parliamentary Inquiry Commission on Population Growth (*Commissão Parlamentar de Inquérito sobre Aumento Populacional*) in March of 1983.

Family planning suddenly attained a prominent place on the floor of the national Senate as well. Former Minister and present PDS Senator Roberto Campos stressed the issue in his first official address to the Senate:

> A negligence of demographic issues is manifested in our timidity, if not our inertia, before the population explosion. In the last decade, we evolved from a position of antipathy to family planning, to a sympathetic apathy, and now, an apathetic sympathy. The last census, of 1980, sets the population growth rate at 2.49 percent, declining in relation to 1960 and 1970, but even so, it is enough to condemn us to relative poverty and to pockets of absolute poverty . . . The country must exorcise the demographic taboo . . . In the south, due to the

> combined effect of education, income growth, and urbanization, there is already spontaneous family planning. All that is called for is to give the poorer classes and regions the opportunity to practice responsible paternity, impossible today due to lack of information and the inaccessibility of preventive instruments.[46]

"Responsible paternity" became the catchword for anti-natalist arguments in 1983. PDS Senator Eunice Michiles, the only woman in the national Senate and president of the *Movimento de Mulheres Democráticas Sociais* (MMDS), the feminine branch of the PDS, made it her primary political banner. In a speech before the Senate on April 28, 1983, the "women's Senator" combined neo-Malthusian arguments with women's rights arguments to propose the creation of an Interministerial Department for Family Planning (*Departamento Interministerial de Planejamento Familiar*) to be directly linked to the Presidency and to be directed by a woman "due to her natural affinity with the program." After blaming a score of national ills on the "population problem," Michiles added that:

> The important fact is that women have been systematically omitted from the discussion of family planning; one cannot omit the fact that it is woman who is the principal agent of human reproduction, the one who spends nine months carrying a child, protecting it with her own body, the one who gives birth with all the joy and suffering that involves . . . the understanding of contraceptive methods opens the doors to feminine independence, in the sense that a woman can decide *how many* children she will have, *when* she will have them, giving her the sensation of control over her destiny, allowing her a greater utilization of opportunities for education, and employment.[47]

The executive branch of the federal government also began its own plans for the institution of a nationally based family-planning program, independently of the deliberations of the national Congress. The Ministry of Health elaborated a new program called the Program for Integral Assistance to Women's Health (*Programa de Assisstência Integral a Saúde da Mulher* or PAISM), which, like Michiles's proposal, also appropriated the "reproductive rights" discourse developed by the Brazilian women's movement, an ideological emphasis notably absent from the pre-1983 State population policies. Elaborated by two feminist-identified women doctors in the Ministry of Health, the Program proclaimed "women's rights to choose" maternity and called for a holistic approach to women's health in contrast to the way in which "traditionally a woman has been attended by the health system almost exclusively during the period in which she crosses the cycle of pregnancy and childbirth, leaving other aspects of phases situated outside that cycle on a secondary plane."[48]

In June of 1983, the Ministry of Health directed all state health departments to begin to discuss state-level implementation of the Program. But in São Paulo, the discussion of a possible family-planning program had already been initiated within the PMDB regime. At the urging of the State Council on the Status of Women, the Department of Health, under the leadership of left-leaning Secretary José Yunes, had initiated a discussion of a comprehensive health program which was to respond to demands of organized women in São Paulo.

Ideologically, the São Paulo Women's Health Program reflects the vertical links established between certain sectors of the women's movement and the post-1983 São Paulo opposition regime. Openly feminist in its discourse, this unprecedented document addressed issues of women's equality, women's sexuality, sex education, reproductive rights, and other issues which were first raised as political claims by the Paulista women's movement.

> Facing, in a general sense, adverse conditions in the workplace (where the worst paid and least gratifying functions are customarily reserved for them), women accumulate domestic obligations in a toilsome double shift which consumes their physical and mental health. As housewives who dedicate themselves integrally to unpaid domestic labor, they also suffer the consequences of carrying out uninterrupted, repetitive, isolated and socially-devalued activities.
>
> Besides the specific conditions of women's work, the role of women in reproduction requires special attention, since pregnancy, childbirth, and lactation are processes that demand their biological, psychological, and social involvement. Women's psycho-social involvement with maternity assumes large proportions in our society, where the sexual division of the labor of childcare determines that this responsibility fall exclusively on the shoulders of women, without the participation of their companions and without the provision of nurseries, day care centers, or other services by the State.[49]

The program proposal surveys and addresses the general health conditions of women in the state of São Paulo and aims to contribute to the "demystification of anti-natalist fallacies . . . and clearly disassociates itself with a demographic policy, that is, does not seek to interfere in fertility, either to reduce it, maintain it, or increase it."[50]

No clear federal or state population guidelines were established in 1983, however, and the implementation stage awaited some sort of consensus within the governing coalition at both the national and state levels. While Figueiredo, Campos, Michiles, the Ministry of Health, and other key government sectors endorsed a national family-planning program, opposition to such a program emerged elsewhere within the State apparatus and among the ranks of the governing party. Key political figures such as then Minister of Social Welfare, Hélio Beltrão, Minister of Education, Ester Figueiredo, Minister Jarbas Passarinho, and some sectors of the ESG and the armed forces, for example, continued to espouse pro-natalist opposition to family planning or advocated a more balanced approach to population control—arguing that increased State outlays in education and health care would lead to a natural reduction of the national birth rate.

The traditional left and sectors of the progressive Church denounced both federal and state family planning as genocidal and imperialist-imposed. The PT in São Paulo, for example, publicly proclaimed its wariness of even the state health department's feminist-inspired Women's Health Program as potentially *controlista*.

Many individual women's movement activists and autonomous women's organizations in São Paulo who are outside the ruling PMDB coalition shared this wariness. They argued that the scarce funds likely to be allocated to these programs (estimated at only Cr$11 million for the federal program as of 1983) would lead to the arbitrary distribution of birth control pills (and not other, more expensive, but safer methods) and to sterilization abuse among low-income populations, rather than to comprehensive health care for women.

Recognizing the cooptative potential of State proposals for family planning which appeared sympathetic to the demands of the women's movement, the autonomous Brazilian women's movement remobilized around reproductive issues. At a June 22, 1983, meeting at *Casa da Mulher* in São Paulo, representatives of several women's groups elaborated a feminist proposal addressing women's health needs as the movement perceived them.

However, this document in fact differed little in its recommendations from the State Health Department's proposal. Points of departure included support for free, publicly funded abortion on demand, the formation of discussion groups for men and women on sexuality and reproduction at public health posts throughout the state, special precautions against sterilization abuse and unnecessary gynecological surgery, and the provision of medical and psychiatric assistance to women victims of rape and battery.[51] Autonomous women's groups were caught in a bind vis-à-vis State initiatives in reproductive politics. Suddenly the State—heretofore perceived as their primary adversary—was espousing a policy that approximated their own. And women's movement activists expressed an urgent need to distinguish their own proposals for reproductive freedom from State attempts to intervene in the control of fertility—a difficult task given that the women's movement had been the first to introduce the notion of State responsibility for the provision of birth control to all women.

Reproductive politics headed the agenda at the III National Feminist Meeting held in Brasília in July of 1983. Attended by women from more than twenty movement-organizations from São Paulo, Rio de Janeiro, Belo Horizonte, Salvador, Recife, Fortaleza, Goiânia, Curitiba, and Florianópolis, the meeting resolved to elaborate a document specifying a national feminist position on birth control, abortion, and family planning to be widely distributed to the press and among opposition leadership. Feminists would form state-based movement commissions to supervise the implementation of state and federal level programs in an attempt to prevent contraceptive abuses such as indiscriminate distribution of pills and forced sterilization.[52]

As of September 1983, State-sponsored family planning remained at the level of ideology and policy formulation. The implementation of the federal program on a national scale was interrupted by the regime's internally-generated succession crisis and the opposition-led mobilizations in support of direct elections, and, later, in support of the indirect candidacy of Tancredo Neves.[53] Nevertheless, throughout 1984, sectors of the national women's movement and of the above-mentioned political opposition consistently pressured the Ministry of Health and other policy-making arenas within which the PAISM was being developed in order to assure that the regime was true to its *anti-controlista* political discourse, a discourse which was made more vulnerable to such organized pressure because it had incorporated or coopted so many of the concrete gender-specific demands raised by women's movement organizations during the past decade.

At the federal level, some State fractions continued to push for the immediate establishment of a "National Council on Population" which would implement a national population control policy. Coopting the most prevalent opposition slogans during the 1984 direct election campaign, which centered on the end to military rule *now* ("Direct Elections Now!," "Tancredo Neves Now!," "Constituent Assembly Now!"), *controlista* segments within the ruling party and the State began advocating "Family Planning Now!" (*Planejamento Familiar Já!*).[54] The Chief of the *Estado Maior das Forças Armadas*, General Waldir de Vasconcelos, made population control one of his principal political concerns during most of 1984—he traveled extensively throughout Brazil, meeting with prominent politicians, industrialists, and community leaders, and arguing that curbing population growth was essential to Brazilian economic growth and prosperity and therefore was a crucial component of "national security." Micheles's MMDS also engaged in a massive, nation-wide propaganda campaign to promote its own neo-Malthusian brand of "feminist" family planning.[55]

The Brazilian women's movement, acutely aware of the cooptative potential of these anti-natalist offensives, launched a counter-offensive of its own. In July of 1984, the IV National Feminist Meeting, held in São Paulo and attended by 97 women from 33 feminist organizations, focused on the issues of women's health and family planning in an attempt to develop a clear feminist position on reproductive rights with which to contest *controlista* policy proposals at the state and national levels. And such a position began to take definitive shape during the remaining months of 1984:

> It would occur to no one to suggest that the federal Program of Integral Assistance to Women's Health is what women aspire to in terms of a public health policy. This program, created by the military regime in recent times, however, contains some of the demands that feminists have been making. In spite of this, we believe that the bureaucratic manipulation of this program could result in few advances [for women] and for this reason, we have decided to participate more actively in its implementation, monitoring it, demanding forums and debates with institutional organisms, participating in conferences and other activities proposed by these organisms.[56]

In November of 1984, several feminine and feminist groups in São Paulo organized the First National Meeting on Women's Health, held in the city of Itapecirica and attended by over 400 women from 19 Brazilian states.[57] The *Carta de Itapecirica*, elaborated by conference participants, called for "the participation of women's groups in the elaboration, execution and monitoring of women's health programs, sex education for all the population, the reclaiming of popular and feminist wisdom against the excessive medicalization [of women's health] and a revalorization of natural forms of life and health."[58]

The feminist movement thus reformulated its historically "anti-State" posture and chose instead to exacerbate the contradictions inherent in State family-planning initiatives. In this sense, then, the Brazilian women's movement and the organized opposition to any attempt at State population control successfully mobilized against those State fractions which promoted *controlista* solutions and "beat" the outgoing authoritarian regime "at its own game." Organized sectors of civil society, spearheaded by women's movement organizations, and of opposition political parties, led by their respective women's divisions, pushed the regime to confront its own internal contradictions regarding the formulation and implementation of population policies.

The Ministry of Health proved to be an unusual "point of access" through which the Brazilian women's movement could influence the authoritarian State apparatus.[59] The team to which the Ministry's PAISM was entrusted was presided by feminist, progressive health professionals who were exceptionally responsive to women's movement demands for "reproductive rights" and who actively combated the efforts of General Vasconcelos, the MMDS, and other sectors of the ruling coalition to wrest the State's would-be family planning program from the purview of the Ministry of Health. To counter Vasconcelos's pro-population control crusade, the Ministry launched its own educational campaign in favor of non-coercive family planning—the women and men responsible for the PAISM within the Ministry also toured Brazil, appearing in public debates and radio and television programs, and even directly debating General Vasconcelos on a few occasions.[60]

Pilot programs of the PAISM were instituted in Sorocaba and Goiás during 1984 and were fairly successful and non-coercive, according to some feminist observers. As Carmen Barroso suggested:

> In the case of women's health, we are in a curious situation of perceiving the desirability of *continuismo*. The government which is now moving out of power had elaborated a relatively innovative program which conquered the respect of sectors of the opposition . . . It would be lamentable if the new Minister of Health wasted the intense work of political negotiations and government financing which has already been undertaken, and did not give the PAISM the importance that it deserves."[61]

Three key systemic variables affected the formulation and implementation of population policies in Brazil over the last decade. The first of these variables was the changing "reproductive ideology" of dominant State fractions within the Brazilian regime over time. By "reproductive ideology" I mean the regime's discursive formulation of the relationship between economic development and population size and relationship between population size and the need to control women's fertility (in a feminist conceptualization, to control women's bodies). As elaborated in the above empirical discussion, the Brazilian regime's dominant reproductive ideology has shifted from pro-natalist (1964 through 1978) to increasingly anti-natalist (1978–83), though as I noted, fractions within the State continued to support pro-natalist arguments. Still other fractions within the State seemed to have developed an ideological discourse which indirectly supported the women's movement's claims for reproductive freedom by arguing that the best strategy for containing population growth was to continue to develop the nation's educational and occupational infra-structure, rather than to enforce population control measures.

This latter position was in the ascendancy after September of 1983, partially as a result of the dramatic crisis of legitimacy suffered by the military regime in its final months. This crisis increased the political opportunity space available to sectors of civil society which both opposed any *controlista* policy prospects, and promoted safe, non-coercive, State-sponsored family planning which would increase women's reproductive options. During the abertura, the State itself became an arena of gender struggle, spearheaded by women's movement organizations.

A second variable affecting the regime's population policy was the shift in political pressures from the international aid community for the adoption of population control measures. Before the 1982 debt crisis, the regime was in a much stronger position to resist the international aid community's historic insistence on population control.

In 1982–83 the economic and political crisis propelled by the largest foreign debt in the Third World tied the Brazilian regime more rigidly to the policy dictates of the international aid community. Pro-*controlista* pressures from international financial agencies, combined with the emergence of anti-natalist ideologies within the ruling coalition, led the Brazilian regime to formulate a policy of State intervention in women's fertility as a means of balancing its key national resource—population—with other, ever-scarcer national resources. Since the Brazilian population growth rate has in fact decreased significantly over the past three decades, one cannot attribute the sudden shift in State population policy (from natalist to anti-natalist) to economic considerations alone—socio-political contradictions arising from the regime's increased dependency on center capitalist countries and their financial institutions must be seen as at least partially responsible for this shift.

The third variable affecting the regime's population policy output is the social bases of support it sought within civil society. This variable was of minimal importance at earlier political conjunctures when the regime relied primarily on repression and coercion in order to enforce the pact of domination, but as the process of political liberalization unfolded, this variable carried increased weight in the formulation of regime policies. The importance of this variable can clearly be seen in the regime's cooptation or appropriation of the reproductive rights discourse of the Brazilian women's movement in its formulation of family planning policy since 1982–83.

The dynamic, dialectical nature of the relationship between women's movements and the State is also illustrated by the data presented above. While State family planning policy discussions did not emanate from a response to women's political claims, the reproductive rights discourse of the Brazilian women's movement figured prominently in those discussions for the first time in Brazilian history. The women's movement, in turn, has been pushed by State initiatives in this area to elaborate a clearer political position on reproductive issues and to mobilize popular support for that position.

These three variables also appear to have been critical in the formulation of population policies at the state level. The Maluf regime's reproductive ideology was more clearly and monolithically anti-natalist than that of the federal government and also contained more openly racist, as well as sexist, overtones. This neo-Malthusian ideology combined with Maluf's attempt to directly procure foreign investment and foreign aid for the state of São Paulo to promote his now amply discredited state development schemes.[62] Thus, Maluf was equally vulnerable to the dictates of the international aid and financial communities concerning population control.

The third critical variable, the social bases of support sought by the Malufista micro-regime, also became more important towards the end of Maluf's tenure as governor when he attempted to garner electoral support for his campaign for a seat in the national Chamber of Deputies. It was at this time that Maluf proposed his *Pró-Família* program, which appealed to more traditional ideologies of gender by couching *controlista* rhetoric in language that, ironically, proclaimed the need for the State to protect the "Brazilian Family." Nevertheless, much of the language of Pro-Family also coopted the local women's movement claim that the State *should* provide free contraception to women of all social classes. Thus, *Pró-Família* appeared as one more neo-populist concession to civil society by a PDS regime which had totally lost its legitimacy within the state of São Paulo by the time of the 1982 gubernatorial elections.

The population policy developed by the Paulista PMDB opposition-led regime, inaugurated in March of 1983, was also strongly influenced by the above-mentioned variables. The dominant reproductive ideology espoused by the new democratic micro-regime was neither pro-natalist nor anti-natalist and apparently was sincerely committed to individuals' right to make their own reproductive decisions. Instead of being subject to political pressures for population control from the international aid community, the PMDB regime in the state of São Paulo was indirectly subject to such pressures from the PDS-controlled federal government. Significantly, the opposition regime's ideological discourse on the relationship between population size and the need to control women's fertility was profoundly influenced by organized, gender-conscious feminist pressure for reproductive freedom from both within and without the governing PMDB coalition. The influence of this organized, gender-specific political pressure on the formulation of family planning policy under the post-1983 São Paulo *governo de oposição* will be further elaborated below.

What remains clear for the present argument is that the third variable, the social basis of support sought by the new regime in order to consolidate political hegemony in the state of São Paulo, was a far more significant variable for the new democratic state government than for its authoritarian predecessors. Given the widely acknowledged role of women's movement organizations in the PMDB's electoral victory in São Paulo in 1982, and of the party's own women's division in mobilizing organized women's support for the PMDB, the new regime was necessarily more responsive to movement demands for safe, non-coercive State-sponsored family planning. And the need to consolidate political support at this key conjuncture in the Brazilian transition to democracy furthered the Paulista regime's commitment to organized women's political demands.

A fourth variable, affecting all of the above gender-based policy outcomes, then, was the existence of gender-conscious political pressure both within and without the state and federal political systems. Without sustained political pressure from feminists both within and without the state, both the federal PAISM and the *Programa da Saúde da Mulher* in São Paulo might have catered more closely to the neo-Malthusian population control dictates of the international aid community, or, given the historic opposition to any form of family planning by the progressive wing of the male-dominant political opposition, not been implemented at all.

A brief history of day-care policy in the municipality of São Paulo will serve further to illustrate how conjunctural variables can influence State intervention in the means of reproduction at the municipal level as well as the national and state levels, and will highlight the dialectical nature of State-civil society relations.

Though the provision of day care by industrial and commercial establishments employing more than thirty women has been written into Brazilian labor law since 1943, the State has rarely enforced this legislation nor has it provided publicly funded alternatives to privately funded day care.[63] Day care only emerged as a political banner in the early 1970s when more women began to enter primary sector production than ever before.

A popular movement for day care emerged among women organized in Church-linked mothers' clubs in São Paulo's Southern Zone in 1973 and spread to other areas of the city in the 1970s. Isolated groups of poor and working-class women began to pressure directly the municipal Family and Social Welfare Department for publicly funded, free, community-based day-care centers for their children—initially getting no response from the local state apparatus. But in March of 1979, at the First São Paulo Women's Congress, those isolated groups of neighborhood women and their middle-class feminist allies launched a unified effort to attain state funds for community-based day care.

In May 1979, a manifesto signed by 46 São Paulo feminine and feminist groups unmasked women's role in the domestic or personal sphere as politically and economically determined, demanding that the State and private capital assume increased responsibility for the reproduction of the labor force:

> We are workers who are a little different than other workers . . . we are different, in the first place, because we are not recognized as workers when we work at home 24 hours a day to create the conditions for everyone to rest and to work. This is not recognized, but our work creates more profit that goes directly into the pocket of the boss.
>
> We are different because when we also work outside the home, we accumulate two jobs—at home and in the factory. And they always pay us *less* for what we do. We work more and are paid less. . . . Women are the ones who most feel the problem of lack of day care . . . even though children, since they are not only children of their mothers, are of interest to all of society. It is society which should create the conditions so that these workers of tomorrow can develop good health and a good education . . . Day care centers are our right.
>
> We want *creches* that function full-time, *entirely* financed by the State and by the companies, close to workplaces and places of residence, with our participation in the orientation given to children and with good conditions for their development—we will not accept mere depositories for our children.[64]

This manifesto was followed by dozens others like it, demanding that "personal" child care become "public day care," pushing the State to assume responsibility for women's domestic burden. The day-care movement quickly expanded throughout São Paulo's *periferia* as women organized in neighborhood women's groups and Church-linked mothers' clubs created *creche* groups or commissions. The movement's political strategies focused on direct revindication vis-à-vis the municipal government ("*batendo na porta da prefeitura*") and popular mobilization in support of public day-care through rallies, assemblies, neighborhood petitions and surveys, and so on.

In late 1979 the day-care movement took its demands to the newly appointed São Paulo mayor, Reynaldo de Barros. The recently inaugurated PDS administration had adopted a legitimacy formula based on neo-populist concessions to the demands of increasingly active popular organizations—a formula inspired by the need to seek electoral support for the PDS in opposition-controlled metropolitan São Paulo. With the "economic miracle" long over and the first gubernatorial elections since 1965 scheduled for 1982, the Malufista-led PDS in São Paulo sought new social bases of support among São Paulo's poor through the provision of meager social services to peripheral, economically depressed neighborhoods.

And *creches* were the social service that featured most prominently in de Barros' neo-populist formula. Despite the fiscal crisis already faced by the municipality of São Paulo in 1979, de Barros promised a commission of thirty women's groups from the city's Southern Zone that he would build 830 day care centers during his administration and that these *creches* would be administered by Community Councils selected and trained by local *Sociedades de Amigos do Bairro* and other neighborhood groups.[65] At the time of his announcement, São Paulo had 123 *creches* serving 17,055 children, but only three of them were publicly funded.[66]

The construction of *creches* during the de Barros administration followed a distinct pattern—*creches* were constructed first where the organized day-care movement was strongest. The first seven public *creches* were built in the neighborhoods in the Southern Zone, where the movement had the longest history and the greatest mobilizational capacity, and *creche* groups in those neighborhoods were given the right to select the day-care workers and administrators for those centers—as stated in the demands of the unified day-care movement. However, de Barros reneged on this latter concession in late 1980, decreeing the mayorship's right to appoint the directors of day-care centers. In 1981 his promise of 830 *creches* in 1979 had dwindled to 330.

Aside from the clearly cooptative nature of the local PDS regime's strategy vis-à-vis the day-care movement, the relationship between the *creche* movement and the municipal government from 1979 to 1982 was observably a dialectical one—movement protests at the *prefeitura* elicited a response from the local State apparatus—the construction of a day-care center in a given neighborhood or a series of centers in a given region of the city, and such concessions in turn fueled the day-care movement on to the mobilization of further protest actions. The movement grew as a consequence, in part, of concrete State responses to women's political demands.

By the end of de Barros' administration, 141 day-care centers, the so-called *creches diretas* entirely financed by the municipal government, had been constructed in the country of São Paulo. De Barros, who became the PDS gubernatorial candidate for São Paulo in 1982, made *creches* a rallying cry of his campaign, stating in numerous campaign materials that "when Reynaldo de Barros came into office São Paulo only had three day-care centers. Now São Paulo has 333!" The 333 were in the planning stages, at best, though de Barros had further serviced his political aspirations by pre-appointing the directors of all unconstructed day-care centers.

The centers that had been constructed but not put into operation during the PDS regime became the primary focus of conflict between the day-care movement and the 1983 PMDB regime. In April of 1983, leaders of the day-care movement met with Governor Franco Montoro, who promised to call on all the state's mayors to give the construction of day care special attention. Movement leaders also met with interim mayor, Altino Lima, and with Municipal Secretary for Family and Social Welfare, Marta Godinho, shortly after the PMDB regime was installed.

The movement's unusual access to PMDB state officials was in part the consequence of the heavy influence of the *Partido Comunista do Brasil* (PC do B), a then-illegal Albanian-inspired party which was part of the PMDB electoral coalition, within the day-care movement. The PC do B-linked

União de Mulheres de São Paulo had assumed much of the political direction of the movement as middle-class autonomous feminists had left the previously cross-class day-care movement over the years. During the first two months of the PMDB administration, the PC do B leadership within the *creche* movement expressed its confidence that the "new, democratic government of São Paulo would support the people's demand for free, community-administered day care."

However, the movement's political posture soon changed as the PMDB regime appeared to place more emphasis on the cost of day care, in light of the economic crisis, than on the political benefits of meeting day-care movement demands. Arguing that municipal coffers had been virtually ransacked by the corruption of the previous regime combined with the 1982–83 debt crisis, the PMDB administration ignored movement demands and decided to make the 68 *creches* built under the de Barros administration indirect, rather than direct, ones. Indirect *creches* have been rejected by the day-care movement as inefficient and pedagogically unsound. Partially financed by the municipal government but administered by private entities (usually charities), indirect *creches* preclude community participation in the administration of community-based day care—a fundamental demand of the movement since its inception.

The result was open conflict between the new democratic regime and the day-care movement as of June 1983—generating protest actions of the sort which had not characterized the movement since the early days of the de Barros administration. PMDB resistance to the further construction of day care and to community administration of existing day-care facilities generated the massive remobilization of the day-care movement in São Paulo. In interviews conducted at the time, movement activists pointed to the irony of the situation—under authoritarian rule, *creches* were conceded by the State; under democratic rule, they were being denied.

Throughout 1984, day-care movement participants argued about what the *Movimento de Luta por Creches's* (MLC) political strategy should be vis-à-vis the new PMDB municipal government. Underlying partisan conflicts, subsumed during 1982 due to the endorsement of day-care movement demands by both the PT and the PMDB electoral platforms, assumed much sharper dimensions within the city-wide movement during 1983 and '84. The PMDB sympathizers within the MLC argued for a new political strategy, one which would be sensitive to the new "democratic" conjuncture and attempt to negotiate with the PMDB regime in spite of its seeming resistance to fundamental movement demands. The *petistas* within the MLC, on the other hand, insisted that the movement had to assume an oppositional or contestatory posture vis-à-vis the local administration and mobilize movement groups throughout the city to delegitimize the new regime's day-care policy by proclaiming it to be "unrepresentative of the popular will."

Unable to arrive at consensus, as feminine and feminist groups had done on the issues of family planning and reproductive rights, the MLC's city-wide and regional organizations had been totally disarticulated by late 1984.

All that remained of the previously massive, united, grassroots movement was a few, dispersed, neighborhood-based day-care groups.[67]

Unlike family-planning policy in the state of São Paulo, State intervention in day care is presently being guided primarily by economic considerations. Since the micro-regime's political legitimacy formula had been thus far based largely on its "opposition" nature, it can politically afford to formulate a day-care policy independently of the support of popular women's organizations which were perceived as key sectors of political support by the neo-populist PDS regime. Also, with unemployment rates skyrocketing in the state of São Paulo, freeing women for paid labor through the provision of public day care is likely to be a low priority for the present regime.

The comparative analysis of two gender-related policy areas under two distinct political regimes in the state of São Paulo reveals that State intervention in the process of reproduction is determined by conjunctural regime variables such as the social bases of support sought by a particular regime, the ideologies of gender and class represented within that regime, the regime's policies for structuring the relationship between State and society, and the regime's ties to particular national and international interests. In the state of São Paulo, the democratically elected, participatory PMDB regime initially proved less pliant to the political demands raised by the day-care movement, whereas the neo-populist, authoritarian PDS regime in fact met women's concrete needs by providing limited accessibility to publicly funded day care, despite its overt manipulation of the issue for electoral purposes.

The data presented thus far suggest that socio-political contradictions emerging from movement-State conflict may lead the State to act relatively autonomously of gender- and class-linked interests.[68] That is, though women's role in the reproduction of the labor force is central to the smooth functioning of the "public" sphere, especially in the context of a dependent capitalist economy, the State may intervene to alter that role at particular political conjunctures, especially those characterized by increased, gender-conscious political pressure from within civil society.

One of the most controversial policy initiatives of the post-1983 PMDB regime in São Paulo may yet prove to be the most significant for promoting concrete improvements in the status of women through public policy. Initially viewed by many non-PMDB women's movement participants as yet another instance of State cooptation and "institutionalization" of the autonomous Paulista women's movement, the policies pursued by the São Paulo Council on the Status of Women during its first two years of legal existence suggest that it may in fact be successfully mediating the structural-historical tendency toward State cooptation and manipulation of gender-specific issues.

The creation of the Council was the result of concerted, gender-conscious political action, through which women's movement activists had gradually conquered a "women's space" within political society. Feminists have been active in the opposition since the '60s and in the PMDB electoral coalition in São Paulo since its inception in 1979. And during the 1982 electoral campaign, the opposition parties as a whole had paid significantly more attention to both the female electorate and organized female constituencies. This fact can be attributed to two principal factors: 1) most women's movement organizations had identified as part of the broad-based opposition to the regime since the early '70s and thus were perceived by the parties as potential (and untapped) sources of political support; and, 2) many feminists and women's movement activists joined the ranks of these parties and were instrumental in pushing male party leadership into including women's issues in the official party platforms.

The PMDB has perhaps the most extensive statement on women's rights of any of the parties, the inspiration for which can be traced to demands raised by São Paulo feminists and women's *Congressos*. As the MDB, the party had been relatively responsive to women's movement organizations. Several candidates and legislators (especially after 1978) had expressed support for gender-specific demands, though the party platform made no specific reference to women's rights. The 1982 PMDB party program, on the other hand, addressed a number of specific demands raised by women's movement organizations:

> the PMDB is opposed to anachronic legislation which impedes the complete exercise of citizenship and civil rights by Brazilian women. The Party defends the guarantee of legal equality in the direction of the marriage union, as well as the legal protection of single mothers, and under any hypothesis, the protection of children. The Party also defends full rights for women workers, is against the discrimination in salaries of which women are victims, is for the abolition of any type of discrimination in employment and hiring of married women, pregnant women, and women with children; will struggle for mandatory day care centers in the workplaces and places of residence, to be maintained by private enterprises and by the State and to be administered under the direct control of those interested.
>
> The PMDB recognizes the existence of discrimination against women, is in solidarity with their struggle for equality of opportunity with men, and supports the demands of women for an equal division of domestic work.[69]

While traditional PMDB female activists focused their energies on "turning out the female vote,"[70] a group of liberal professionals and academic women (mostly former participants of feminist groups and research centers) constituted themselves as a "Study Group on the Situation of Women" within the Montoro campaign. The activities of this group were not directed at mobilization and recruitment, but rather at the formulation of a specific "Proposal of PMDB Women for the Montoro/Quercia Government" which was widely distributed in pamphlet form during the final months of the campaign. The proposal included four basic areas of action to be implemented and supervised by a *Conselho da Condição Feminina* (literally, the Council on the Feminine Condition) which would serve as an instrument for a *global politics* destined to eliminate the discriminations suffered by women."[71] The four proposed areas of action included: 1) women's work and the elimination of salary and employment discrimination; 2) women's health and reproductive rights; 3) *creches*; and, 4) the protection of women against violence. The women who composed this study group and those who supported their work within the party tended to come from the ranks of the more professional-based feminist groups in São Paulo—such as the *Frente de Mulheres Feministas, Pró-Mulher*, and the *Casa da Mulher Paulista*—which concentrated their feminist politics around raising women's issues in the mainstream media, academia, and the arts.

Following the overwhelming victory of the PMDB in São Paulo, the *Grupo de Estudos da Situação da Mulher* and other PMDB women militants began to instrumentalize their proposal for the creation of a *Conselho Estadual da Condição Feminina* which would implement the *Programa Estadual em Defesa dos Direitos da Mulher* (State Program in Defense of Women's Rights). And the divided women's movement began a process of reconciliation and remobilization, the symbolic expression of which was the planning of the 1983 International Women's Day celebration.[72]

Discussions of how the *Conselho* would in fact be constituted, whether a *Secretaria da Mulher* would be preferable to a *Conselho*, and which areas of government policy toward women would be prioritized under the new administration, remained a closed partisan discussion within the PMDB. The fact was that some sectors of the women's movement (those which had supported the PMDB during the campaign) were now in "power" and the "opposition" women (members or sympathizers of the Partido dos Trabalhadores) and non-partisan women's movement activists saw themselves as marginalized from the decision-making process which was to define the new administration's policy toward women.

The result was that even before the PMDB had officially announced its plan of action vis-à-vis women, strong opposition to the *Conselho da Condição Femenina* was already emerging among the ranks of the more grassroots-based feminist groups, the more politicized neighborhood women's groups, and the *creche* movement. This incipient opposition viewed the plan for the *Conselho* as a possible instance of institutionalization of the women's movement by the State and viewed the sectors of the movement involved in policy planning as "unrepresentative" of the women's movement as a whole. However, the women's movement "as a whole" was so disarticulated after the electoral conjuncture that it was unable to present a concrete plan of action for confronting the more organized and cohesive PMDB plan for women.

Created by Governor Montoro on April 4, 1983, the *Conselho Estadual da Condição Femenina* is granted broad "advisory" powers, but has no executive or implementation powers of its own, nor does it have an independent budget, being totally dependent on the *Gabinete Civil* for financial and technical assistance. The decree makes no mention of the *Programa Estadual em Defesa dos Direitos da Mulher*, developed and proposed by PMDB feminists. Instead, the *Conselho* is granted the power to:

> *propose* measures and activities which aim at the defense of the rights of women, the elimination of discrimination which affects women, and the full insertion of women into socio-economic, political and cultural life . . . incorporate preoccupations and suggestions manifested by society and *opine* about denunciations which are brought before it . . . *support* projects developed by organs, governmental or not, concerning women, and *promote* agreements with similar organizations and institutions . . . [italics mine].[73]

The avowed purpose of this *Conselho*, according to both the governor and the PMDB women who proposed it, is to give women influence over several areas of the State administration rather than isolate women in a separate women's "department." However, since the *Conselho* is given *no* executive power, its influence of policy is necessarily somewhat limited—though its very creation unquestionably represented an advance for women relative to the previous PDS regime.[74]

The Council's official tenure began in September of 1983, and since that time many of the original fears of the women's movement as to institutionalization and cooptation have proven partially unfounded. The Council openly admits to its presently "partisan" composition, stating that "the criterion for supra-partisan representation [within the Council] is still a dream, unviable at the time of the Council's creation."[75] Furthermore, the women who made up the original Council were aware of the problematic nature of the Council's relationship both to the male-dominated State apparatus and the autonomous women's movement. In one of its earliest official publications, Eva Blay, then president of the Council, addressed the precariousness of its position within the regime:

> Another question to be profoundly considered refers to the political-administrative form chosen [to represent women's interests within the new government]; a Council. Social movements, and among them, women's movements, desire and should guarantee their autonomy vis-à-vis the State. To be part of the State apparatus in order to be able to utilize it from within but at the same time maintain the freedom to criticize it is an extremely complex question. Nevertheless, this difficulty must not constitute an obstacle which paralyzes the participatory process. The [political-administrative] form devised to avoid the reproduction of vices typical of the traditional [political] structure is the creation of a Council which has a majority representation of sectors of civil society. The mechanisms of selection [of said representatives], have yet to be defined as it is hoped that organized groups or independent feminists will pronounce their opinions on the subject . . . Created within an opposition party, due to the initiative of a study group, the Council on the Status of Women now belongs to all women, who can and should manifest themselves as to its future direction.[76]

But the problem posed by the Council itself as to which women's organizations were to be represented within the Council and which women were to represent them immediately generated a great deal of conflict between the Paulista women's movement and the new regime. The July 1983 appointment to the Council of fifteen PMDB party women,[77] predominantly from the more liberal, professionally based feminist organizations in São Paulo, led some PT-linked women and independent women's groups, under the leadership of *petista* municipal councilwoman Irede Cardoso, to accuse the Council of being "undemocratic" and "unrepresentative" of the women's movement as a whole. The Council in turn argued that its political strength and effectiveness depended on the strength and effectiveness of women's movement organizations, that it needed the support and collaboration of those organizations in order to gain clout within the state administration. Some autonomous women's organizations indeed recognized the Council as a legitimate political conquest of the women's movement (and particularly of organized women within the PMDB) and chose to approach the new institution from a cautious, but supportive perspective.

The creation of the Council in and of itself served to remobilize and strengthen the women's movement in São Paulo. Confronted with a new "women's institution" created by a regime theoretically sympathetic to women's claims, São Paulo feminist and feminine movement organizations had to concretize and redefine their political priorities so that they could be effectively channelled through the Council or else risk having those priorities manipulated by the new regime for its own political gain.

A few autonomous women's groups called a public forum in May of 1983 to discuss the movement's potential relationship to the Council and this forum was established as a monthly event to discuss relevant political developments. These monthly forums were the first instance of on-going, unified feminist and feminine political action since the 1981 International Women's Day celebrations. This movement response to State policy initiatives is another clear illustration of the dialectical nature of the relationship between social movements and the State.

The most relevant of the Council's policy initiatives for purposes of the present discussion is its intervention in the formulation and implementation of day-care and population policies promoted by the PMDB regime since 1983. The Council elaborated the following position on day-care policy under the PMDB administration:

1. The right to day care should be considered an extension of the universal right to education.
2. Official policy on day care should integrate existing initiatives, guarantee minimal coherence in the actions of the various organs involved [in policy implementation], avoid the dispersion of resources to innumerable programs and organs, and be open to new proposals.
3. In this integrational effort, it is necessary to be clear that the State and society must assume responsibility for the child, guaranteeing it care, protection, and education, principally for the children of working women who constitute that portion of the population most in need of places where they can safely leave their children during working hours.
4. The popular movements, and especially the women's movement, have revindicated day care as a right of working mothers and have already

accomplished, through their struggles, the recognition of that right through the installation of "direct *creches*" in several cities.

5. Day-care policies to be elaborated should integrate the community, parents and professionals, in the process of implementation of the day-care center and in the guarantee of popular participation in its functioning.
6. The day-care centers should function during periods that correspond to the working hours of mothers.[78]

This statement indicates the Council's need to juggle the policy constraints imposed by the local regime which clearly favors "indirect *creches*," *and* the demands of the popular-based day-care movement which constitute an important segment of the Council's constituency within civil society.

After considerable deliberation, the Council decided to support the regime's contention that the present economic crisis precludes the creation of more direct *creches*, which represent greater state investment in the means of reproduction than *indirect* or privately administered *creches*. At the same time, the Council sided with the day-care movement in supporting its claim that those *creches* which were designated as *direct* by the previous PDS Paulista regime should be maintained as such, as these represented a "right of working mothers" legitimately acquired through "popular struggle."

The Council also participated in the preparation, supervision, and coordination of the Special Inquiry Commission on Day Care created in the São Paulo municipal legislature in October of 1984. The Commission, presided by PMDB councilwoman Ida Maria, was created partially as a response to the conflicts that had emerged between the local PMDB administration and the day-care movement since 1983.[79] The Council refined its position on day care in its testimony before this Commission:

> The municipal chain of direct *creches* constitutes a conquest of the popular movements and the women's movement and as should be *discussed*. An effort in the sense of *discussion* and adopting a *creche* policy which perfects the existing chain of direct *creches* is urgent, as [the present system] aside from its high cost (due to inefficiencies and excessive centralization) offers low quality care.[80]

Again, this statement reflects the precarious *structural* position of a "women's space" within the State apparatus which claims to represent the interests of women *outside* the pact of domination. On the issue of day care, the Council, as part of the State apparatus, supported the regime yet appears also to have pushed the local administration to acknowledge the legitimate claims of the popular-based day-care movement.

In 1985, the Council redefined its day-care policy—focusing now on the *private* provision of day care by enterprises employing more than thirty women, an aspect of the 1943 labor laws (CLT) which had never been enforced by the State.[81] The Council encouraged unions and government agencies to increase their monitoring of the CLT provision, thus resolving the conflict between the day-care movement and the State by deflecting the revindications for, and the costs of, day care onto the private sector.[82] The Council has also been pressuring the new Ministry of Labor to revise the CLT's current provisions for day care in the work place. At the state level, it also hopes to promote closer Department of Labor supervision of commercial and industrial establishments in order to ensure that they are providing pedagogically sound and conveniently located day-care services for their female employees.

Interestingly, the day-care issue has been "depoliticized," in a genderic sense, by the Council. Whereas day-care movement organizations had insisted that day care was a "woman's right" or a "mother's right," given women's socially ascribed responsibility for domestic labor and childrearing, the Council now argues that "the *creche* is a right of children, of workers . . . the right to day care should be seen as an extension of the universal right to education."[83] This political "renaming" of the day-care issue seems to have been strategically warranted given the clear resistance of some sectors of the state and local administration to the idea and feasibility of direct *creches*.

Christian Democratic factions within the ruling PMDB coalition pushed for alternative approaches to the "problem" of day care. Since 1983, Dona Lucy Montoro, the governor's wife, in particular has been promoting the idea of "*mães crecheiras*" or day-care mothers—neighborhood women who would be employed and trained by the State to care for the children of other working women and provide these children within a "familial environment"—as an alternative to more expensive, "impersonal," and "inefficient" public day-care centers.

Opponents to this alternative, both within the day-care movement and the State apparatus, argue that given the precarious conditions of life in urban, working-class neighborhoods, *mães crecheiras* could not possibly meet children's pedagogical, recreational, and nutritional needs adequately. Just as importantly, they suggest, transferring the responsibility for child care from one low-income woman to another and paying the second substandard wages and granting her no workers' benefits is glaringly exploitative. Moreover, opponents argue, a *mãe crecheira* policy would reinforce existing patterns of gender inequality by creating *more* low-paying, low-status jobs for women, jobs which merely commodify women's socially ascribed mothering roles.[84]

Confronted with this traditionalist tendency within the local and state administrations, the Council began arguing for day care in broader terms, appealing to "universal rights" rather than "women's rights" in order to widen political support for Council day-care proposals within the ruling coalition. The Council's "generic" discourse on day care, however, has also had the indirect effect of depoliticizing the *genderic* content of the Struggle for Day Care Movement's historic political claims.

Significantly, there has been little gender-conscious political pressure for day care from without the PMDB ruling coalition since mid-1984. And in the absence of such pressure, day-care policy is at present being determined primarily by intra-State conflicts rather than by movement-State interaction.

While divergences within the PMDB regime and civil society delayed the implementation of the state's Women's Health Program, the Council consistently supported the inclusion of "conception and contraception, with information about and access to all of the contraceptive methods" within the Program. The Council has sponsored several public forums about this issue and one of its sub-committees, the Women's Health Commission, has worked closely with the state's Department of Health in the further elaboration of the Program.

The Council had to contend with a rather different "reproductive ideology" in the case of family-planning policy. The left or *autêntico* wing of the PMDB ruling coalition predominated within the State Department of Health and, as we saw above, the left in Brazil as a whole has historically equated family planning with imperialist population control initiatives. Though this position is certainly warranted given the neo-Malthusian content of family-planning policies proposed by the right and the international aid community, the Council has insisted that the *Programa da Saúde da Mulher* is also a legitimate need of all Paulista women. The Council's Commission on Women's Health argued that if the program was properly administered and accompanied by popular education, it would advance the status of women by enabling women of all social classes to make informed reproductive choices. In short, as Margaret Arilha put it, "the Council had to combat the ghosts of population control" among left-wing sectors of the PMDB "micro-regime" and other progressive sectors of civil society.[85]

The Council exerted constant pressure to get the Health Department to begin the implementation phase of the program. Its access to the department's policy process was facilitated by the fact that a representative from the Health Department sat on the Council (whereas, for example, there was no such access to the municipal agencies in charge of day-care policy). The Council's structure and membership, combining representatives from both civil society and the State, proved especially effective in the area of family planning policy. Council members directly lobbyed the Secretary of Health, the directors of the department's seventeen regional subdivisions, and other key policymakers throughout 1984 in order to convince them of the worthiness and the urgency of implementing the Program for Women's Health. It also launched an extensive propaganda campaign to dispel fears of population control among progressive sectors of civil society. As a direct consequence of the Council's sustained efforts, the Health Department made the Women's Health Program one of its programmatic and budgetary priorities for 1985.

After months of discussion and deliberation between the Council and the Health Department, it was agreed that family planning representatives, trained and supervised by the Department of Health in conjunction with the Council's Commission on Women's Health, will supervise the implementation of the Program for Women's Health in all seventeen subdivisions in the state of São Paulo.[86] The Council also facilitated the women's movements' access to the policy implementation process, urging the Department of Health to hold monthly public "Forums to Accompany the Implementation of the Program for Women's Health."[87] The São Paulo Council was also instrumental in generating new access points through which women's movement organizations could impact the federal PAISM, pressuring the Ministry of Health to create a "Commission on Reproductive Rights," which, like the state forums, bring policy-makers, health professionals, and women's movement activists together to ensure a safe, accessible, and non-coercive federal family-planning policy.

What differentiates the São Paulo Council's actions within the regime, then, from similar administrative organs created throughout Latin America (largely as a consequence of the U.N.-sponsored Women's Decade), is the pronounced *feminist* presence within this "women's space" within the State. Though non-feminist women and PMDB party militants are also members of the Council, a feminist position on family planning, largely derived from the contributions of the Brazilian feminist movement to political discourse on population policy, has emerged firmly within Council policy.

In undertaking the supervision of the implementation of family planning policy at the state level, the Council has accomplished what the autonomous women's movement could never accomplish on its own due to its structural position "outside" the pact of domination. And the continued mobilization of women's movement organizations on a national, state and local level around reproductive rights and family planning has unquestionably contributed to the success of this gender-conscious family planning policy.

The Council has formulated and monitored the implementation of a number of innovative public policies which address the special needs and concerns of Paulista women, maintaining consistent, organized, and gender-conscious political pressure within the local state apparatus. In addition to the policy initiatives discussed above, for example, the São Paulo Council on the Status of Women's Commission on Violence against Women was instrumental in persuading the mayor in August of 1985 to create the *Delegacia da Mulher*, a police precinct, staffed entirely by specially trained female officers, which processes cases of rape, sexual abuse and domestic violence.[88] The ground-breaking recognition of this gender-specific aspect of "crime" by the State is unprecedented in Brazil, and indeed the "women's precinct" structure is unparalleled elsewhere in the world. Again, on this issue, the Council's effectiveness in this policy area is partially attributable to the São Paulo women's movements' constant protest actions and public education campaigns around the issue of violence against women.

But the danger of cooptation or institutionalization of the women's movements and the private issues they have politicized remains in that the activities of the Council may absorb or preempt such gender-conscious political pressure emanating from the autonomous women's movement.

Since 1983, the São Paulo Council on the Status of Women has assumed many of the activities which were previously orchestrated by women's movement organizations. In 1984 and 1985 the International Women's Day celebrations in São Paulo, which had previously been peak mobilizational moments for the autonomous women's movements, were transformed into peak mobilizational strategies in support of the Council's conception of "women's interests" and in support of the PMDB's local and national "political projects." In 1984, the theme of International Women's Day was "Direitos, Diretas e Paz" (Rights, Direct Elections, and Peace), a theme reflecting not only the concerns of the autonomous women's movement (couched in more "acceptable" rhetoric) but also the interests of the nation-wide political project of the PMDB in promoting direct elections for 1984.

The danger in the Council's strategy is that, however unintentionally, it has to some extent pre-empted the mobilization of gender-conscious political pressure from outside the PMDB regime. Some of the Council's activities have at least temporarily absorbed or pre-empted the political initiatives of organized female constituencies within civil society.

A further source of concern among movement participants "outside" the State apparatus has been that the institutionalization of some of the women's movements' gender-specific demands within the new PMDB regime has threatened to over-absorb the dynamism of feminism, and women's mobilization in general, as forces for social change within civil society. After all, what distinguishes social movements from political parties is that movements have cultural and social, as well as political goals. They seek normative, as well as structural, transformations of society more actively than do traditional political parties.

The exclusion of the more grassroots-oriented sectors of the women's movement from the Council on the Status of Women, its overwhelmingly white and middle-class composition, and its assumption of mobilizational initiatives among women in São Paulo thus, have indirectly served to suppress or delegitimize some of the more radical or strategic demands which have spurred the Brazilian women's movement since the mid-1970s. As David Bouchier notes, in the United States the early radical core of the feminist movement was "written out of history when [the] movement gain[ed] momentum and wider political and commercial considerations [came] into play."[89] In the case of the United States, "Liberal feminism [was] more quickly and thoroughly integrated into the upper middle class status quo than any other protest movement in history . . ."[90]

This same "quick" and "thorough" integration of the more "acceptable" or moderate demands of the Brazilian women's movement seems to have been under way in São Paulo since 1983. And such institutionalization can be viewed as a more subtle form of State cooptation of women's political claims, one typical of liberal democratic regimes, such as the post-1983 PMDB micro-regime in São Paulo.

The indirect suppression or delegitimation of the more radical core of the women's movement's political claims, through the institutionalization of its more acceptable claims (i.e., those which do not directly conflict with larger political or developmental objectives), potentially weakens feminism as a force for change within civil society—even as progress toward women's emancipation is being made within the State and political society. Furthermore, the likelihood of more thorough cooptation of movement demands is also increased, for as Bouchier suggests:

> a radical core produces a fruitful and dynamic tension in a social movement . . . a mass following of people with moderate views is needed to influence elites. But a radical wing, constantly raising unresolved issues and generating new ones, constantly on the alert for cooptation and retreat, is essential to preserve the oppositional movement from a gentle slide into the prevailing hegemony. They also, quite unintentionally, help give the more moderate wings of the movement an appearance of relative acceptability which they might not otherwise achieve.[91]

In the case of São Paulo, the institutionalization of genderic political claims thus far has not resulted in the deradicalization or depoliticization of the more radical genderic content of those claims.[92] As we have seen, the continued dynamism of autonomous women's movement organizations within civil society and the successful reorientation of movement strategies, in response to the new democratic political conjuncture have been key factors in preventing State family planning policy from becoming cooptative or coercive State population control initiatives.

The day-care issue *has* to some extent been "depoliticized," in terms of its genderic content. The blame for this, however, cannot be placed solely on the State Council on the Status of Women. Rather, the responsibility for the genderic depoliticization of day-care policy is shared by day-care movement organizations which were unable to articulate a united position in response to the new political conjuncture. In failing to sustain gender-conscious political pressure within civil society, the divided day-care movement indirectly weakened the bargaining position of the Council within the local administration and failed to pressure either the Council or relevant government agencies to implement a day-care policy which reflected the historic, gender-specific demands of the Struggle for Day-Care Movement.

The potential "institutionalization" and deradicalization of the women's movement is of even greater concern at the present political conjuncture in Brazil. The PMDB women who mobilized organized female constituencies in support of Neves's candidacy also sought his support for their gender-specific political claims. In June of 1984, São Paulo State Assemblywoman Ruth Escobar, elected on a feminist platform in 1982, organized a "supra-partisan" commission of over sixty women to present the candidate with a *Carta das Mulheres*, specifying the genderic issues which they felt must be included in the candidate's platform and asking that he consider the creation of a Women's Council or Ministry at the federal level. And in January of 1985, Neves officially endorsed the creation of such an institution and appointed a multi-partisan commission of women legislators, presided over by Escobar, to elaborate a legislative proposal for the creation of a National Council on the Status of Women.[93]

From the experience of the São Paulo Council, one can surmise that unless the National Council is composed of a minimally "feminist" majority, it might well become merely another democratic ornament, like so many similar bodies presently in place in other Latin American regimes. The PMDB women who formulated the proposal seem to be aware of this danger, suggesting that the Council be composed of "women representative of civil society whose political social trajectory has been linked to the struggle for equal rights between the sexes in various areas of political action."[94]

Our discussion of the São Paulo Council on the Status of Women suggests that successful institutional mediation *is* possible, *if* the following conditions are met: 1) a gender-conscious, feminist-identified majority must prevail within the Council; 2) the Council must facilitate movement access to the State apparatus and policy-making arenas; 3) Council activities must not pre-empt autonomous women's movement activities; and, 4) the autonomous women's movements must sustain gender-conscious political pressure from without the regime in order to prevent the deradicalization or depoliticization of the genderic content of their political claims.

The *Conselho Nacional dos Direitos da Mulher* (National Council on Women's Rights or CNDN) was officially installed on September 11, 1985, thanks to the continual lobbying of Escobar and other women legislators who pressured for rapid congressional approval of the executive proposal developed by the women's parliamentary commission earlier in 1985.[95]

Escobar, who presided over the original Council, secured an initial budget of $Cr6 billion for the CNDM's first three months of operation alone—an allocation greater than that of many of the smaller executive departments or ministries. The CNDM does have a fairly solid feminist majority, in spite of its need to accommodate the multitude of political tendencies within the ruling PMDB-PFL coalition and within the national women's movement itself.

In an interview conducted in late August of 1985, Escobar outlined some of the principal politcal and organizational goals of the National Council: "The Council will be a more dynamic organ than a Women's Ministry could possibly be. It will open new spaces, new channels . . . it will create new groups within the various ministries which will enable us to influence policy more broadly, more effectively." Escobar suggested that the Council structure is a preliminary step toward the creation of a full-scale Ministry on the Status of Women. She argued that women must first "conquer greater political space and strength" within the State apparatus, within policy-making arenas (hence the presence of women legislators on the Council), and must prove their mobilizational capacity vis-à-vis organized female constituencies in civil society before such a Ministry can be created—"without such a power base, the Ministry would be ineffective . . . like in Venezuela, where a Ministry on the Status of Women was created in the mid-'70s and dissolved less than a year later" due to lack of organized feminist pressure both within and without the State.

Escobar was confident that partisan rivalries and ideological differences could be overcome within the CNDM—the multi-partisan commission which planned the Council had reconciled many partisan tensions, she argued, as she insisted that "women's issues ultrapass the parties even if they [the parties] don't want them to." However, the ruling coalition of the *Nôva República* is extremely disparate and the ascension of center-right factions or the increased hegemony of the *Partido da Frente Liberal* (whose platform on women's issues is little more progressive than that of its "parent party," the PDS) might compromise the CNDM's ability to promote women's interests at the national level.[96]

The danger, ultimately, is that this new "woman's space" within the post-authoritarian State apparatus, at the national, state, and local levels, might become a kind of female political ghetto, a dependent capitalist version of what Philippe Schmitter has termed "societal corporatism." Under a liberal, democratic regime, such a ghetto might preclude those calling for a more radical restructuring of the politics of gender in Brazil from participating in the new regime's policy process, while legitimating those women whose demands for gender-specific reforms are in keeping with the priorities of the still bourgeois—and still male-dominant—Brazilian New Republic.

NOTES

1. Women's movement organizations and women's divisions of the major opposition parties organized several women's demonstrations in major Brazilian cities during the nation-wide campaign for direct elections in 1984. The State Councils on the Status of Women in São Paulo and Minas Gerais and the Women's Division of the PMDB provided the initiative and the organizational infrastructure for these demonstrations. After the opposition shifted to a strategy of defeating the regime at its own game by organizing a "Democratic Alliance" within the Electoral College in support of the candidacy of Tancredo Neves, the women's division of the PMDB organized a national campaign of "*Mulheres com Tancredo.*"

2. For a comparative perspective on the relationship between female political participation, women's movements, and various types of Latin American political regimes, see Elsa Chaney, *Supermadre: Women in Politics in Latin America* (Austin: Univ. of Texas Press, 1980), and "Old and New Feminists in Latin America: The Case of Chile and Peru," *Journal of Marriage and the Family* 35 (1973): 333–43; Norma Stoltz Chinchilla, "Women in Revolutionary Movements: The Case of Nicaragua," in *Revolution in Central America*, edited by Stanford Central American Action Network (Boulder: Westview Press, 1983), and "Mobilizing Women: Revolution in the Revolution," *Latin American Perspectives* 4, 4 (1977): 83–102; Jane Deighton et al., *Sweet Ramparts: Women in Revolutionary Nicaragua* (London: War on Want and Nicaraguan Solidarity Campaign, 1983); Patricia Flynn, "Women Challenge the Myth," *NACLA Report on the Americas* 14, 5 (1980): 20–35; Julia Silvia Guivant, "O Sufrágio Feminino na Argentina, 1900–1947," *Boletim das*

Ciências Sociais 17 (May–July 1980); Cynthia Jeffress Little, "Moral Reforms and Feminism: A Case Study," *Journal of Interamerican Studies and World Affairs 17, 4* (1975): 386–97; Vivian M. Mota, "Politics and Feminism in the Dominican Republic: 1931–1945 and 1966–74," in *Sex and Class* in Latin America, edited by June Nash and Helen I. Safa (New York: Praeger, 1976); Muriel Nazzari, "The Woman Question in Cuba: An Analysis of the Constraints to its Solution," *Signs* (1983); Nancy Caro Hollander, "Si Evita Viviera," *Latin American Perspectives* 4, 4 (1977) and "Women: The Forgotten Half of Argentine History," in *Female and Male in Latin America*, edited by Ann Pescatello (Pittsburgh: Univ. of Pittsburgh Press, 1973); and Jane Jacquette, "Female Political Participation in Latin America," in *Sex and Class in Latin America*, edited by June Nash and Helen I. Safa (New York: Praeger, 1976) and "Women in Revolutionary Movements in Latin America," *Journal of Marriage and the Family* 35, 2 (1973): 344–54. Susan C. Bourque and Kay B. Warren situate much of this literature within four major analytic/conceptual frameworks, in *Women of the Andes: Patriarchy and Social Change in Two Andean Communities* (Ann Arbor: University of Michigan Press, 1981), see especially chaps. 2 and 3. Much of this literature is reviewed in my own doctoral dissertation, "The Politics of Gender in Latin America: Comparative Perspectives on Women in the Brazilian Transition to Democracy" (Ph.D. dissertation, Yale University, 1986).

3. For a comparative analysis of women's movements, gender politics, and regime transitions in Peru, Chile, Argentina, and Brazil, see Jane Jaquette, ed., *Feminism, Women's Movements and Transitions to Democracy in South America* (Boston: Allen & Unwin, forthcoming).

4. For a discussion of the gendered content of State power, see Zillah R. Eisenstein, *Feminism and Sexual Equality: Crisis of Liberal America* (New York: Monthly Review Press, 1983); Irene Diamond (ed.), *Families, Politics and Public Policy: A Feminist Dialogue on the State* (New York: Longman, 1983); and Mary MacIntosh, "The State and the Oppression of Women," in *Feminism and Materialism*, edited by A. Kuhn and A. M. Wolpe (London: Routledge and Kegan Paul, 1979). These and other Western feminist theories of the State, however, rarely consider the relationship between gender, class, and imperialism within the context of dependent capitalism.

5. For a discussion of the relationship between the "public" and the "private," and "production" and "reproduction," see Mariarosa Dalla Costa and Selma James, *The Power of Women and the Subversion of the Community* (London: Falling Wall, 1973); Michele Barrett, *Women's Oppression Today: Problems on Marxist Feminist Analysis* (London: Verso, 1980); Susan Moller Okin, *Women in Western Political Thought* (Princeton: Princeton Univ. Press, 1979); Jean Bethke Elshtain, *Public Man/Private Woman* (Princeton: Princeton Univ. Press, 1981); and Mary O'Brien, *The Politics of Reproduction* (London: Routledge and Kegan Paul, 1983). For a Brazilian perspective on this "split," see Madel T. Luz, "O Lar e a Maternidade," in *O Lugar da Mulher*, edited by M. T. Luz (Rio de Janeiro; Graal, 1982); and Maria Lygia Quartim de Moraes Nehring, *Família e Feminismo: Reflexões sobre Papéis Femininos na Imprensa para Mulheres* (Ph.D. dissertation, Univ. of São Paulo, 1981), part I.

6. See, for example, Heidi Hartmann, "The Unhappy Marriage of Marxism and Feminism: Towards a More Progressive Union," in *Women and Revolution*, edited by L. Sargeant (Boston: South End Press, 1981), and "Capitalism, Patriarchy and Job Segregation by Sex," *Signs* 1, 3 (1976); George Chauncey, "The Locus of Reproduction: Women's Labour in the Zambian Copperbelt, 1927–1953," *Journal of Southern African Studies* 7, 2 (1981); Eileen Boris and Peter Bardaglio, "The

Transformation of Patriarchy: The Historic Role of the State," in *Families, Politics, and Public Policy*, edited by I. Diamond (New York: Longman, 1983); and Lisa Peattie and Martin Rein, *Women's Claims: A Study in Political Economy* (Oxford: Oxford Univ. Press, 1983).

7. The term "gender" is preferred to "sex" in my analysis. As O'Brien has explained, "The word 'sex' is avoided simply because it has too many levels of meaning. Sex can be instinct, drive, an act in response to that drive, a gender, a role, an emotional bomb or a causal variable . . . For the social relations between men and women and for the differentiation of male and female the word 'gender' is preferred (1983:13). Gender, genderic, gender-based, gender-specific are used throughout the text as a means of reinforcing the idea that women and men are social and political, not biological, categories, produced historically.

8. For an opposing argument, see Catherine A. MacKinnon, "Feminism, Marxism, Method, and the State: Toward a Feminist Jurisprudence," *Signs* 8, 4 (1983): 635–58, and "Feminism, Marxism, Method and the State: An Agenda for Theory," *Signs* 7, 3 (1982): 515–42. MacKinnon argues that "however autonomous of class the liberal state may appear, it is not autonomous of sex" (1983), p. 658. This perspective is both ahistorical and deterministic and leaves a crucial question of political *praxis* unanswered—if the State *is, and* is always male, then what are people concerned about gender-specific social change to do in the meantime? In my view, the State is *male-dominant*, rather than male (an argument parallel to that of post-structuralist, Marxian theories which suggest that the State is bourgeois-dominant rather than bourgeois).

9. Zillah Eisenstein suggests that the relative autonomy of the State from male interests in late capitalist society derives from the conflicting and even contradictory needs of capitalism and patriarchy. See *Feminism and Sexual Equality*, especially pp. 87–113.

10. Martin Carnoy, *The State and Political Theory* (Princeton: Princeton Univ. Press, 1984), 259.

11. In "Movimentos Sociais Urbanos: Balanço Crítico," in *Sociedade e Política no Brasil Pós-64*, edited by Bernardo Sorj and Maria Hermínia Tavares de Almeida (São Paulo: Brasiliense, 1983), Ruth C.L. Cardoso suggests that the expansion of State services itself contributed to the generation of revindicatory movements: "The popular classes always revindicated decent housing and urban services, but the pressure mechanism was different. The existence of public policies with social ends makes contemporary States, regardless of how antipopular they may be (and frequently are), implement global social policies which themselves create expectations of demand," p. 229, my translation. For a similar perspective on the relationship between the State and the "new social movements" in Brazil, see Renato Paul Boschi and Lícia do Prado Vallardes, "Movimentos Associativos de Camadas Populares Urbanas: Análise Comparativa de Seis Casos," and Pedro Jacobi, "Movimentos Populares Urbanos e Resposta do Estado: Autonomia e Controle vs. Cooptação e Clientelismo," both in *Movimentos Coletivos no Brasil Urbano*, edited by Renato Paul Boschi (Rio de Janeiro: Zahar, 1983). Studies of social movements in the U.S. also point to a dynamic interrelationship between State and civil society. See Luther P. Gerlach and Virginia H. Hine, *People, Power and Change: Movements of Social Transformation* (New York: Bobbs-Merrill, 1970); Frances Fox Piven and Richard A. Cloward, *Poor People's Movements* (New York: Vintage, 1977); and Doug McAdam, *Political Process and the Development of Black Insurgency, 1930–1970* (Chicago: Univ. of Chicago Press, 1982).

12. However, it is interesting to note that many of the women involved in the day-care movement in contemporary Brazil are working-class housewives, not wage-earning women. Cynthia Sarti's research on the day-care movement suggests that many movement participants see the creation of *creches* in their neighborhood as a possible source of employment close to home.

13. For a theoretical perspective on this dimension of women's consciousness, see Temma Kaplan, "Female Consciousness and Collective Action: The Case of Barcelona, 1910–1918," *Signs: A Journal of Women in Culture and Society* 7, 3 (Spring 1982). Lourdes Beneria and Marta Roldán provide a compelling empirical account of working-class women's consciousness in *The Crossroads of Class and Gender: Industrial Homework, Subcontracting and Household Dynamics in Mexico City* (Chicago: Univ. of Chicago Press, 1987). See also Carol Andreas, *When Women Rebel: The Rise of Popular Feminism in Peru* (Westport, CT: Lawrence Hill, 1985).

14. See Ralph Della Cava, "The People's Church, the Vatican, and Abertura," this volume.

15. Current estimates place the number of CEBs in contemporary Brazil at over 80,000. These are primarily concentrated in the periphery of Brazil's urban centers. The CEBs function alongside thousands of non-Church-linked popular organizations which predate the authoritarian regime, but which have become increasingly politicized over the course of the last two decades. And parallel women's associations often exist alongside both of these types of community organizations, perhaps making them the most numerous among the various new forms of associability found in Brazil today. For an analysis of Church-linked Mothers' Clubs and the limitations of such organizations, see Jany Chiriac and Solange Padilha, "Características e Limites das Organizações de Base Femininas" in *Trabalhadoras do Brasil*, edited by Fundação Carlos Chagas (São Paulo: Brasilense, 1982). For a more positive assessment of this type of feminine organization, see Cora Ferro, "The Latin American Woman: The Praxis and Theology of Liberation," in *The Challenge of Basic Christian Communities*, edited by S. Torres and J. Eagleson (Maryknoll, N.Y.: Orbis, 1981). On the Church's "new" discourse on women and community participation, see Elsa Tamez, ed., *Against Machismo: Rubem Alves, Leonardo Boff, Gustavo Gutiérrez, José Marquez Bonino, Juan Luis Segundo and Others Talk about the Struggle of Women* (Oak Park, Ill.: Meyer Stone Books, 1987).

16. On the ideological and organizational constraints on women's participation in Church-linked neighborhood associations, see Sonia E. Alvarez, "Women's Participation in the 'People's Church': A Critical Appraisal," paper presented at the XIV International Congress of the Latin American Studies Association, New Orleans, March 17–19, 1988.

17. For a general discussion of this phenomenon, see Eva Alterman Blay, "Movimentos Sociais: Autonomia e Estado-Uma Análise Teórica dos Movimentos de Mulheres entre 1964–1983," paper presented at the VI Reunião Anual da Associação Nacional de Pós-Graduação em Ciências Sociais, Aguas de São Pedro, Brasil, Oct. 24–27, 1983, and "Mulheres e Movimentos Sociais Urbanos: Anistía, Custo da Vida e Creches," *Encontros com a Civilização Brasileira-Mulher Hoje*, no. especial (1980). See also "A luta das mães por um Brasil melhor," *Cadernos do CEAS* 58 (Nov./Dec. 1978): 19–27. For a discussion of changes in the domestic political economy of the working classes under authoritarianism, see Ana Maria Q. Fausto Neto, *Família Operária e Produção da Força de Trabalho* (Petròpolis: Vozes, 1982).

18. On the *Movimento Feminino pela Anistia*, see Therezinha Godoy Zerbini, *Anistia: Semente da Liberdade* (São Paulo: Escolas Profissionias Salesianas, 1979).

On the *Movimento Custo da Vida*, see Tilman Evers, "Os Movimentos Sociais Urbanos: O Caso do Brasil," in *Alternativas Populares da Democracia*, edited by J.A. Moisés, et al. (Petrópolis: Vozes, 1982). On the *Movimento de Luta por Creches*, see Maria da Glória Marcondes Gohn, "O movimento de Luta por Creches em São Paulo: Reconstitução Histórica e Algumas Considerações Teóricas," paper presented at the V Encontro Anual da Associação Nacional de Pós-Graduação e Pesquisa em Ciências Sociais, Nova Friburgo, Rio de Janeiro, October 1981; Carmen Barroso, *Mulher, Sociedade e Estado no Brasil* (São Paulo: Brasilense, 1982), esp. 151–54 and 167–68; and "Creche," *Suplemento dos Cadernos de Pesquisa* 43 (Fundação Carlos Chagas, Nov. 1982).

19. A distinction between "feminine" and "feminist" women's movement organizations is commonly made by both movement participants and social scientists in Latin America. Paul Singer clarifies the usage of these concepts: "The struggles against *carestia* or for schools, day-care centers, etc., as well as specific measures to protect women who work interest women closely and it is possible then to consider them *feminine* revindications. But they are not *feminist* to the extent that they do not question the way in which women are inserted into the social context." "O Feminino e O Feminismo," in *São Paulo: O Povo em Movimento*, edited by P. Singer and V.C. Brant (Petrópolis: Vozês, 1980), 116–117, my translation, emphasis in the original.

20. For an analysis of the dynamics of feminist movements in other Latin American nations during the 1970s and '80s, see Nancy Saporta Sternbach, Patricia Chuchryk, Marysa Navarro, and Sonia E. Alvarez, "Latin American Feminisms: From Bogotá to Taxco," *Signs* (forthcoming). See also Cornelia Butler Flora, "Socialist Feminism in Latin America," *Women and Politics* (1984); and Patricia Chuchryk, "Protest, Politics and Personal Life: The Emergence of Feminism in a Military Dictatorship, Chile 1973–1983" (Ph.D. dissertation, York University, 1984).

21. Data from Renato Paul Boschi, "The Art of Associating: Social Movements, the Middle Class and Grassroots Politics in Urban Brazil," Final Report to the Tinker Foundation, Dec. 1984 (mimeo), pp. 83–84. For further information on women's work and education, see Carmen Barroso, *Mulher, Sociedade e Estado*, pp. 13–86; and Fúlvia Rosemberg, Regina P. Pinto and Esmeraldo V. Negrão, *A Educação da Mulher no Brasil* (São Paulo: Global, 1982).

22. Jo Freeman highlights the role of *networking* in the emergence of social movements in "On the Origins of Social Movements," in *Social Movements of the Sixties and Seventies*, edited by J. Freeman (New York: Longman, 1983).

23. For a more detailed analysis of the origins and dynamics of second-wave feminism in Brazil, see Sonia E. Alvarez, "A Latin American Feminist Success Story?," in *Feminism, Women's Movements and Transitions to Democracy in South America*, edited by Jane Jaquette (Boston: Allen & Unwin, forthcoming).

24. For testimonials of women's experiences within these organizations and in exile, see Albertina de Oliveira Costa, et al., *Memórias das Mulheres do Exílio* (Rio de Janeiro: Paz e Terra, 1980).

25. For a discussion of the Brazilian feminist movement and the issues it has politicized, see Anette Goldberg, "Feminismo em Regime Autoritário: A Experiência do Movimento de Mulheres no Rio de Janeiro," paper presented at the XII Congress of the International Political Science Association, Rio de Janeiro, Aug. 1982, and "Os Movimentos de Liberação da Mulher na França e na Itália (1970–80): Primeiros Elementos para Um Estudo Comparativo do Novo Feminismo na Europa e no Brasil," in *O Lugar da Mulher*, edited by M.T. Luz (Rio de Janeiro: Graal,

1982); Ana Alice Costa Pinheiro, "Avances y Definiciones del Movimiento Feminista en el Brasil" (Master's thesis, Colégio de México, 1981); Branca Moriera Alves and Jacqueline Pitanguy, *O Quê é O Feminismo?* (São Paulo: Brasiliense, 1981); Marianne Schmink, "Women in Brazilian 'Abertura' Politics," paper presented at the Latin American Studies Association Conference, Bloomington, Ind., 1980; Maria Lygia Quartim de Moraes Nehring, part II; and Carmen Barroso, *Mulher, Sociedade e Estado no Brasil,* 163–66. For a discussion of the sanctioning of violence against women in Brazilian law and feminist mobilization around this issue, see Mariza Corrêa, *Morte em Família: Representações Jurídicas de Papéis Sexuais* (Rio de Janeiro: Graal, 1983) and Heloisa Pontes, "Práticas Feministas no Brasil Contemporâneo—um Estudo de Caso do SOS-Mulher," March 1983 (mimeo). For an overview of feminist initiatives in other areas of women's rights, see Comba Marques Porto e Leonor Nunes de Paiva, "Direito: Diferentes mas Não Desiguais," in *Mulheres em Movimento,* edited by Equipe Projeto—Mulher do Instituto de Ação Cultural (Rio de Janeiro: Marco Zero, 1983).

26. On the relationship between women's movements and political parties, see Iara Maria Ilgenfritz da Silva, "Movimentos de Mulheres e Partidos Políticos: Antagonismos e Contradições," paper presented at the V Encontro Anual da Associação de Pesquisa em Ciências Sociais, Nova Friburgo, Rio de Janeiro, 1981; Fanny Tabak and Sílvia Sánchez, "Movimentos Feministas e Partidos Políticos," paper presented at same Encontro, 1981; Sílvia Pimentel, "A Necessária Participação Política da Mulher," mimeo, n.d.; and Maria Teresa Miceli Kerbauy, "A Questão Feminina: Mulher, Partido, e Representação Política no Brasil," paper presented at the XII International Conference of the Latin American Studies Association, Albuquerque, N. M., April 1985.

27. On the critical nature of the 1982 electoral conjuncture, see Bolivar Lamounier, "'Authoritarian Brazil' Revisited: The Impact of Elections on the Abertura," this volume.

28. As Ruth Cardoso points out, "Instead of the movements strengthening the political parties, as many had hoped, party militancy often weakened the movements . . . 'Taking party' implies breaking the consensus created by a common *vivência,*" from "Movimentos Sociais Urbanos: Balanço Crítico," pp. 237–38. I would add that in fact these partisan tensions, deriving from women's "larger" world view and encompassing their perspective on strategies for general social change, have been underlying tensions in Brazilian feminism since its inception and were merely aggravated by the 1982 electoral conjuncture.

29. June Hahner, in *A Mulher Brasileira e Suas Lutas Sociais e Políticas: 1850–1937* (São Paulo: Brasiliense, 1981), provides the most comprehensive discussion of the early feminist movement in Brazil. An English language version of the above is "Feminism, Women's Rights and the Suffrage Movement in Brazil," *American Journal of Sociology* 78, 3 (1980): 792–811. See also Branca Moreira Alves, *Ideologia e Feminismo: a Luta da Mulher pelo Voto no Brasil* (Petrópolis: Vozes, 1980); and Susan Besse, "Freedom and Bondage: The Impact of Capitalism on Women in São Paulo, Brazil, 1917–1937" (Ph.D. dissertation, Yale University, 1983).

30. For a more complete discussion of the history of population politics in Brazil, see Mário Victor de Assis Pacheco, *Racismo, Machismo e "Planejamento Familiar"* (Petrópolis: Vozes, 1981); and Peter McDonough and Amaury de Souza, *The Politics of Population in Brazil: Elite Ambivalence and Public Demand* (Austin: Univ. of Texas Press, 1981).

31. A recent study by feminist social scientist Hildette Pereira de Melo, for example, estimated that there are 1.5 to 2 million provoked, illegal abortions in Brazil per year—one of the highest rates in the world (*Istoé*, Sept. 21, 1983, p. 38). In 1980 alone, Melo counted 201,597 women interned in public hospitals (40 percent of the Brazilian health care network) due to post-abortive complications such as hemorrhaging, uterine perforations, and infection. The prohibition of safe, legal, inexpensive abortion, then, cost the Brazilian State an estimated average of Cr$2,157,000 per post-abortion medical sequel—making the total estimated cost of the non-availability of safe, effective contraception Cr$434,844,689,000 in 1980—enough to amply finance a State family planning program which would give women the concrete means with which to avoid unwanted pregnancies.

32. For an analysis of the role of gender in right-wing mobilizations of women against Allende in Chile, for example, see Maria de los Angeles Crummett, "El Poder Feminino: The Mobilization of Women against Socialism in Chile," *Latin American Perspectives* 4, 4 (1977): 103–13; and Michele Mattelart, "Chile: The Feminine Side of the Coup *d'État*," in *Sex and Class in Latin American*, edited by J. Nash and H. I. Safa (New York: Praeger, 1976).

33. Maria Ignácia d'Avila Neto explores some of the psychosocial dimensions of the relationship between authoritarianism and the subordination of women in *O Autoritarismo e A Mulher: O Jogo da Dominação Macho-Fêmea no Brasil* (Rio de Janeiro: Achiame, 1980). For a discussion of militarism as the quintessential expression of mysogonistic patriarchy, see Ximena Bunster-Burroto, "Surviving Beyond Fear: Women and Torture in Latin America," in *Women and Change in Latin America*, pp. 297–325, edited by J. Nash and H. Safa (Massachusetts: Bergin and Garvey, 1985). Maria Elena Valenzuela provides an incisive account of the patriarchal foundations of Pinochet's Chile in *Todas Ibamos a Ser Reinas: La Mujer en Chile Militar* (Santiago: Ediciones Chile y América, 1987).

34. For a discussion of neo-Malthusian ideology and population policy in Latin America, see Bonnie Mass, *Population Target: The Political Economy of Population Control in Latin America* (Toronto: L.A. Working Group and Women's Educational Press, 1976); and Karen L. Michaelson (ed.), *And the Poor Get Children: Radical Perspectives on Population Dynamics* (New York: Monthly Review, 1981).

35. For a concise summary of State population policies and programs since 1964, see Conselho Estadual da Condição Feminina, "O Direito de Ter ou não Ter Filhos no Brasil," *Cadernos* 1 (1986).

36. Walter Rodrigues, *Painel sobre Planejamento Familiar, presentado na Escola Superior de Guerra* (Rio de Janeiro: BEMFAM, 1978), p. 60, my translation.

37. Walter Rodrigues, p. 64.

38. *Folha de São Paulo*, Aug. 11, 1982. The extent to which Pró-Família was actually implemented during the final months of the Maluf administration is still unknown. Relevant public documents are "missing." My own extensive search uncovered little information regarding policy implementation. I could not track down any related documents in any of the several state government organs involved in either population policy formulation of implementation. The present opposition government suspects that the *Pró-Família* program may have been the victim of corruption and health-related abuses and has established a special legislative commission (*Commissão Especial de Inquérito*) to investigate these and other Malufista abuses of State power.

39. Significantly, Resk's candidacy in 1978 had been endorsed by feminist groups in São Paulo.

40. Commissão de Estudos sobre Planejamento Familiar das Entidades Feminis-

tas de São Paulo, "Controle da natalidade e Planejamento Familiar," São Paulo, Brasil, 1980 (mimeographed), my translation.

41. Commissão, p. 4.

42. Ibid., p. 8.

43. The rationale is that controlling the "exaggerated" population growth rate in underdeveloped nations is a quicker, more cost-efficient solution to the inequitable distribution of national resources. This neo-Malthusian idea is to decrease the number of people who need to be housed, fed, schooled, employed, and so on, rather than increase State investments in housing, social welfare, education, and so on.

44. For an analysis of the crisis, see Fishlow and Bacha and Malan, this volume.

45. *Folha de São Paulo*, March 2, 1983.

46. Ibid., June 9, 1983.

47. Brasil. Senado Federal. "Palestra da Senadora Eunice Michiles sobre Planejamento Familiar," April 28, 1983 (mimeo), emphasis in the original, my translation.

48. Brasil. Ministério da Saúde. "Assistência Integral à Saúde da Mulher. Subsídios para uma Ação Programática," June 1983, (mimeo), p. 4.

49. Estado de São Paulo. Secretaría de Saúde. "Programa da Saúde da Mulher," 1983 (mimeographed), p. 5.

50. Ibid., p. 6.

51. Entidades Feministas de São Paulo, "Carta ao Conselho Estadual da Condição Feminina," May 25, 1983 (mimeographed).

52. Rio feminists also brought their proposal for a National Pro-Abortion Campaign to be initiated on Sept. 22, 1983, throughout Brazil. The result of a national meeting on Women's Health, Sexuality, Contraception, and Abortion, held in Rio on March 4–6, 1983, and financed by the Pathfinder Fund, the campaign was not endorsed by many of the women present as they feared that a stress on abortion would be problematic at a time when the government had anti-natalist intentions which could compromise this demand politically—did the feminist movement want abortion to be legalized as a further State measure to control population size?

53. For a comprehensive, journalistic account of the 1984 nation-wide mobilizations in favor of direct elections, see Ricardo Kotscho, *Explode um Novo Brasil: Diário da Campanha das Diretas* (São Paulo: Brasiliense, 1984). On the PMDB shift to a strategy of "beating the regime at its own game" by seeking to defeat the PDS's candidate within the Electoral College, see Gilberto Dimenstein et al., *Complô que Elegeu Tancredo* (Rio de Janeiro: Editora JB, 1985). For more analytical treatments of strategic realignments both within the government and within the opposition, see Bolivar Lamounier and Marcus Faria Figueiredo, "A Crise e a Transição para a Democracia no Brasil," *IDESP, Textos* 5 (1984). Francisco Weffort's *Por Quê Democracia?* provides the most incisive and compelling analysis of why progressive sectors within the political opposition aligned themselves behind the seemingly "bourgeois" cause of reestablishing political democracy in Brazil (São Paulo: Editora Brasiliense, 1984). See also Skidmore, this volume.

54. Interview with Margaret Arilha, Coordinator of the São Paulo State Council on the Status of Women's Commission on Women's Health and member of the Council's Executive Committee, São Paulo, Aug. 22, 1985.

55. Movimento de Mulheres Democráticas Sociais, "Planejamento Familiar Já!," 1984. (mimeo.).

56. Coletivo Feminista Sexualidade e Saúde, "Brasil: Mujeres y Salud," in *La Salúd de las Mujeres: La Experiencia de Brasil, Reflexiones y Acciones Internacionales*, edited by ISIS International (Santiago, Chile: ISIS, 1985), 11.

57. Organizing groups included Casa da Mulher de Bela Vista, Casa da Mulher de Grajau, Coletivo Feminista Sexualidade e Saúde de São Paulo, Serviço Orientação Família, and Centro Informação Mulher.

58. Coletivo Feminista Sexualidade e Saúde, "Brasil: Mujeres y Salud," p. 13.

59. I am indebted to Margit Mayer for having drawn my attention to differential "access points" within the State apparatus. See her "Urban Social Movements and Beyond: New Linkages between Movement Sectors and the State in West Germany and the United States," paper presented at the Fifth International Conference of Europeanists, Washington, D.C., Oct. 18–20, 1985.

60. Interview with Margaret Arilha, São Paulo, Aug. 22, 1985.

61. Carmen Barroso, "O Continuismo, a Saúde e o Sexo Forte," *Folha de São Paulo*, March 3, 1985.

62. For a discussion of Malufista abuses of State power, see José Yunes, *Uma Lufada Que Abalou São Paulo* (São Paulo: Paz e Terra, 1982).

63. A study conducted in 1979 by the *Delegacia do Trabalho* in the state of Santa Catarina, for example, revealed that out of 593 companies required to provide day care for their female employees by law, only three did so. An enforcement drive led to compliance by 32.8 percent of the firms by 1981.

64. Movimento de Luta pela Creches, "Manifesto," March 1979.

65. De Barros requested and was granted a budget of Cr$9 billion from the Banco Nacional de Habitação (BNH) for the construction of day-care centers in the country.

66. *Estado de São Paulo*, Oct. 21, 1979.

67. Much of the above discussion is based on interviews with Maria Amélia de Almeida Teles, Coordinator of the Conselho Esadual da Condição Feminina's Day Care Commission and former member of the Executive Committee of the MLC, 19 Aug. 1985; and, Isabel, the Director of the Jardim Miriam day-care center and former member of the regional and citywide coordinations, 23 Aug. 1985.

68. It should be noted that the lack of State intervention in the *physical* reproduction of the labor force in Brazil in the pre-1983 period did not have a neutral effect on gender power imbalances. By denying women the means with which to make informed, effective choices about their fertility, the Brazilian State had actively contributed to women's continued confinement in their roles as mothers and nurturers. This "non-decision making process," then, had direct political implications for women's lives.

69. Partido do Movimento Democrático Brasileiro, "Programa Nacional," 1983.

70. The sectors within the PMDB which had wanted a women's branch of the party implemented their strategy by creating the *Comitê Feminino Pró-Montoro* in Aug. 1982, headed by the candidate's wife, Dona Lucy, and his daughter, Gilda. This "women's auxiliary" did not raise gender-specific issues but rather focused on mobilizing relatively "apolitical" middle- and working-class housewives for work in the Montoro campaign—a common strategy for female recruitment and not unlike that implemented by MMDS. The *donas de casa* of São Paulo were specifically targeted by Montoro's political discourse during the last months of the campaign and he made frequent appeals to "the housewives of São Paulo who have their feet firmly on the ground and will therefore vote for the PMDB since they know better than anyone that what São Paulo needs is change." In a speech reproduced in *Folha de São Paulo* during the final days of the campaign, Montoro expressed confidence that the PMDB would receive the immense majority of female votes throughout Brazil—"One of the reasons for this," according to the candidate, was the "fact that

donas de casa know how to evaluate the consequences of the dramatic unemployment index and of the constant rise in the price of food, clothing and education, with a great deal of realism. The PMDB is strong among *donas de casa* because they are the ones who feel the tragedies that the federal government imposes on the Brazilian people, who are tired of seeing prices rise because of our leaders and members of the PDS." *Folha de São Paulo*, Nov. 6, 1982.

71. Proposta das Mulheres do PMDB para o Governo," 1982, emphasis in the original.

72. International Women's Day has been a central rallying date for unified action for the Paulista women's movement since 1975. In 1979, 1980, and 1981, the various types of women's movement organizations coordinated their efforts to sponsor the I, II, and III Congressos da Mulher Paulista which brought together approximately 800, 4,000, and 6,000 women militants, respectively. Incidents during the II Congresso surrounding the attempt by massista sectors of the movement (those organizations linked to sectarian political tendencies on the left) to alter the agreed-upon agenda of the Congresso to include a discussion of "larger" political issues, such as the Assembléia Nacional Constituinte, led to the expulsion of the *massista* sectors from the Coordination of the movement (which had been established after the I Congresso). The *massistas* have since organized their own IWD celebrations—in 1981 there were *two* III Congressos in São Paulo, and in 1982 and 1983 two centralized celebrations of IWD—with the massistas and some sectors of the neighborhood women's movement which constitute their base, on the one hand, and feminists and other sectors of the neighborhood women's movement on the other.

73. Governo do Estado de São Paulo, Decree no. 20892, *Diário Oficial*, April 5, 1983, my emphasis.

74. Montoro appointed *no* women to any of the 24 posts in his state cabinet and only two were appointed to the municipal cabinet—in posts which have traditionally been held by women in liberal democracies, the *Secretaria Municipal da Família e Bem-Estar Social* and the *Secretaria Municipal de Educação.*

75. Governo do Estado de São Paulo. Conselho da Condição Feminina, untitled newsletter, 1984.

76. Governo do Estado de São Paulo. Conselho da Condição Feminina, untitled document, Fall 1983.

77. Due to the lobbying efforts of the São Paulo Black Women's Collective and the Unified Black Movement, the Council later added a black woman to its group of "representatives" of civil society. It is important to note that people of color are extremely poorly represented in the New Republic and that race-specific political claims are among the most difficult to articulate within institutional arenas, due in large part to the continuing myth of "racial democracy" upheld by even the most progressive white Brazilians. It is also worth noting that black women have been dissatisfied with the extent to which the predominantly white feminist movement has confronted race issues in women's lives and have formed their own, separate feminist organizations in recent years.

78. Governo do Estado de São Paulo. Conselho da Condição Feminina, *Mulher*, I, Sept. 1984.

79. Interview with Ana Maria Wilheim, legislative staff of Vereadora Ida Maria Jansco, 20 Aug. 1985. See also São Paulo. Camara Municipal. Commissão Especial de Inquérito sobre Creches. "Relatório Final. Nossos Filhos, Nosso Futuro. Vamos Melhorar Nossas Creches," 1984.

80. Ibid., emphasis in the original.

81. Interview with Maria Amélia de Almeida Teles, São Paulo, 19 Aug. 1985.

82. Governo do Estado de São Paulo, Conselho Estadual da Condição Feminina, Commissão de Creche, "Levantamento e Características dos Berçários/Creches no Local do Trabalho", 1985. This study found that virtually none of the establishments obligated by law (those that employ more than 30 women) to provide day care for infant children of their employees do so. In the entire state of São Paulo, only 38 businesses maintain day care services for their employees.

83. Governo do Estado de São Paulo, Conselho da Condição Feminina, "Creche no Local do Trabalho," 1985 (pamphlet).

84. Interview with Ana Maria Wilheim, São Paulo, 20 Aug. 1985.

85. Interview with Margaret Arilha, São Paulo, 22 Aug. 1985.

86. The historical tendency in Latin America and elsewhere in the Third World has been for State-sponsored "family planning" to consist of the arbitrary distribution of birth control pills and the encouragement of sterilization, especially among women of color and poor women. Thus, the existence of a gender-conscious mechanism, within the State apparatus, to supervise the implementation of State-sponsored family planning is critical in mediating this historical tendency.

87. Informal interview with Ana Maria P. Pluciennik, Coordinator of the Women's Health Program, Secretaria da Saúde, São Paulo, 21 Aug. 1985.

88. Since its creation, the Delegacia has been reportedly receiving 200–300 complaints per day. Similar "women's precincts" have been installed elsewhere in Greater São Paulo and in other Brazilian cities.

89. David Bouchier, "The Deradicalization of Feminism: Ideology and Utopia in Action," *Sociology* 13, 3 (1979): 392.

90. Ibid., 394.

91. Ibid., 397.

92. In other states where the PMDB ruling coalition was politically narrower than in São Paulo, as in Minas Gerais where center and center-right elements of the PMDB predominated, the range of genderic policy initiatives and reproductive ideologies represented within the regime was far narrower as well. In 1983, Governor Tancredo Neves had also created a Council on the Status of Women, but unlike its São Paulo counterpart, the Council in Minas was composed primarily of women party members with few, if any, links to or experience within autonomous women's movement organizations. The women's movement in Minas, in fact, has remained aloof from the Council, viewing it merely as a "bureaucratic ornament," directly subservient to the PMDB's more conservative elements—"a place where deputies' wives can be appointed to the government."

93. Interview with Ruth Escobar, President of the National Council on Women's Rights, São Paulo, 26 Aug. 1985.

94. Partido Movimento Democrático Brasileiro, "Terceira Proposta—Minuta, Conselho Nacional da Condição Feminina," March 15, 1985 (mimeo.).

95. Informal interview with Beatriz Schumacher, member of Ruth Escobar's legislative staff, 25 Aug. 1985.

96. For a more detailed analysis of the political role of the National Council on the Status of Women and post-1985 policy developments, see my "Contradictions of a Woman's Space within a Male-Dominant State: The Political Role of the Commissions on the Status of Women in Post-Authoritarian Brazil," in *The Bureaucratic Mire: Women's Programs in Comparative Perspective*, edited by Kathleen Staudt (forthcoming).

8

The New Unionism in the Brazilian Transition

MARGARET E. KECK

In his first official act as the Minister of Labor of Brazil's "New Republic" in 1985, São Paulo lawyer Almir Pazzianotto[1] declared an amnesty for all trade unionists removed from their posts by previous governments. He then asked for the repeal of the law which institutionalized work on Sundays and holidays in commercial establishments, and proposed that workers have the right to make the rules which regulate union elections at the local, federation, and confederation levels. The following week he canceled the 1978 amendment to the Labor Code which made the establishment of central labor organizations illegal. The Minister also announced his intention of raising the minimum wage by more than 100 percent prior to May 1.*

Pazzianotto's initiatives were intended to demonstrate good will on the part of the new government, which hoped to negotiate a "social pact" with unions in order to gain time for dealing with Brazil's pressing economic problems—in particular the foreign debt. Such a pact was not immediately forthcoming, however, and Pazzianotto could not develop a genuinely pro-labor program and at the same time maintain his position in a highly conservative government. The positive accomplishments of the first year and a half of the New Republic vis-à-vis labor were thus accomplishments of

*I want to thank the Columbia University Institute for Latin American and Iberian Studies, which funded one stage of the research for this paper, Alfred Stepan, Paulo Sérgio Pinheiro, Scott Mainwaring, Caroline Domingo, and Larry Wright for their useful comments on different drafts, and the researchers in the CEDEC labor project in São Paulo for the many discussions which helped to clarify my ideas. Much of the information on the 1981–84 period was gathered in collaboration with the Brazil Labor Information and Resource Center. An earlier version of this paper was presented at the XII International Convention of the Latin American Studies Association, Albuquerque, New Mexico, April 18–20, 1985.

omission: the easing of repression of strikes, and the refusal of the Minister to intervene in striking unions in spite of considerable pressure to do so.

It is significant that Pazzianotto's early remarks referred to a reformulation of the labor relations system, which had gone virtually unchanged since the Consolidation of Labor Laws (CLT) went into effect under Vargas in 1943. Reformulation of the labor code was the subject of extensive discussion among Brazilian union leaders during the late seventies and early eighties, and as early as 1970 President Geisel established a government commission to study the question. The *abertura* period opened up some additional space for union activity, mainly an increase in direct bargaining between unions and employers and the toleration of inter-union organization which was illegal under the CLT. A proposed revision of the labor code was published by the government in 1979, with concessions on internal union organization, collective bargaining, wage policy, and union representation.[2] Unions rejected the proposal due to continuing limitations on the right to strike and other government controls over union autonomy.

The elements of the Brazilian labor code were formulated under Vargas in the 1930s and integrated into the Consolidated Labor Laws (CLT) in 1943. One of the keystones of the Brazilian corporatist model, the legislation remained virtually intact up to the end of the military regime. The CLT intended that unions be organs of collaboration with the government for the promotion of social peace. The code established a monopoly of representation for State-recognized unions in each occupational category. It gave the Labor Ministry broad powers over union by-laws, leadership, and finances, with the right to remove union leaders from their posts, veto slates in union elections, and intervene in unions, replacing union officers with government functionaries. Financing was provided by a union tax (*contribuição sindical*), representing one day's pay per year, deducted from each worker's paycheck whether or not the worker was a union member, and distributed according to criteria determined by the government, mainly for use in social protection programs. Union members paid dues over and above the tax. A system of labor courts was to oversee contracts, providing compulsory arbitration in case of disputes. Strikes were only legal in rare instances and after a complex bureaucratic procedure had been followed. Labor contracts were mainly individual contracts between the worker and the employer; collective contracts, while legally allowed, were rare.[3] The law made no provision for union representation at the plant level. Federations and Confederations were vertically organized by occupational category, with officers chosen in elections where each union had one vote, regardless of size. Central organizations were preventively outlawed in 1978.[4]

While the Labor Code gave the government powerful mechanisms with which to control unions, it also established what were at least in principle some of the most advanced social welfare programs in existence at the time. These welfare provisions, as well as other incentives for labor leaders to work within the system (such as the prospect of social mobility through the State bureaucracy, either in the social welfare institutes or by being appointed to a labor judgeship on the Labor Courts), were important cooptive elements in the labor legislation. Influence in the bureaucracy, in turn, gave union leaders the ability to win benefits for their members, thus reinforcing their own legitimacy, and diminishing the incentive to demand structural changes.

Maintaining social and economic control over the working class through the labor code was part of a political model whose underlying values had remarkable staying power in Brazil. Wanderley Guilherme dos Santos's definition of "regulated citizenship"

> . . . the concept of citizenship whose roots lie not in a code of political values, but in a system of occupational stratification, and moreover that system of stratification is defined by a legal norm. In other words, all those members of the community who are located in any of the occupations recognized and defined in the law are citizens.[5]

describes the essentially functional role which workers were expected to play within the "organic" nation. The politics of elite conciliation was historically successful in coopting individuals and excluding workers in general from the public sphere. The labor code regulated workers' activities in the private one.

By the beginning of the 1980s, the code was coming under increasing attack both by employers and by workers' organizations. For the former, particularly those in the modern industrial sectors, the "control" aspects of the labor code did not compensate for the bureaucratic encumbrances placed in the way of negotiating agreements; direct bargaining procedures would provide a much quicker way of resolving problems. Some entrepreneurs even went so far as to say that the strike law should be revised;[6] few, however, came out in favor of union organization which was completely autonomous from State control.

For labor, the call for autonomy became the watchword of a broad mobilization and the development of new kinds of demands and organization in the 1970s. For many unions involved in the revitalization of the labor movement, autonomy implied a great deal more than the revision of the legislation; it meant the right of workers to determine for themselves the forms of organization and strategies appropriate for them. The question of the rights of labor, therefore, moved into the broader arena of political relations in the whole society. We should note, however, that as the political conjuncture moved closer to one in which real reforms seemed possible, the definition of autonomy began to vary from union to union, ranging from one which merely involved an end to the State's power of intervention in union affairs to full independence of unions from the State, with a gradual and programmed phase-out of State financing through the union tax.[7]

The rise and politicization of new forces in the labor movement in the 1970s took place in a situation where there was very little space available for labor to play a political role in national decision-making. In this narrow space, the unions themselves became politically contested terrain among tendencies with different visions as to the appropriate path for the future of the movement.

With Brazil's return to civilian rule, it was reasonable to expect that a number of changes would take place, transforming both the political environment in which labor organizes and the rules which govern its organization. One of the key questions concerned the extent to which the labor movement itself would be able to take the initiative for changes affecting its future, or whether the initiative would continue, as in the past, to come from the State. Which of these outcomes occurred would in large part be determined by the developments which took place in the last ten years of the military regime. This paper will explore the kinds of demands and organizational forms which arose in the Brazilian labor movement during the final years of the military regime, and the strategies which different tendencies within the labor movement claimed would lead to the strengthening of unions and the empowerment of workers.

The Legacy of Vargas

Prior to 1964, the application of the more restrictive elements of the labor code varied according to the political conjuncture. Different political relations between unions and the State provided possibilities for unions and their leaders to win benefits sometimes in spite of and sometimes precisely because of the stipulations of the code.[8] With the second Vargas government, the pattern of what has been called populist unionism began in which labor exchanged support for government policy—particularly nationalist economic policies—for a relative relaxation of more stringent controls over unions; thus, for example, unions were able to mobilize large-scale strikes in 1953 and 1957 for economic gains. This kind of relationship between labor and the State reached its zenith under Goulart; in the early 1960s, however, cycles of mobilization and lack of resources on the part of the government led to increasing radicalization on both sides, frightening the middle class and business communities and helping to establish the psychological backdrop for the military coup in 1964. In addition, during the rapid polarization of that period, the radicalization of union leaders around national policy issues increasingly left the union rank and file behind, which helps to explain why in spite of the appearance of so much strength, there was no large-scale labor response to the military takeover.

The kind of unionism characteristic of the pre-1964 period was thus one in which union leadership traded the support and mobilization of their memberships with the State in exchange for benefits for members and often for themselves as well. Their ability to play the role of political brokers was to some extent facilitated by the mandatory structuring of unions: the union tax provided a financial cushion for union leaders regardless of active membership recruitment. Any effort to organize unions outside the official structure, in addition to being illegal, was likely to fail; as the Communist

Party found when attempting to create parallel unions in 1947–51, it was essentially impossible to compete with the financial and social resources available to the official unions.

The union tax and other benefits were not, however, enough to secure the positions of union leaders at all times. Particularly in periods of significant strike mobilization, the union did have to maintain an active presence on the shop floor. Nonetheless, for the most militant union leaders, rank and file organization took second place to participation in national policy debates, while less politically active labor bureaucrats had little interest in organizing an active rank and file at all.

Union activity from the mid-'40s on was closely linked to left party activity, particularly the PCB (Brazilian Communist Party) and sectors of the PTB (Brazilian Labor Party). Parallel central organizations, beginning with the MUT (Movement for Unification of Workers) in the 1940s and ending with the CGT (General Workers' Command) in the 1960s, served as coordinating organs for militant union leaders, bypassing the confederation structure, in which the large unions were outnumbered and could be outvoted by smaller ones which sometimes existed little more than on paper.[9]

Thus while the State maintained significant legal and structural control over unions, its need for labor's political support required some degree of consent to that control. As Timothy Harding pointed out,

> Workers supported populism not because they were conservative or satisfied with the status quo, but because they hoped for real solutions to their pressing problems. Once they participated in mobilization through their unions in campaigns initiated by the government, they became increasingly conscious of their own needs and their own power.[10]

Dismantling Populism: Labor Under the Military

The military regime brought about a dramatic change in the relationship between the labor movement and the State. Determined to purge populist and left influences in both State and society, the military moved to ensure the demobilization and exclusion of the labor movement from political life. Existing labor legislation provided a ready-made tool for bringing unions back under tight control, and was applied to its fullest. Nonetheless, the new regime went further, enacting new legal mechanisms which reduced the arenas in which labor had been able to exercise some bargaining power prior to 1964. The military's determination to crush the political power of unions was particularly significant when we consider that it was precisely labor's role as a political actor which had been crucial to its economic bargaining position.

Central union organizations like the CGT were abolished. Between 1964 and 1970 the Labor Ministry carried out 536 interventions in union organizations, removing the elected leaders from office and appointing replacements.[11] After purging the unions, the new regime handed over to them greater responsibility for social programs, forcing union officers to devote more of their time to administrative tasks. At the same time, a wage squeeze policy which continued throughout the "economic miracle" period[12] and the replacement of seniority and job tenure guarantees with the Time in Service Guarantee Fund (*Fundo de Garantia do Tempo de Serviço*—FGTS)[13] brought real hardship to workers. The FGTS made it easy for companies to lay off workers at will, producing high rates of turnover in some sectors, and facilitating the dismissal of union activists.[14] A move to allow revitalization of the unions in 1967–68 (the *renovação sindical*) came to an abrupt end with the repression of the Contagem and Osasco strikes in 1968.[15] The advent of Institutional Act No. 5 at the end of that year ushered in the most repressive period of military rule and eliminated the likelihood of any recurrence of such events.

Considering wage increases to be the main cause of inflation—and recognizing that union wage campaigns were important periods of mobilization and politicization—the military regime instituted a new wage policy designed to bring both elements under strict control.[16] The determination of wage increases according to a set formula using data provided by the government eliminated the unions' main material function, i.e. the ability to win real wage increases for their memberships. In May 1964, a new strike law (Lei No. 4330) set out the bureaucratic process to be followed in order to call a legal strike, a process so complicated and drawn-out as to render legal strikes virtually impossible.[17] Real wages declined; with the elimination both of the issue (wages) and the weapon (the strike) through which they had mobilized their members and exercised bargaining power, unions were severely weakened.

Social security reform was another key aspect of the new regime's effort to reduce the power of unions. Prior to 1964, social security institutes had become a fiefdom of PTB labor leaders and an essential element in their power base. The administrative unification of the institutes under the National Institute for Social Security (*Instituto Nacional da Previdência Social*—INPS) placed social protection squarely in the hands of State technocrats. At the same time, the regime used the expansion of social protection in the seventies to increase its legitimacy. Sectors previously not covered—rural workers, domestic workers, and the self-employed—were brought under the INPS system in the '70s and rudimentary forms of social assistance were created. Unions were expected to co-administer State policy, but had no voice in its formulation.[18]

These changes meant that the forms of union mobilization and organizing which had previously won benefits for labor were no longer tenable. Not only had leadership itself been purged, but also its main organizing foci had been destroyed. With wage increases controlled by the State, the annual wage campaign was deprived of its meaning. With job security provisions abolished, any plant organizer could be fired immediately. And the technocratization of the social security institutes removed the main area of labor penetration in the State bureaucracy. Under the military, strikes would be repressed, and the authoritarian State was not dependent upon or even particularly interested in the support of workers. Nor could labor win benefits through links to a political party. The MDB, which had never paid much attention to labor in any case, was too weak prior to 1974 to be worth attention, and its sphere of activity, Congress, had been robbed of any power of initiative.

For leaders who wanted unions to be more than administrators of social protection programs, the logical shift would eventually be to try to win concessions directly from employers. This change in strategy, however, did not begin to emerge until the second half of the '70s and, even then, union leaders were at first unable to obtain a response from employers.[19] Collective bargaining implies that workers possess weapons of coercion with which to confront employers—specifically, the right to strike. Thus until this right began to be won *de facto* with the strikes in 1978, collective bargaining was more talked about than practiced. The few gains won by unions prior to 1978 were through the action of lawyers in the labor courts, which effectively meant delegation of union power and a complete lack of involvement of the union rank and file.[20]

To win concessions directly from employers required, in addition to the right to strike, a very different kind of relationship between the union leadership and the rank and file, in which the latter would have to be organized for on-going shop-floor or local activity, instead of being occasionally mobilized for mass action as in the pre-1964 period. This possibility involved a dramatic change in focus for the Brazilian labor movement, and as such did not emerge all at once, but in pieces, involving the conjunction of a number of elements which evolved simultaneously.

An increase in plant-level organization had been part of the rhetoric of a large number of unions, but its meaning and implementation differed. The Second National Congress of Union Leaders in 1967, for example, called for the organization of factory commissions; for most unions present these were conceived as temporary committees which would carry out specific tasks—like gathering signatures for a petition. For more radical unionists, factory groups were meant to be the cornerstone of a more authentic and militant kind of trade unionism.[21]

While the repression of the Osasco and Contagem strikes in 1968 put an end to this kind of public debate among union leaders, workers did carry out a number of plant-level actions in the early '70s. Unions were generally not the organizers of these actions, but they were usually called in to negotiate an agreement with the employers. In the auto sector in the ABC region, located in greater São Paulo, a number of slowdowns, work to rule actions, and simultaneous refusals to work overtime took place in the second half of 1973 in which the union was not called in even to negotiate.[22]

The scattered militancy of workers at the level of particular plants and the bypassing of legally mandated State intervention in settlements went virtually unnoticed and unmentioned in the early part of the '70s. With the change in national conjuncture which began in 1974 with Geisel's *distensão*, the political space for labor activity began to increase. Some of the bureaucratic controls over unions were relaxed—for example, Ministerial audits of union budgets. In 1976, a change in the law allowed unions to use up to 20 percent of the proceeds from the union tax for administrative expenses, not subject to authorization by the Labor Ministry.[23] This relaxation of State controls coincided with the visible growth of political opposition to the regime nationally, expressed, for example, in the new electoral strength of the MDB. Outside the electoral arena, the Bar Association was stepping up its campaign for a return to the rule of law; the press was exploring the limits of a gradual relaxation of prior censorship. Within this context, new allies were available for labor leaders wishing to break out of the relative paralysis of the late 1960s and early 1970s. Thus when labor unrest began to take the form of a mass movement in the late '70s, led by union leaders elected within the official structure who were publically challenging the structure and its rules, the "new unionism" became a public phenomenon.

The Workers of the Miracle

At the same time as tighter control over wage policy and union activity limited the material benefits accruing to workers from the period of rapid economic expansion in the late 1960s and early 1970s in Brazil, the industrial working class was undergoing major changes. Between 1960 and 1980 the number of people employed in the secondary sector (including manufacturing, construction, and "other industrial activities") grew from 2,940,242 to 10,674,977; in other words, the number employed in industry grew by 3.6 times in only 20 years.[24] If we compare it with 1950, the number had almost quintupled.[25] Over the same period, the urban population was also growing at a rate of around 5.65 percent per year;[26] nonetheless, during the 1970s, unlike previous periods, urban employment grew at a higher rate than did the employable urban population of ten years and above—faster, in fact, than did the urban population as a whole. The service sector also expanded

Table 1. Geographical Distribution of Workers: Extractive and Manufacturing Industry (in %)

Region	*1970*	*1980*
North	1.43	2.50
Northeast	9.95	10.30
Southeast	70.58	65.62
South	16.92	19.64
Center-West	1.12	1.87

Source: IBGE, *Censo Industrial*, 1970, cited in Duarte Pereira, "Um perfil da classe operária," *Movimento* (28 de abril-4 de maio de 1980), p. 13; IBGE, *Censo Industrial*, 1980. Calculations made on the basis of "pessoal ligado à produção" in extractive and manufacturing industries.

considerably, particularly in the State and social service sectors rather than in the more marginal personal service sector.[27]

Industry remained heavily concentrated in the southeast of the country, particularly in the state of São Paulo, which alone accounted for 49 percent of secondary sector employment in 1970[28] and around 47 percent in 1980.[29] Nonetheless, as we can see from Table 1, between 1970 and 1980 some regional diversification of industry did take place.

Brazilian workers were also young: data from 1976 show that some 49 percent of workers in extractive and manufacturing industry were between 18 and 30 years old, and 34 percent were between 18 and 21 years old.[30] These last figures show that for a very large percentage of the Brazilian working class, the pre-1964 period was at most a childhood memory. Thus, with the exception of particular areas with a very strong historical tradition and memory of labor organizing,[31] these young workers were building their organizations on the basis of experience gained under the authoritarian regime.

The Rise of the New Unionism

The "new unionism" in Brazil refers to more than the strike waves in the late 1970s. It involved diverse sectors of the labor movement in an organizationally heterogeneous movement, and the goals of its component unions were not always the same. In its initial stages from 1977 to 1979, the "new unionism" was the expression of a combative attitude towards union activity more than anything else, differentiated from traditional trade union practices in Brazil in several ways.

First, the leaders of the new unionism, who placed greater emphasis on rank-and-file organization, promoted contact between union leadership and the rank and file. Secondly, many of them were demanding substantial revision of the existing labor legislation, with the goal of creating unions which would be autonomous from the State; this involved recognition of the right to strike, and the right to bargain with employers without State interference. Third, for the first time in a decade they were willing to take the risks involved in militant action—to strike, for example—even in the face of expected government repression.

It is important to emphasize that the "new unionism" was not a development "parallel" to the existing union structure. Leaders like Lula in São Bernardo were often in fact quite hostile to attempts to create parallel labor organizations outside the unions, insisting instead on the need to use those democratic mechanisms which existed—the sovereignty of the union assembly, for example—to transform the *Estado Nôvo* structure from within.

> If there is a legally constituted union to represent the workers, what must we do? We must bring into the unions the best that there is within the factories. (. . .) there can be as many commissions as there are groups of workers, but that they

> should work within the union, either to get rid of the union leadership or to make the union leaders work, whatever. But, I repeat, within the union, if what they want is to change unionism. Because in union assemblies it's the union members who make decisions. If there are 1000 workers from different groups and these people decide to change the rules of the game, then these people can come and change the rules of the game. What has to be done is to create conditions for these people to begin to participate.[32]

The unions involved in the revitalization of the labor movement covered a broad spectrum of occupational categories in the modern sector, as well as important segments of the rural labor movement. The growth of the modern sector of industry and services during the period of the "economic miracle" had meant the rapid expansion of these unions. The number of rural unions also increased from 625 in 1968 to 1154 in 1972; 1,745 in 1976; and 2,144 in 1980. The exceptionally rapid growth of rural unions after 1973 (from 2,930,692 members in 1974 to 5,139,566 in 1979)[33] was spurred by the State's delegation of the administration of rural social protection programs to rural unions;[34] at the same time, the concentration in land ownership, and the expansion of capitalist agriculture and emphasis on export and industrial crops over food crops in rural areas, led to an increase in land struggles. The innately radical nature of the land struggles and the involvement of the Catholic Church's land pastoral (*Comissão Pastoral da Terra*—CPT) were important in promoting a combative leadership in the rural unions.[35] The Pernambuco sugar-cane cutters' strikes, beginning in 1979, directed national attention to workers in the rural sector. The agricultural unions, and in particular the president of CONTAG (Confederação Nacional dos Trabalhadores na Agricultura), José Francisco da Silva, began to play a much greater role in national labor politics than was previously the case.

During the late 1970s, the "new unionism" evolved a distinctive discourse and set of aims in a number of informal and formal meetings and conferences. In 1978, for example, combative unionists (known as *autênticos*—literally "authentics," a name intended to differentiate them from labor leaders under the thumb of the military government) forced a debate during the Fifth Congress of the National Confederation of Workers in Industry, and issued a "Statement of Principles" signed by leaders of thirty-seven unions.[36] The document urged political democratization, an economic development policy with emphasis on raising the standard of living of the population, and union autonomy, the right to strike, collective bargaining, the right to plant-level union representation, and freedom to join international labor organizations.[37] Horizontal organizations of the most combative unions began to form in a number of cities and states.[38]

The Role of the Metalworkers

The Wage Recovery Campaign in 1977, and the unprecedented strike waves in 1978–79 and to a lesser extent in 1980 clearly established the metalworkers, particularly the auto workers and other metalworkers' unionists from the ABC industrial belt of São Paulo, as the leaders of the new labor militancy. In 1973, metalworkers made up some 32.79 percent of the industrial labor force in the southeastern part of Brazil;[39] in 1980, 34.1 percent of workers in manufacturing in Brazil were metalworkers, a proportion which rose to 43.6 percent in the state of São Paulo.[40] Luís Inácio Lula da Silva, president of the Metalworkers' Union of São Bernardo and Diadema, became an important media figure from 1977 on. His charismatic command of giant assemblies of workers during the strikes gave him an almost mythical aura. Up until the end of 1979, when the Workers' Party was formed under his leadership despite opposition from more traditional sectors of the labor movement to forming a party, Lula was almost universally recognized as the most "authentic" new Brazilian labor leader.

The Wage Recovery Campaign grew out of the discovery by DIEESE, an independent research association supported and financed by the unions, of a discrepancy in the government's calculation of the cost of living figures for 1973. DIEESE's claims caught the attention of a group of World Bank economists, who confirmed the DIEESE figures in a secret report to the Brazilian government. Summaries of large sections of the report were published in *Folha de São Paulo* on July 31, 1977, and within a month the Getúlio Vargas Foundation published a "revision in its accounts" for 1973, with the inflation figure rising from 15.5 percent to 20.5 percent.[41] Upon learning of this, the São Bernardo Metalworkers' Union asked DIEESE for a study of how much the metalworkers had lost because of the manipulation of the figures. DIEESE returned the figure 34.1 percent.

With the results of the DIEESE study, the São Bernardo Metalworkers, together with other metalworkers' unions from the region (Sto. André, Mauá, Ribeirão Pires, and Rio Grande da Serra) launched a campaign for the recovery of the 34.1 percent. While the government refused to consider the question, and the labor court refused to convoke management to discuss negotiation, the Wage Recovery Campaign was nonetheless an important step forward. It brought the labor movement national attention which it had not enjoyed since the defeat of the 1968 strikes, and made Lula into a spokesperson for workers' grievances. In 1977 he was interviewed on national television and in major magazines, and appeared on the cover of the newsweekly *Istoé*. Justice Minister Petrônio Portella invited him to Brasília in January 1978 to discuss the problems of labor, as part of an initiative to open a dialogue with the opposition.[42]

In addition, the campaign served as a mechanism for making workers aware that the union was more than a dispenser of social services, and pulled together under the demand for wage recovery a number of localized struggles occurring in particular plants. Prior to the Wage Recovery Campaign, there was a tendency for union demands to be made at the juridical level, requiring good relations with union lawyers more than mobilization and organization of workers in the plants.[43] As a result of the campaign, which mobilized tens of thousands of workers in rallies supporting the demand, the unions came to appreciate the importance of organizing in the factory, and workers in the factories began to see the unions as organisms which supported their demands and as the main instruments they had for making those demands felt.[44]

Thus while it was not the union that told the Scania workers on the 7:00 shift to sit down in front of their machines on May 12, 1978, the union was called in to do the negotiating an hour after the strike began. Within days, workers from Ford and other auto plants in the region had followed the Scania example. Negotiations by the union won an 11 percent raise, almost double the amount Scania had initially been willing to offer.[45] For the first important strike in Brazil in ten years, this was an impressive outcome.

The 1978 strikes spread from the São Bernardo Metalworkers to include at least 24 professional categories and over 500,000 workers in six states and the Federal District. The following year, strikes broke out all over Brazil. More than 3,000,000 workers from some 113 categories went out that year, some in sectors where unions had the capacity to lead them and others not.[46] Lula, Olívio Dutra from the Rio Grande do Sul bank workers, and João Paulo Pires Vasconcelos from the João Monlevade, MG Metalworkers became a sort of consulting squad, helping in some cases to negotiate between union leaders and their rebellious memberships. The chaotic nature of the 1979 strikes led labor sociologist Maria Hermínia Tavares de Almeida to comment that many of them seemed inspired more by the need to bear witness to the aspirations of workers for freedom, autonomy, and the right to full citizenship, than by any concrete demand.[47] There is no question that by 1979, the question of the citizenship and participation of workers was no longer an abstract discussion among intellectuals, but rather had been placed on the agenda of the debate about democracy by the actions of the labor movement itself.

In 1979, the São Bernardo Metalworkers were much better prepared for a strike than they had been the year before, but so, on the other hand, were the employers. The major firms in the area were completely stopped. Huge strike assemblies were held in the stadium of São Bernardo. With government intervention in the union and employer resistance, however, the workers agreed to return to work and allow a 45-day cooling-off period. During this period, while negotiations were going on with the employers, tension was building up to a fever pitch in São Bernardo, with workers prepared to resume the strike at any moment. Finally, after offering a 6 percent increase, the employers refused to go any further. It was clear that they were prepared to resist a strike for longer than the union was able to maintain one; without a strike fund, a long strike appeared effectively impossible. Lula and the other directors of the union were confronted with a situation in which they had to take to the strike assembly a proposal which they did not support, but which they were sure was all they would get.

Lula has described that strike assembly as the hardest day of his life. There had been a May Day rally in the stadium with 150,000 people; the whole city was mobilized, ready to go out again. The assembly to which Lula was to speak was expecting him to either bring in a favorable agreement or lead them out. Instead, after a speech in favor of the agreement by another director of the union, he asked for and won a vote of confidence in the union leadership.

The 1979 strikes reached fifteen states and spread far beyond the metalworkers, to include urban service workers, textile workers, miners, bank workers, construction workers, teachers, sugar plantation workers, and many others. While most concentrated on wage demands, some of the strikes went beyond these to challenge aspects of the union legislation, asking for factory level union representation and job security provisions; in no case did employers make concessions on the latter questions. On the other hand, the government's decision to alter the wage law, making the annual wage increase twice yearly, can be counted as a victory for the labor movement as a whole.

The Legacy of the Strikes

The 1978–1979 strikes helped to increase workers' consciousness of their importance as political actors. But it also made some union leaders aware that industrial action alone was unlikely to win their demands, as the labor ministry and repressive apparatus could be counted upon to intervene on the side of the employers. The 1978 strikes had caught both business and government by surprise, but they were surprised no longer. Particularly among metalworkers, whose actions remained at the core of the upsurge of the "new union movement," tentative discussions began about the idea of forming their own political party.

The labor upsurge beginning in 1978 captured the imagination of the country and helped to give a social dimension to the debate about democracy by demonstrating that organized opposition to authoritarian rule went far beyond intellectuals and political elites. There was considerable disagreement among intellectuals and politicians over the political implications of new labor activity. First, did the movement among metalworkers represent a new phase of development for the labor movement as a whole, or did it only reflect the capacity of a labor aristocracy to win more for itself? Secondly, did the metalworkers' demands, tactics, and form of organization provide examples and leadership for less well-organized segments of the labor movement, or did their growing militancy risk increasing divisions within the working class? Finally, these questions were important for assessing the role of unions and the appropriate strategy for them to adopt in the process of political change under way. Far from being a purely academic debate, the differing assessments of developments in the labor movement reflected and contributed to debates on the shape of the Brazilian transition, the degree to which fundamental social change was possible, and the kinds of institutions which would best further the movement towards democracy.

The automobile workers whose actions set off the strike waves in 1978–79 are at the heart of Brazil's industrial working class. In relation to the majority of the labor force (in which 36.4 percent earns one minimum wage or under and 60.8 percent earns at most two minimum wages),[48] they are relatively well paid.[49] Other workers saw them as an elite in the 1960s—the first to have access to durable consumer goods like televisions, refrigerators, and in some cases even cars.[50] While such consumer luxuries were in fact limited to a tiny minority of the most skilled metalworkers, they represented an aspiration which, particularly during the "economic miracle" period, had a powerful impact on other workers. Nonetheless, within the auto plants themselves the situation was less than ideal. In a study of shop floor conditions and organization in the early 1970s, John Humphrey found widespread discontent among workers not only about wages, but particularly about working conditions. A high degree of turnover, used as an employer tactic to avoid paying higher salaries to workers with more seniority in the firm, contributed to a generalized insecurity about employment.[51] This study, and the conclusions which Humphrey drew from it, played a part in the debate over the significance of the "new unionism." The *perceptions* which these workers had of the hardships they faced contrasted sharply with the relative privilege which auto workers enjoyed in relation to the rest of the working class.[52]

In her analyses of the developments in the 1970s, Maria Hermínia Tavares de Almeida concentrated more on the structure of the labor market than on workers' perceptions of their situation. She argued that the development of the modern industrial sector and technological changes in the labor process led to a segmentation of the working class between workers and unions in the traditional and modern sectors of the economy. New demands emphasizing collective bargaining and shop floor representation, and the pursuit of radical changes in the labor legislation, came from these new sectors, where workers had higher wages, better working conditions, and a higher level of mobility than in traditional industry. The demand for direct collective bargaining, for example, made sense in the modern sector where workers were more easily mobilizable due to their concentration in a few large plants, and employers were often more flexible. In addition, the frequent demand that wage hikes reflect productivity increases in the firm created an even greater distinction between the two sectors.

This segmentation of the working class, claimed Tavares de Almeida, made it difficult to generate class-wide demands and alternatives to the current system. The greater capacity of unions in the modern sector to bargain and their emphasis on demands specific to the sector led to an apolitical attitude, in which industrial action rather than class-wide political organization was seen as the central mechanism for change. As a result, she argued, the initiative regarding the system as a whole remained in the hands of the State, and while some changes could be expected in the labor legislation to provide more flexibility, the basic corporatist structure was likely to stand. Unions had shown an ability to bypass existing rules, but not to create and impose new ones.[53]

Based on the data from his study, Humphrey disputed what he claimed were Almeida's underlying assumptions, that is, the substantially greater degree of opportunity and job satisfaction on the part of workers in the modern sector. On the contrary, he says, high turnover rates and restrictive wage policies prevented the isolation of auto workers from the rest of the class. Improvements in wages and working conditions won by the modern sector would become an example to workers in the traditional sector, rather than a means of permanent separation. Attention to the shop floor level of organization constituted a very different kind of challenge to state control over labor relations than did the populist union politics prior to 1964, but the possibility of resolving conflicts at this level was not restricted to the modern sector. Finally, the development of inter-union horizontal organization expressed a desire for solidarity and support among unions, and represented a potential source of initiative for change.[54] Thus, for Humphrey, the kinds of actions and demands of the metalworkers did not demonstrate an apolitical attitude. Because shop floor relations exemplified larger structures of social control, changes in these relations constituted challenges to the more general configuration of social relations.[55]

The questions involved in the Humphrey-Tavares de Almeida debate began to be clarified in the 1980s. For reasons which will be discussed in the next sections of the paper, Humphrey's argument contains important elements for understanding the kinds of changes in labor practices which began to spread throughout the Brazilian working class in the 1980s. This was not, however, due only to the generalization of shop floor militancy. The *political* conjuncture of the late '70s and early '80s produced opportunities for action which went beyond what one might have expected from an analysis of labor market structure; the growing relation between union leadership and shop floor organization in this period made it possible for labor to make use of these opportunities.

The tendencies which began to develop in the '70s did so in a political environment in which the State was willing to allow a *de facto* growth in union activity. The strikes were of crucial importance, both in terms of infusing extraordinary new energy into the labor movement and as a massive demonstration of unrest within civil society in Brazil. At the grassroots level, they helped to generate mass support for the opposition from various kinds of social movements. Progressive politicians and the press heralded the developments in São Bernardo as a new kind of grassroots democracy in Brazil, pointing to the massive participation of workers in strike assemblies, the responsibility of leaders to those assemblies, and the proliferation of shop floor initiatives. Labor mobilization fed the image of an increasingly powerful opposition within civil society to continued military rule. The appearance of consensus, however, masked real differences over the ends at stake: for many of the "new unionist" labor leaders, the establishment of a civilian democratic regime signified mainly a change in the context of their struggle, rather than its goal.

With the founding of new political parties in 1979 and the appearance of inter-union organizations, the labor upsurge became more than just another manifestation of opposition to authoritarian rule within civil society. The Workers' Party, formed in São Bernardo do Campo in October 1979, insisted both on the *specificity* of workers' demands within the democratic struggle and on the need for workers to have an independent political organization. Many leaders of the elite opposition (and many labor leaders as well) considered this attitude as at best a naively utopian view of the possibilities available and at worst as destructively divisive at a moment when opposition unity was paramount. Differences over opposition strategy in the political sphere had their counterparts in political differences among union leaders as to how best to proceed, how far to push, and to what extent workers and unions *on their own* could expect to win major improvements in their lot. Divisions, latent in the 1970s, between those who espoused reliance on industrial action and those who favored a more moderate approach, gradually crystallized in the 1980s into different organizations advocating distinct approaches to the relations between labor and the State.

Labor Action in the Eighties

Strikes in 1980 met with a more determined response from the government than had those of the two preceding years. During the metalworkers' strike that year, São Bernardo was occupied by troops, and the union was placed under intervention. Its leaders were jailed, purged from union office, and charged with violations of the National Security Law. This tough stance on the part of the government, together with the economic downturn of the early 1980s, meant that the labor movement temporarily had to abandon mass strike actions as the major focus of labor activity. The main trends in the early '80s, however, in spite of the lower incidence of strikes, were in many other ways a continuation of the developments of the '70s, and established the context for labor action after the return to civilian rule in 1985. An important new factor was the increased salience of political parties both at the level of union elections and at the level of national organizations.

Labor activity in the early '80s took place at two organizational levels. First, an increase in plant-level organization fed the growing tendency towards collective bargaining, both at the plant and at the industry level, and away from the settlement of disputes by the labor courts. Secondly, while during 1980–84 there was a decrease in the number of large-scale industry-wide strikes, the strengthening of links between union leadership and rank-and-file organization was reflected in 1984 in a significant increase in the number of short strikes in single plants (out of a total of 626 strikes, 500 were in one plant).[56] Large-scale strikes returned to the forefront in 1985, with the inauguration of the new civilian government. Finally, at the level of union leadership, there was increasing attention to the creation of national inter-union horizontal organizations.

Union Organization and Demands

With the difficulty in winning significant wage concessions due to the government's wage legislation and the less favorable economic conjuncture, demands in the 1980–84 period focused on: a) job security questions; b) frequency of wage adjustments (due to the enormous increase in inflation); and c) the recognition of shop floor union representation. These were all issues which played a part in the demands of the strikes in the late '70s, but which took on added importance with the change in conjuncture. In the '80s, they increasingly became part of a process of direct bargaining between unions and employers. The direct bargaining strategy, and the increased attention paid to shop floor issues during those years, paid off in greater organizational capacity by the time the new civilian government was inaugurated in 1985.

Job security, for example, had always been an issue in the metalworking sector in particular, as employers used a high turnover rate as a way of keeping wages down. As unemployment rose in 1981, it became even more important. According to IBGE figures, in mid-1981 more than 900,000 people lost jobs in the six major metropolitan areas of Brazil, and by August, unemployment in those cities was estimated at 2,000,000.[57] A DIEESE study completed in June 1981 showed 12.8 percent unemployment in the metropolitan area of São Paulo alone, and still more dramatic, 18.4 percent underemployment among those who had jobs.[58] The FGTS provided little protection in a situation of widespread and protracted unemployment.

The demand for more frequent wage adjustments grew out of the extreme hardship which the sharp rise in inflation posed for workers. The Figueiredo government's decision in 1979 to make wage adjustments biannual had been an attempt to pull the rug out from under the impulse behind the strike waves of 1978–79, and the measure did contribute to a sharp fall in the number of strikes in 1980. Nonetheless, over the next few years the rate of inflation rose from 110.2 in 1980 to 211.0 percent in 1984,[59] and the prices on basic goods, primarily foodstuffs, rose even faster. According to DIEESE figures, the amount of labor time necessary to earn a basic basket of goods at the minimum wage had risen from 138 hours 3 minutes in 1978 to 163 hours 44 minutes in 1981. In 1983, for the first time since the study began, the price of a basic basket of goods *exceeded* the monthly minimum wage. (See Table 2.) This situation was exacerbated by the removal of government subsidies on basic goods, as well as a more restrictive wage law passed in 1983. Thus in 1984 a central demand in union bargaining was an increase from biannual to quarterly wage adjustments, or else an anticipation of the biannual adjustment by several months. Many unions, including the largest in the country, the São Paulo Metalworkers' Union, were successful in winning more frequent adjustments in direct bargaining with employers.

The demand for shop floor union representation took several forms, and was particularly characteristic of the union tendency eventually identified

Table 2. Minimum Wage and Minimum Essential Ration

Year	*Minimum Wage at Time (CR$)*	*Value of Ration*	*Hrs. Work/Ration*
1959	5.90	1.65	67 hrs. 7 min.
1960	9.44	2.36	96 hrs. 0 min.
1961	13.22	3.23	82 hrs. 7 min.
1962	13.22	5.77	104 hrs. 45 min.
1963	21.00	9.26	105 hrs. 50 min.
1964	42.00	N.A.	N.A.
1965	66.00	4.35	88 hrs. 33 min.
1966	84.00	37.93	108 hrs. 22 min.
1967	105.00	44.27	101 hrs. 11 min.
1968	129.60	53.52	99 hrs. 7 min.
1969	156.00	72.67	111 hrs. 48 min.
1970	187.20	84.13	107 hrs. 52 min.
1971	225.60	101.18	107 hrs. 38 min.
1972	268.80	133.99	119 hrs. 38 min.
1973	312.00	216.90	166 hrs. 51 min.
1974	376.80	252.60	160 hrs. 54 min.
1975	532.80	348.10	156 hrs. 48 min.
1976	768.00	494.29	154 hrs. 28 min.
1977	1106.40	582.56	126 hrs. 22 min.
1978	1560.00	897.33	138 hrs. 3 min.
1979	2268.00	1586.17	167 hrs. 51 min.
1980	4149.60	3004.63	173 hrs. 47 min.
1981	8464.80	5774.83	163 hrs. 44 min.
1982	16608.00	10205.96	147 hrs. 29 min.
1983	34776.00	35349.85	243 hrs. 58 min.
1984	97176.00	92468.23	228 hrs. 22 min.
1985	333120.00	281575.00	202 hrs. 52 min.
1986*	804.00	569.82	170 hrs. 06 min.
1987*	2400.06	2056.83	205 hrs. 41 min.

* 1986–87 figures are in cruzados. Note that 1,000 cruzeiros = 1 cruzado.

Source: *Boletim do Diesse,* Ano II: Sept. 1983, p. 6; Ano III: Sept. 1984, p. 45; Ano IV: Oct. 1987, pp. 74-75. Data refer to price of minimum ration in city of São Paulo for month of September.

with the Workers' Party and the CUT (*Central Única dos Trabalhadores*). In some cases, shop floor representation took the form of factory commissions elected at the plant level and organically associated with the union; an agreement of this type was made first in 1981 with Ford, and later with a number of the large auto companies. In 1982 factory commissions began to be established in middle-sized firms as well.[60] A number of other corporations established Quality Control Circles on the Japanese model, in an attempt to short circuit the unions' push for shop floor organization linked with the union. Companies also argued that the existence of *Comissões Internas de Prevenção de Acidentes* (Accident Prevention Commissions—CIPAs) in the plants already provided for worker organization on the shop floor; a number of unions responded to this argument by demanding that CIPA elections be taken seriously and announced in advance.[61]

Table 3. Real Minimum Wage and Index of Real Minimum Wage (July 1940 = 100)

Year	*Minimum Wage*	*Index of Real Minimum Wage*
1940	44,658.67	98
1941	40,705.97	89
1942	36,547.64	80
1943	35,894.18	79
1944	37,900.24	83
1945	30,538.62	67
1946	26,796.12	59
1947	20,476.36	45
1948	18,913.56	41
1949	19,219.72	42
1950	18,150.44	40
1951	16,765.85	37
1952	45,001.39	98
1953	37,064.00	81
1954	45,051.66	99
1955	50,590.01	111
1956	51,394.26	112
1957	55,963.85	122
1958	48,611.37	106
1959	54,419.33	119
1960	45,695.97	100
1961	50,809.35	111
1962	46,390.35	102
1963	40,779.08	89
1964	42,136.25	92
1965	40,632.86	89
1966	34,637.55	76
1967	32,768.58	72
1968	32,069.43	70
1969	30,858.49	68
1970	31,406.84	69
1971	30,054.24	66
1972	29,515.03	65
1973	27,047.44	59
1974	24,730.66	54
1975	25,927.89	57
1976	25,758.82	56
1977	26,841.81	59
1978	27,655.20	61
1979	27,290.24	61
1980	28,144.15	62
1981	28,861.57	63
1982	25,215.04	55
March 1983	23,568.00	52

Source: *Boletim do DIEESE* (Abril de 1983), pp. 14–15.

In other cases the demand for shop floor union representation was for the right to have shop stewards, or sometimes simply the right of union officials to visit the shop floor without being accompanied by an official of the company. In a number of cases companies began in practice to recognize shop floor union representation in spite of the fact that this was rarely written into the contract. Other demands won have included monthly meetings between the union and the firm to discuss workers' problems, the union's right to verify job safety measures taken by the firm, and the right to have a union bulletin board in the plant.[62] The advantage of legal recognition, obviously, given that provision is not made in the CLT for shop floor representation, was the need for the guarantee of the jobs of shop floor representatives; the broad prerogatives of companies to dismiss workers at will allowed for a large amount of arbitrariness in the acceptance or not of *de facto* arrangements.

While it would be an exaggeration to contend that the above trends demonstrated that rank-and-file organization had become effective overall in Brazil, the increase in plant-level action did indicate the growing importance of plant-level questions for unions, and growing contacts between the leadership of unions and an intermediate level of cadre in the factories who were capable of mobilizing the rank and file on the shop floor. Collective bargaining, particularly at the plant level, presupposes an intimate knowledge of local conditions which requires an intensification of these contacts.

In addition, demands based on shop floor issues often represented a demand for change in social relations in the workplace in a much more direct way than do more general kinds of industry-wide issues. Negotiations in the 1984 plant-level strikes often included problems related to working conditions, inadequacy of facilities, coffee and lunch breaks, health and safety measures, and other questions bearing on the quality of work relations in that plant. In strikes in non-industrial sectors, for example the numerous strikes of bus drivers and bus ticket collectors in 1984, the most common demand was for an end to the practice of discounting from the worker's pay losses from assaults and repairs due to wear and tear on the bus. This kind of demand had important implications as a challenge to social relations of domination in the work situation.

With the end of military rule, many workers believed that there was a chance to improve their concrete situations, and unions took advantage of a significantly lower likelihood of repression to resume large-scale strikes. Hundreds of the strikes which took place in 1985, in addition to wage and shop floor issues, were for the first time part of a coordinated campaign for a reduction of the work week from 48 to 40 hours. While few unions succeeded in winning the full reduction, many won partial reductions of from one to three hours per week through collective bargaining. Perhaps the most dramatic strike of the year was that of the bank workers; in the first national bank strike in Brazil's history, some 700,000 bank workers remained out for three days in early September, after having carried on a national educational campaign in preparation for the strike. The campaign turned what might have been a popular rejection of the strike—typical of strikes of service sector employees—into considerable popular support.

Of great importance in the climate of labor militancy in 1985 was the change in government. The likelihood that strikes would be met with force was significantly less; Labor Minister Pazzianotto played an active role in mediating strikes, but resisted pressure from other members of the government to use repression or intervene in unions. At the same time, however, conservative ministers in key economic posts held out little hope for an improvement in workers' situation. Indeed, they called for additional belt-tightening. Proposals for a "social pact" abounded, but seemed to the unions to represent little more than yet another attempt to make workers carry the burden of sacrifices for the whole economy.

The move towards collective bargaining over both workplace and wage issues also placed a greater emphasis on the work of technical cadre able to assess the position of a company and its bargaining strategy. The union research organization, DIEESE, played an important role in this process, as did lawyers and technical aides employed by particular unions. In the early 1980s, DIEESE, headquartered in São Paulo, established state-level offices in eight other cities, in order to deal more specifically with problems in those areas, and had a number of sub-offices in specific unions. Use of DIEESE information as the basis for developing bargaining strategies had grown significantly since the time when it played a major role in the Wage Recovery Campaign of 1977.

Two developments seem likely as a result of a shift in emphasis to plant-level bargaining, shop floor activism, and concern with working conditions. One is a change in the functions of the union leadership, away from bureaucratic tasks related to the administration of social assistance programs and towards the coordination of shop floor demands and activities, and the development of bargaining strategies. The other is likely to be a closer relationship between the articulation of demands at the national level and the demands of local unions than there was in the pre-1964 period. This second development is evidently more difficult to predict, as national leadership must in principle act on the basis of a double legitimacy: recognition by its interlocutors outside the unions (the State, employers' associations), and recognition by the lower ranking union leadership. During most of the pre-1964 period the significant relation was the one with the State. Developments in the '70s and '80s indicate that at the very least national leadership is likely to be held more accountable than in the past, and the possibility exists that a more representative kind of relationship might evolve. The next section will explore these issues in the formation of national organizations in the eighties.

National Organizations

At the same time as many unions began to devote more attention to specific local problems and forms of organization, there was also an increase in organization-building at the national level. This process was highly politicized. Between 1977, when the idea of holding a National Conference of the Working Class (CONCLAT)[63] was first proposed, and 1981, when it was finally held, the first informal groupings of trade union leaders had given way to increasingly well-organized tendencies with different strategic approaches to union organization and policy.

By the end of 1978, there were three visible tendencies within "combative" unionism. The first, which called itself the *oposições sindicais* (union opposition) was composed of rank-and-file unionists who favored the organization of factory commissions and action outside of the official union structure. This tendency, important during the periods from 1966 to 1968 and from 1977 to 1979, lost some of its vitality with the growing activism of union leaders within the official structure. Because of its continued emphasis on the illegitimacy of the existing labor legislation, many unionists of this tendency were unwilling to become involved in national organizations until basic changes in union structure had taken place. The second, called the *autênticos,* worked within the union structure, supported factory-level organization and participation by the rank and file, and emphasized union independence in relation to the state and to employers. This tendency was led by Lula and the Metalworkers' Union of São Bernardo and Diadema. The third tendency emphasized organizing to win leadership positions within the labor movement, particularly at the federation and confederation level, and promoted the creation of a *Unidade Sindical* group to coordinate demands and activities at state and national levels. Union leaders close to the Brazilian Communist Party played an important role in this group.[64]

Differing positions became more clearly defined in 1979. For the *Unidade Sindical* group, workers had not yet accumulated enough force to challenge the structure; thus the best tactic was still to concentrate energies on winning positions in the official labor hierarchy. For the *autênticos*, this was insufficient, and the crucial task was to organize workers to participate more in both union and political life. While emphasizing the importance of winning control of unions, the *autênticos* considered the federations and confederations as too unrepresentative to be worth bothering with. The former position thus stressed institutional pressure (from the union hierarchy) as the potential means of winning union demands; for the latter the solution lay in direct action at the union and plant levels. By implication, thus, for *Unidade Sindical*, labor's demands would be met via direct interaction with State institutions (mediated and supported by political parties) in much the same way as they had been prior to 1964; for the *autênticos*, the struggle was more directly focused on economic power within society, i.e. the firms. While the State role was acknowledged to be important, it was not expected to "grant" rights which had not already been won in practice. In October 1979, when many of the *autênticos* were involved in founding the Workers' Party, the major tendencies in the union movement began to be identified with divergent political parties as well. Most of the *Unidade Sindical*, including trade unionists who were members of the Communist parties, elected to join the PMDB, the broad opposition front constituted as the successor party to the MDB.

The founders of the Workers' Party (PT) included *autêntico* labor leaders, intellectuals, Catholic activists, and members of small, extra-legal, mainly Trotskyist parties. The trade unionists involved were attempting to provide the basis for an authentic class politics, arguing that no system could be democratic without full recognition of the rights of labor to organize autonomously and to strike. Multi-class coalitions would thus be undertaken only around specific issues, rather than as the institutional form of political participation.

In spite of its legalization and participation in elections, the PT has remained essentially society-centered in its orientation. Its emphasis on working-class self-development and mistrust of State and parliamentary institutions were a direct inheritance from both the *autêntico* unions and from grassroots-oriented Catholic activism. For the *Unidade Sindical* and many other trade unionists, as well as many progressive politicians involved in other parties, this was considered a naive position, given their assessment of the overall weakness of the working class. Pragmatic considerations mandated a close alliance with the elite opposition and reliance on the leaders of the post-authoritarian regime to bring about change.

Tensions increased during 1980 between the *autênticos* and the *Unidade Sindical* over the Metalworkers' strike in São Bernardo. In spite of its importance in terms of demands, a much higher degree of rank-and-file strike organization and the degree of community and Church solidarity which it generated, the strike ended without having met any of the economic demands with which it had set out to challenge the government's new wage policy. Unlike previous strikes, this one did not enjoy the support of a broad range of unions; *Unidade Sindical* saw it as adventurist, weakening the labor movement and threatening to close the space which the government's *abertura* had thus far allowed unions.[65]

1981, the year in which the CONCLAT was finally to be held, brought a new situation for most industrial unions. The aggravation of the economic recession—in large part a reflection of the recession in advanced capitalist countries—produced a drastic increase in lay-offs and unemployment. Strikes were fewer and more defensively oriented, usually local actions to defend jobs and gains already won. Strikes against lay-offs were common, and the strike at Ford, in spite of its failure to win immediate readmission of the 400 workers laid off, did win a historic agreement in which the company agreed to recognize an elected Factory Commission whose first task would be to negotiate criteria for readmission of those laid off. The Ford agreement established a precedent for direct bargaining with companies over forms of shop-floor representation.[66] By the time the CONCLAT was to be held, however, the possibility of winning significant economic gains through mass action had been essentially eliminated, both by the display of government repression in 1980, indicating that the political space for such actions had narrowed, and by the worsening of the economic conjuncture.

There was clearly a need to discuss union strategy for confronting a new situation.

The National Conference of the Working Class (CONCLAT) was preceded by state-level preparatory conferences (ENCLATs) in seventeen states. The intensely partisan debate at the ENCLATs, particularly in São Paulo and Rio de Janeiro, led many unionists to fear that the CONCLAT would be stillborn. To prevent this from happening, a number of the main union leaders involved met for twenty hours on August 20–21 to produce a document which would serve as a basis for the debate in the CONCLAT, and to reach a compromise which would allow the deliberations to proceed to the end.

The CONCLAT was finally held on August 21–23, 1981, at Praia Grande, SP, with 5,247 delegates from 1,126 unions and professional associations. Discussion covered a wide range of issues: social security policy, employment and job stability, wage policy, agrarian reform, union unity, freedom, autonomy, and organization. At the insistence of the unions led by Lula, the plenary approved a diluted motion calling for the discussion of a general strike. The major problem at the CONCLAT arose over the composition of the National Pró-CUT Commission, the body which was to continue the work of the CONCLAT on an interim basis, to study the issues involved in the formation of a national organization, and to call the next CONCLAT. The Executive Commission's attempt to present a unitary slate failed because a majority of the slots were filled by partisans of the *Unidade Sindical*. Two alternatives were eventually presented, one by Lula and one by Arnaldo Gonçãlves, president of the metalworkers of Santos. Both slates contained the names of compromise candidates. When neither slate won a decisive majority, the leaders were forced to hammer out a compromise, conceived mainly by José Francisco da Silva of CONTAG, in which the rural unions would fill 23 out of 54 seats on the commission, and each of the major blocs present at the conference would fill 50 percent of the remainder.[67]

Once established, the Pró-CUT was seriously divided between union leaders led by the São Bernardo Metalworkers, who favored a move towards rank-and-file unionism and who emphasized direct action (particularly strike action), and those who favored a more moderate approach to union action and the creation of a national organization which would function more from above in the policy arena than from below as a coordinator of new forms of labor initiative at the grassroots and union level.

Complicating the degree of conflict already present was the importance of the upcoming national elections in November 1982. Electoral competition between the Workers' Party and the PMDB (in which members of the *Unidade Sindical* participated) for the votes of workers, especially in the state of São Paulo, sharpened the existing polarization. Some members of the Pró-CUT Commission argued that it was impossible to imagine holding a unitary trade union conference in the face of the widespread politicization around the elections, and suggested the postponement of the next conference until 1983. José Francisco da Silva of CONTAG defended this position, arguing also that in spite of advances by many individual unions since the CONCLAT in 1981, the inter-union organizations had not made progress towards unifying the struggle. Better than another CONCLAT which might try to form an unrepresentative central organization before the subject had been debated enough among workers would be to strengthen inter-union organizations at the state level, and to promote more debate.[68] The São Bernardo tendency argued in response that the mandate of the commission extended only to 1982, and that the conference should be held in any case. The former position carried the day, and the conference was postponed until August 1983.[69]

In spite of the exacerbation of conflict between the *Unidade Sindical* and the *autênticos* over the question of CONCLAT, the government's IMF-mandated wage austerity policy in 1983 provided an incentive for joint action. The wage policy was embodied in a series of decree laws, intended to contain wage increases well under the rate of inflation. They also attempted to eliminate the redistributive aspect of wage policy, instituted in 1979, whereby the lowest paid workers received raises 10 percent over the rise in the official cost of living index (INPC).

Strikes to protest the austerity measures broke out in July, at first including only unions linked to the *autêntico* tendency—oil workers, metalworkers of ABC, São Paulo subway workers, and bank workers. On July 21 a broader one day "general strike" was called, and was fairly successful in São Paulo and Rio Grande do Sul.[70] This strike was significant as the first explicitly political strike since 1964, and as a demonstration that in spite of mistrust and division among tendencies in the labor movement, joint action was still possible.

Government response to the strike, on the other hand, reinforced divisions. Police repression of strikers in São Paulo was concentrated in the ABC region, provoking protests from state congressmen and the acting president of the PMDB, Teotónio Vilela. At the federal level, in addition to the oil workers and the São Bernardo metalworkers unions from the earlier strike, the government intervened in the bank workers' and subway workers' unions of São Paulo. The only unions which suffered intervention as a result of the strike were those whose leaders were affiliated with the PT, indicating on the one hand that these unions were perceived as more of a threat to the status quo than were those closer to the *Unidade Sindical* tendency, and on the other that the government still saw repression by removal of the union officers as an effective means of reducing the influence of the *autênticos*. The São Paulo metalworkers, whose president Joaquim dos Santos Andrade[71] was the self-proclaimed leader of the July 21 strike, and other unions which had played a leading role, were untouched.

Joint action in response to the government's wage initiatives did not, however, prevent the battle which was shaping up over the next CONCLAT. The dispute took the form of a fight over representation, with the

Unidade Sindical arguing for enlarged delegations at the Confederation and Federation levels and for the exclusion of most associations not recognized by the CLT (which included many of the public employees associations as well as others whose base of representation was sometimes dubious). The *autênticos* argued that given the structure of Brazilian unionism the federations and confederations were not representative of the workers they covered,[72] and that representation should be on the basis of unions and of rank-and-file delegates elected by workers proportional to the size of their base. They also called for the immediate creation of a central union organization, while the *Unidade Sindical* still considered such a move precipitous.

While the battle over organizational questions formally precipitated the split in the Pró-CUT, the deeper disputes discussed above had in fact made the prospects for reconciliation more and more difficult to imagine. The opposing tendencies had begun to devote ever more attention to winning control over unions whose officers were up for election, with the explicit objective no longer being only the defeat of *pelego* leadership, but increasingly an expression of the rivalry between the two activist tendencies. The Pró-CUT formally split apart in July 1983, and the *autênticos*, calling for the formation of a *CUT Pela Base* (whose English translation would be essentially a CUT from the Rank and File), held an organizing convention in São Bernardo in August 1983, with 5,059 delegates from 665 unions and 247 other labor organizations. The convention established a central labor organization called the CUT (*Central Única dos Trabalhadores*—Central Workers' Organization). The opposing tendency in turn held a convention at Praia Grande, São Paulo, in November, with 4,254 delegates from 1,258 unions, federations, and confederations, and formed an organization which it called CONCLAT (*Coordenação Nacional da Classe Trabalhadora*), with the word "Coordination" implying a rejection of the immediate creation of a "Central Organization."

While it was not considered out of the question that the CUT and the CONCLAT might at some future date decide to form a single organization, the questions which divided them were not easily soluble. Different strategic approaches were grounded in different visions of society, which in turn were heavily influenced by the way in which different union leaders experienced the authoritarian period. In an analysis of interviews with leaders of metalworkers' unions from both the CUT and the CONCLAT, Roque Aparecido da Silva found that the fact that the former had as a rule lived most of the authoritarian period as factory workers and the latter as union officers had produced profoundly divergent visions of society. For CONCLAT leaders, the solution to problems of labor relations was to be found within broader social and political institutions, provided that the rules of the game were changed in such a way as to give workers a fair chance. For leaders of the CUT, on the other hand, who had experienced the difficult conditions on the shop floor during the authoritarian regime firsthand, the problem was structural. The solution could only lie in broad social transformation; given that workers could not rely on allies in other social sectors, workers themselves were the only possible agents of that transformation.[73]

The success of the CUT relative to the CONCLAT over the next two years should not be interpreted as a conscious choice on the part of workers of one vision of broad social change over another. The explanation for the growth of the CUT lies in the fact that its confrontational strategy, combined with its emphasis on direct bargaining, was highly successful in winning concrete gains for the membership of its affiliates. Emphasis on shop floor organization and closer relationships between union leaders and rank-and-file workers provided a basis for the success of many of the plant-level strikes in 1984; the greater degree of unity among the CUT leadership facilitated the coordination of strikes in 1985, enabling stronger unions to reinforce the claims of weaker ones. While unions which were members of the CONCLAT won victories during this period as well, the heterogeneity of the CONCLAT, combined with its generally conciliatory approach, made it less effective in consolidating the fruits of those victories.

The role of the Labor Ministry was also key to the consolidation of the CUT's position. While victories in the 1984 strikes, mainly won at the level of individual plants, often occurred in spite of the efforts of the Ministry, Pazzianotto's encouragement of the direct bargaining process and his refusal to intervene in strikes provided a more favorable conjuncture for coordinated action in 1985. In addition, the new Labor Minister removed the legal restriction on the formation of central organizations. The decrease in the likelihood of repression clearly benefited that sector of the labor movement most able to mobilize its resources.

By late 1985, particularly after the exceptionally well-coordinated bank workers' strike in September, the CUT began to be recognized as the predominant organization in the labor movement. Its membership included around 1,250 unions, representing around 15,000,000 workers. In absolute numbers of unions, CONCLAT was still ahead, but those numbers were deceptive; while more bank workers' unions may have belonged to the CONCLAT, for example, the four bank workers' unions which belonged to the CUT represented over 70 percent of the bank workers in the country. Of the 6,112,000 workers estimated by the Labor Ministry to have been on strike during the first eleven months of 1985, some 60 percent were led by CUT unions, and most of the other 40 percent received some support from the CUT.

By early 1986, many CONCLAT leaders were sufficiently nervous about the CUT advances that they decided that it was time to form a genuine central organization; a "coordination" was no longer enough. By choosing to call themselves the CGT, these leaders were attempting to demonstrate the historical continuity of their movement. The CGT's president, Joaquim dos Santos Andrade, declared his intention to combat actively the CUT for influence in the unions, disputing union elections with all possible resources, and displaying a new militant rhetoric.

Political Parties and Labor Organization

As should be apparent from the above discussion, divergent and finally opposing tendencies in the active sector of the labor movement were already implicit prior to the establishment of new political parties under the 1979 party law. Nonetheless, with the establishment of the parties, union positions became increasingly identified with political party positions.

This was especially evident in the case of the *autêntico* group, as there was significant overlap between the union leadership and the leadership of the Workers' Party. While the CUT included a number of unions whose officers were not involved in the party, as did its leadership, it was unquestionably dominated by unions whose leaders were PT members.

Party identification in the *Unidade Sindical* and the CONCLAT formed at Praia Grande in November 1983 was somewhat more complicated, because of the illegality of the Brazilian Communist Party (PCB), the Communist Party of Brazil (PC do B), and the 8th of October Revolutionary Movement (MR-8), of which the first was by far the most important. Illegal Trotskyist parties similarly included under the umbrellas of the PT and the CUT played a substantially smaller, though quite vocal, role. While the *Unidade Sindical* tendency and the CONCLAT coordination formed in 1983 were considered in partisan terms to be associated with these parties and with the PMDB, there were significant internal differences among components of the organization.

Because CONCLAT organizers chose to give a prominant place in the organization to federations and confederations, it included many unionists who were not part of the combative wave in the 1970s. In addition, the first president of CONCLAT was José Francisco da Silva, who prior to the split in the pró-CUT was not definitively identified with either the Unidade Sindical or the *autèntico* fraction, in spite of stronger political links with the former. Possessing an indisputable power base of his own (rural union membership outnumbers the membership of all urban unions combined), his decision not to attend the São Bernardo conference in August 1983 was of decisive importance for the division between the national organizations along partisan lines. The heterogeneity of the CONCLAT, however, made it difficult for the organization to formulate positive policies; its significance derived more from the prominence of some of its members than from actions it proposed.

Because developments in the relations between unions and political parties during the transitional period took place in a closed political situation, analysis of their dynamics is particularly complex. While we can point to partisan competition for leadership of unions and of national organizations, its intensity came more from issues internal to the labor movement than from disputes which developed in the broader political sphere and parachuted into the unions to gain support from labor for goals formed outside the labor movement. There is thus an important difference between the partisan struggle "over" the unions and partisan struggle "in" the unions in

Table 4. Total Number of Unions by State and Union Membership by State: Urban and Rural (as of December 31, 1979)

State	*Tot. No. of Unions*	*Memb. Urban Unions*	*Memb. Rural Unions*
Rondonia	3	905	———
Acre	8	560	21,617
Amazonas	54	35,744	22,985
Roraima	1	189	———
Pará	102	41,364	147,588
Amapá	6	1864	———
Maranhão	169	22,588	277,590
Piauí	109	21,293	147,041
Ceará	218	69,341	509,848
Rio Grande do Norte	149	57,110	179,406
Paraiba	167	72,575	292,741
Pernambuco	183	197,823	376,198
Alagoas	68	51,641	122,633
Sergipe	84	22,987	67,036
Bahia	233	172,454	309,705
Minas Gerais	445	325,425	416,541
Espirito Santo	70	48,866	127,466
Rio de Janeiro	234	1,387,871	58,431
São Paulo	546	1,536,358	429,144
Paraná	290	221,773	511,679
Santa Catarina	342	193,425	315,184
Rio Grande do Sul	486	424,002	599,780
Mato Grosso	32	9854	65,552
Mato Grosso do Sul	25	2513	20,032
Goiás	83	154,650	79,222
Distrito Federal	17	66,391	1103
TOTAL:	4124	5,139,566	5,098,522

Source: IBGE, *Anuário Estatístico do Brasil* (1983), page 782.

the Brazilian case. In spite of the fact that struggles over union leadership were referred to (particularly outside the unions) in partisan terms, the players had not changed, and the terrain had not really shifted. Even the excursion of workers into electoral politics in the 1982 elections did not fundamentally change, for the PT, the idea that the political and the union spheres should be separate—in spite of the considerable overlap in personnel between the two.

From the time of the 1978 electoral campaign, the São Paulo (P)MDB attempted to broaden its base of support by including some union and popular leaders in its slate. Fernando Henrique Cardoso, in his senatorial campaign, consulted regularly with labor leaders and chose Maurício Soares, a lawyer for the São Bernardo Metalworkers' Union, as his alternate. While the MDB did win widespread support from workers in this election, it did not establish an ongoing relationship with unions, and the popular leaders elected were not represented in the leadership of the party.

Opposition politics in the traditional political sphere was still very much a matter for elites.

The Brazilian Communist Party approached this situation much as it always had, attempting to approximate itself to power—not to the State, as during the populist period, but to the future State, in the form of the leading opposition party. PCB members ran on MDB and later PMDB tickets for Congress, and campaigned actively for the party. As the transition to civilian rule drew near, the party began a campaign for its legalization, a situation which it had last enjoyed in the mid-'40s. The party's approach to political power, therefore, was to work together with that sector of the opposition which could be expected to occupy the State after the demise of the military regime. The PCB's overall assessment of the Brazilian political situation remained much the same as it had been in the '50s and '60s: that Brazil needed to experience a liberal democratic period and national economic development before the conditions were ripe for the working class to come to power.

Union leaders close to the PCB took a comparable position with regard to union politics. Given that the working class was not yet strong enough to impose its will forcefully in society and in the political sphere, the proper strategy was to win hegemony by gaining access to leadership positions within the existing class organizations, and expect to win substantial gains through political alliances with the opposition in government. This was consistent with the party's approach to labor and social questions prior to 1964. Communist trade union cadre may well have been unwilling to risk hard won positions by adopting what they often considered the naive and hotheaded militancy of their younger colleagues. With the legalization of the party in 1985, and with the formation of the CGT, this position may begin to change. While the party can be expected to take moderate reformist positions at the level of national politics, it is likely to be a serious contender in organizational battles within the unions, and may be forced to adopt a more militant rhetoric in order to develop a stronger rank-and-file constituency. In addition, during the first years of the New Republic PCB union leaders have faced an increasingly well-financed and organized challenge from unionists linked to the American Institute for Free Labor Development (AIFLD), who are intent upon breaking apart the moderate-PCB alliance within the CGT and winning control of the organization.[74]

For labor leaders involved in the creation of the Workers' Party, the focus of the new unionism on shop floor issues and on winning new rights from employers involved a separation of industrial action from the political role of representing and advocating the rights of labor in the political sphere. The PT was to be an extension of and at the same time separate from labor organized institutionally in unions, and was—as a party—to respect the autonomy of those unions.

At the same time as it was to remain separate, however, its role was seen as complementary: the 1979 strike, according to Lula, had demonstrated the impossibility of winning major gains for workers through purely industrial action; to win these gains, workers needed a political organization of their own, founded and headed by and for workers themselves. The party was not so much to *lead* workers, as to *express* in the political sphere needs and demands which workers already felt and which arose at the level of social and union organizations. For the labor leaders involved, the creation of the party was thus a strategic response by a sector of the labor movement for the achievement of goals which had already been articulated elsewhere.

As a legal party competing in the electoral arena with other parties, the PT had to evolve a broader appeal. Its focus, however, remained largely class-based, making the claim that the rights of workers (and by derivation, the rights of everyone) to participate were central to the process of democratization, and that elite politics, characteristic of other parties in the electoral game, relegated workers' rights to a subsidiary place.

Throughout the early '80s, the PT had a great deal of difficulty in reconciling its emphasis on encouraging self-organization within the working class and its role as a party in relation to state institutions. The anti-statism of *autêntico* unionism and of Catholic activists in the party led to an ambivalent attitude about action at that level, evident, for example, in the many conflicts between party leaders and PT members of Congress. In addition, in their concentration on social organization party leaders sometimes miscalculated the consciousness of their constituency, as in the failure to adequately explain the party's decision not to participate in the indirect presidential elections. Once the decision had been made through a series of party congresses, there seemed to be an assumption that the reasons would be clear to workers simply because they were workers.

The PT's determination not to pre-empt the positions of the labor movement, ironically enough, had a similar effect to the position taken by others not to form a labor party. No *party* organization was prepared to formulate political positions for labor and fight for them in national political arenas. Instead, political parties, including the PT, waited for the initiative to come from union organizations.

Labor and Democratization in Brazil

For almost a decade before a civilian government was finally installed in 1985, the labor movement, like many other sectors of Brazilian society, was in a state of ferment. New forms of organization responded to changes in the political climate, but also reflected deeper changes in the structure of the Brazilian working class. Unions sought new strategies to protect their members and strengthen their organizations. The labor movement which mounted record numbers of strikes in the first months of the New Republic was a far cry from the one which circulated the first petitions in favor of wage recovery. What then are the prospects for democratization of labor relations? And what have the changes discussed here contributed to democratizing Brazil?

Labor and the State

One important set of changes is taking place at the level of the relations between unions and the State, in the reform of labor legislation. While some small reforms took place soon after the inauguration of the civilian government, major changes would have to await the Constituent Assembly convened in 1987.

The *de facto* changes which took place in the preceding decade helped to create conditions for new kinds of institutional relationships to arise. While unions remained legally tied to the State, they had won in practice a broader arena of action. The legal restrictions on strikes proved increasingly ineffective as unions induced employers to bypass the mediating institutions of the State's labor relations apparatus and deal directly with the unions.

In addition, some of the military regime's attempts at repression of labor militancy in the 1980s produced the opposite of the intended effect: while intervention in *autêntico* unions may have had short-term demobilizing effects, it also contributed to the formation of new leadership. Again and again, the new union leaders elected after these interventions were those endorsed by the purged officers. In addition, the willingness of employers in some instances to negotiate over conflicts with the expelled leadership, rather than with the government-appointed intervention team, further undermined and delegitimized the government's attempts.

Legal changes are important in a system where the legal system has structured the relations among social groups as thoroughly as in Brazil. The fact that labor mobilization prior to 1964 did not lead to changes in the legal relations between labor and the State facilitated use of the existing corporatist legislation by the military to discipline the labor movement.

Nonetheless, as the debate over new labor legislation heated up in mid-1985, institutional autonomy was still a complex question, and one on which there was substantial disagreement even among activist union leaders. While a broad spectrum of union leaders wanted an end to State interference in union affairs and an end to the requirement that unions be recognized by the State, they were virtually unanimous in seeking to maintain the principle of only one union representing the same category of workers in a same territorial base. The financial question was a serious stumbling block. In spite of gains made in practice, most unions continue to depend heavily on the proceeds of the union tax for their functioning.[75] A voluntary dues system could not immediately replace this funding reservoir. Thus few union leaders favored the immediate abolition of the union tax, but rather a gradual phase-out in which it would increasingly be replaced by voluntary dues. The financing question was further complicated by the relationship between State administration of the union tax and the welfare functions of unions to which large portions of this funding were directed. While unions' role in administering social protection programs contributed to their bureaucratization, it was also an important incentive for workers to join unions. Given the inadequacy of the public health system in Brazil, unions were under some pressure from their memberships to maintain extensive medical facilities.

In spite of a widespread recognition of the need for legal changes, union organizations had difficulty in formulating concrete proposals which could be counterposed to those emanating from the Labor Ministry. The task of promoting labor's interests in the Congress and the Constituent Assembly was coordinated not by a union organization or a party, but by a lobbying organization called DIAP (the Inter-Union Department for Parliamentary Action). Formed at the instigation of a group of unions in 1983, DIAP is a voluntary organization run by labor lawyers, which has attempted to fill the gap left by the unions' inability to develop a unitary approach to labor legislation.

The Constituent Assembly has passed a number of procedural and substantive measures which, if not altered in the final voting, could potentially signal significant changes in union life in Brazil.[76] It is interesting to note, however, that many of these changes have already been won in practice in many areas. The new Constitution should include a much broader right to strike than has hitherto been enjoyed in Brazil, a right which, moreover, will extend for the first time to public employees. Public employees will also be granted the right to unionize. These are important steps, yet public employees did not wait for legal sanction to mobilize or organize; indeed, they made up almost 44 percent of the 55,000,000 workers who went on strike in 1987.[77] The reduction of the work week from 48 to 44 hours mandated by the Constitution had already been won by workers in many industries, and some had gone even further. Similarly, the provision for shop floor representation in firms with over 200 employees responds to the struggle which has been under way for the last decade.

The new Constitution also removes a number of the Labor Ministry's prerogatives regarding unions. It stipulates that workers are free to form unions without authorization from the Ministry, and to manage their own affairs without State intervention. These provisions, together with the right to shop floor representation, could bring major advances along the road to union autonomy. On the other hand, the Constitution also mandates that no more than one union may represent the same occupational category within the same territorial base (thus maintaining the principle of *unicidade sindical*) and preserves the *contribuição sindical.* The coexistence of these last two cornerstones of corporatist labor legislation, on the one hand, and a more liberalized organizational framework, on the other, is likely to be an uneasy one, and there are a number of institutional and procedural questions which will have to be resolved in ordinary legislation and/or in practice.

While the practical import of the legal changes in Brazilian labor relations will take some time to assess, the attention paid to labor issues in the Constituent Assembly, and the level of polarization they provoked, speak to the change in labor's political clout in Brazil. Aside from the length of the presidential term, which was perhaps the most divisive issue in the Assembly's deliberations, the debate over job security (*estabilidade no emprego*) for workers may have produced the most heated discussion. The importance of the issue for workers placed it high on DIAP's agenda; for employers, on the other hand, the demand for employment guarantees seemed to challenge the most basic aspects of social relations. The defeat of the provision, and its replacement with an increase in FGTS benefits accruing to workers dismissed without just cause, presented the issue as a technical matter and clouded some of the deeper questions involved. To understand these questions, we need to look beyond the structural relations between unions and the State.

Labor and the Democratization of Society

Another process, less visible and more difficult to predict, is taking place within civil society and is symbolized by the growth in the practice of direct bargaining. The development of organizational requisites for bargaining are only one side of its significance for social relations overall. Negotiation between workers and employers as a direct and determinant manner of resolving problems in industrial relations involves a new departure for Brazil. Far from apolitical, it challenges the core of the exclusionary system, whereby relations of domination were enforced by vertical State structures which prevented the face-to-face confrontation of class interests, denying, in fact, the very existence of those interests.

First, while bargaining is evidently a form of conciliation, its basis is the recognition of legitimate conflict. The adversary relation has never been accepted in Brazilian political culture.[78] Focused on ideas of consensus, cordiality, and solidarity, the rejection of conflict as a social phenomenon has been reinforced by an exceptionally stratified social system in which lower orders "know their place."[79]

The conciliation and bargaining among elites, which have taken place on the supposition that differences are more apparent than real, and that dissidents can be coopted into the system, have not been reflected in the relation between elites and the great mass of the population—rural and urban workers, marginals, the poor. While "the nation" was embodied in "the people," the latter served a symbolic purpose only, to be convoked en masse to support State initiatives at times but mainly to serve as a backdrop—a repository where cordiality and deference reigned as benevolent symbols of the Brazilian character. Conflict, within this system of values, could only be personalized; it could not be a "state of things"—a feature of a relationship defined not by persons but by social roles.

Thus not only in the context of the resolution of plant problems, but also in the context of Brazilian social relations as a whole, bargaining takes on a different kind of importance. The *recognition of conflict*, and the possibility of negotiating differences between workers and employers, presupposes a degree of formal *equality* which the legal structures were not designed to incorporate. In addition, the act of negotiation in a bargaining relationship is an *impersonal* act, in which the bearers of two socially distinct roles attempt to resolve conflicts regardless of their relations and assets as persons—it is as citizens and representatives that they meet at the bargaining table. In a system like Brazil's, where social and political relations have tended to be highly personalized, this too is unusual.

It is also an eminently political phenomenon. The mediation of labor relations by the State placed workers in a straitjacket at the workplace which was reflected in their action in the political sphere. The practice of workers as citizens in political society was limited by their position as less than citizens in social relations. The demand for autonomy of workers' organizations has appeared sporadically since the 1940s; the practice of autonomy, however, is potentially transformative. It involves a process of class formation in which workers begin to identify themselves through horizontal relations in the workplace and in the community, in shared conditions and struggles, rather than through a vertical pattern of categories established by the State. This process has been reinforced by the growth of community activism, such that shared conditions are identified both with relations of domination in the workplace and with shared patterns of social (and individual) consumption in the community. The spread of unemployment has also blurred the lines between worker and non-worker, forcing the definition of class identity to move outside State-structured relations and into the sphere of common experience.

Changes in labor relations, through strikes, growing workplace organization, and a more responsive relationship between leadership and rank and file, as well as the generalization of face-to-face negotiating between unions and employers, have contributed to this process. The turn towards direct bargaining and collective organization, begun in the large automobile plants in the São Paulo industrial belt, spread to smaller plants and to other sectors of the population. In 1984, when close to 2,000,000 workers went out on strike over wage demands, job stability, and working conditions, the majority of strikes were settled through negotiation. Strikes took place at the factory level rather than at the industry level, many beginning inside the factory and some going so far as to involve occupations. Aside from the industrial sector, teachers, professors, and public employees challenged bans on strikes in essential sectors. Rural day-laborers (*bóias-frias*) struck all over Brazil, in revolt against new work rules and miserable pay.

The relatively better position of workers in industries in the modern sector does not seem to have led to a division of the working class along the traditional/modern axis. Analysis of sectoral data on Brazilian industry suggests that the sharp division which she posited between two clearly defined sectors of the industrial economy was somewhat exaggerated. Taking as an indicator the ratio of wages to value added in different industrial sectors as portrayed in the Industrial Census of 1980, the differences are smaller than we would expect. (See Table 6.) While this can be no more than indicative, when considered together with the average sectoral wage it suggests that at the very least, the diversity within the so-called modern sector means that we cannot draw conclusions about a union constituency

Table 5. Occupational Categories, Median No. Workers, and Average Wage (1980)

Occupational Category	*No. Workers*	*Average Wage*	*% Variation from Av. Wage*	*% of Ind. Work Force*	*% of Ind. Wage Bill*
Mining	64,521	12,088	18.81	1.69	2.01
Transform. of non-metal. mineral products	295,555	7,641	−24.89	7.75	5.82
Metallurgy	420,024	11,776	15.74	11.02	12.75
Machinery	427,317	16,417	61.36	11.21	18.09
Elect. & Communic. Mats.	196,710	10,849	6.63	5.16	5.50
Transportation Materials	232,142	14,854	45.99	6.09	8.89
Wood Products	188,754	6,125	−39.79	4.95	2.98
Furniture Products	126,007	7,332	−27.93	3.30	2.38
Paper and Paper Products	85,575	10,663	4.80	2.24	2.35
Rubber Products	45,570	12,699	24.81	1.19	1.49
Leather and Fur Products (not incl. shoes & clothing)	34,400	7,180	−29.42	0.90	0.63
Chemicals	112,691	14,877	46.22	2.95	4.32
Pharm. & Veterinary Prod.	23,045	9,587	−5.77	0.60	0.56
Perfumes, Soaps, Candles	18,009	8,704	−14.44	0.47	0.40
Plastics	95,952	8,763	−13.86	2.51	2.16
Textiles	325,703	8,438	−17.06	8.54	7.08
Clothing, Shoes, Fabric	385,239	6,201	−39.05	10.10	6.16
Food Products	417,223	6,490	−36.21	10.94	6.98
Beverages	38,191	9,191	−9.66	1.00	0.90
Tabacco	15,057	10,853	6.67	0.39	0.42
Editorial and Graphics	91,364	11,794	15.92	2.39	2.77
Various	84,262	8,311	−18.31	2.21	1.80
Aux. Support & Indust. Serv.	87,721	15,268	50.06	2.30	3.45
TOTAL	3,811,032	38,773,838,530	100.00		

Source: Calculated from figures in *Censo Industrial* 1980, pp. 264–67.

Table 6. Metalworkers' Strikes—1984—Size and Location*

	Sao Paulo	*Minas Gerais*	*Rio de Janeiro*	*Pernambuco*	*Rio Grande do Sul*
100,000 +	—	—	—	—	—
10,000–99,999	2	1	1	—	—
5000–9999	5	1	—	—	—
1000–4999	36	1	5	—	1
500–999	34	—	2	—	—
1–499	68	3	2	2	3
No Data on Strikers	4	—	2	—	—
TOTAL	149 (**)	6	12	2	4

*Labor Ministry statistics give the total number of strikes during 1984 as 626. The data included in this table is based on information about strikes provided in the monthly *Boletim do DIEESE*, out of which a data base was created which includes information about 392 strikes. Data on the number of strikers is given for 356 of those strikes.

Thus the data included in this table is intended to be indicative rather than comprehensive, and should not be taken to cover all strikes which occurred in 1984.

**Out of the 149 Metalworkers' strikes in the state of São Paulo, 106 were in the São Paulo metropolitan area: 53 took place in the ABCD region, and 53 in São Paulo itself.

in any given jurisdiction from the best paid workers in the most technologically advanced factories in that area. The average wage of metalworkers, while almost twice that of clothing workers, still did not reach three minimum wages in this period. Thus most metalworkers' unions are not negotiating for a privileged stratum of workers with interests different from the rest of the working class, but rather for a more heterogeneous base whose common interests are those of the class as a whole.

Division in the labor movement, instead of being grounded in a distinction between traditional and modern sectors, is grounded in a political struggle among union leaders over the strategic options available for change in the position of labor in Brazil. The different experiences of labor leaders in prominent positions in the two central organizations under the authoritarian regime produced very different visions of State and society, and of the role of unions in relation to other institutions.

The orientation of union leaders in the CONCLAT towards political bargaining and towards global approaches to resolving the problems of workers seemed likely to place them in a prominent position during the initial reform period of the transitional civilian regime. Nonetheless, the strategic position of the unions in the CUT within the modern industrial sectors meant that a social pact to which they did not agree would be impossible to enforce. The strikes of April–May, 1985, coordinated by the CUT, sent a clear signal to the new government. Demanding a reduction in the work week from 48 to 40 hours and quarterly wage adjustments, the strikes included over forty occupational categories, and were largely successful in winning gains through direct negotiations with employers. The message was that the CUT unions were not prepared to wait for global solutions at the level of political institutions. The success of this strategy led

Table 7. Occupational Categories and Strikes—1984**

	No. Strikes	*Total Strikers*	*Avg. Days*
Professors/Teachers	29	492,331	7
Rural Day Laborers	29 (*)	478,920	3
Public Employees	24 (*)	227,150	6
Metalworkers	172 (*)	234,148	3
Bankworkers	7 (*)	38,300	1
Bus Drivers & Conductors	16 (*)	37,810	3
Construction Workers	13	32,490	4
Chemical Workers	20 (*)	12,599	4
Security Guards	5 (*)	6486	3
Textile Workers	3	3520	2.5
Transport Workers	12 (*)	3031	2.5
Cabinet Makers	6	2295	5
Other	56	115,015	
Total	392	1,684,095	

*Data on strikers is not available for 6 Professors/Teachers' strikes; 3 strikes of rural day laborers, 6 Public Employees strikes, 4 Metalworkers' strikes, 1 Bus Drivers' strike, 1 Chemical Workers strike, and 1 Security Guards' strike. Thus the total number of strikers in those instances would be higher.

**Labor Ministry statistics give the total number of strikes during 1984 as 626. The data included in this table is based on information about strikes provided in the monthly *Boletim do DIEESE*, out of which a data base was created which includes information about 392 strikes. Data on the number of strikers is given for 356 of those strikes.

Thus the data included in this table is intended to be indicative rather than comprehensive, and should not be taken to cover all strikes which occurred in 1984.

to the doubling of CUT membership between the time of its formation and the end of 1985. By 1986, faced with the slow and continuous erosion of their base, CONCLAT leaders were divided about what to do. Some favored reunification into a single central organization, but the faction led by Joaquim dos Santos Andrade and the PCB decided to fight back and transform the immobilized CONCLAT into a new central organization, the CGT.[80]

The kinds of changes taking place from below are much slower to emerge. Seen from the point of view of social relations as a whole, their direction implies an assertion of rights rather than a demand for concessions. This in turn leads one to expect that from some sectors of the labor movement, particularly those identified with the CUT, confrontational positions will continue to coexist with the development of new negotiation practices. Nonetheless, the opening up of new legal and institutional space for union organization is a hopeful sign. The success of Brazilian democracy depends on its ability to accept confrontation and conflict as part of "normal politics" among free citizens. Grafting liberal institutions atop an authoritarian social structure provides a weak foundation; changes in social relations must provide a civil society strong enough to "re-present"[81] itself in political relations and political institutions for a democratic system to survive. The role of the "new unionism" and the future of Brazilian labor relations stands at the center of that process.

NOTES

1. Pazzianotto came to prominence as the lawyer for the Metalworkers' Union of São Bernardo and Diadema during the strikes of the late '70s; he was chosen for the post of Labor Minister because he was respected by different sectors of the labor movement, as well as by business leaders.

2. For a discussion of the relations between unions and the authoritarian State and the creation of a potential space for negotiation, see Amaury de Souza and Bolivar Lamounier, "Governo e Sindicatos no Brasil: A Perspectiva dos Anos 80," *Dados* 24: 2 (1981): 139–60.

3. Collective bargaining was possible under the CLT, and where they existed collective contracts took legal precedence over individual ones. There is some evidence that, at least in São Paulo, bargaining over wage increases was becoming more common in the early '60s. In a survey of 23 contract disputes in São Paulo between January and March 1964, Mericle found that 47.8% were resolved by collective bargaining. Nonetheless, the lack of a "duty to bargain" provision in the labor code meant that employers' only incentive to bargain was the desire to avoid compulsory arbitration in the labor courts, an incentive which was rarely operative. See Kenneth Scott Mericle, "Conflict Regulation in the Brazilian Industrial Relations System" (Ph.D. dissertation, University of Wisconsin, 1974), 200–207.

4. The complete annotated text of the labor code can be found in Adriano Campanhole and Hilton Lobo Campanhole (eds.), *Consolidação das Leis do Trabalho e Legislação Complementar* (São Paulo: Editora Atlas, 62nd ed., 1983). Detailed discussions of the establishment of the Brazilian Labor Code are in José Albertino Rodrigues, *Sindicato e Desenvolvimento no Brasil* 2nd ed. (São Paulo: Símbolo, 1979); and Kenneth Paul Erickson, *The Brazilian Corporative State and Working Class Politics* (Berkeley: Univ. of California Press, 1977). For an article providing an excellent summary of the structure and functioning of the system, see Kenneth S. Mericle, "Corporatist Control of the Working Class: Authoritarian Brazil since 1964," in James M. Malloy (ed.), *Authoritarianism and Corporatism in Latin America* (Pittsburgh: Univ. of Pittsburgh Press, 1977).

5. Wanderley Guilherme dos Santos, *Cidadania e Justiça* (Rio de Janeiro: Editora Campus, 1979), 75.

6. "Lei de Greve deve ser revista," *Folha de São Paulo*, 1 May 1982, p. 16, contains statements by Roberto Della Mana, Director of the Department of Labor Coordination of FIESP, the São Paulo Federation of Industrialists.

7. A good examination of these questions can be found in a dossier published in *Folha de São Paulo*, 2 June 1985, pp. 35–37.

8. For a conjunctural approach to the relations between labor, the State, and other political actors, particularly in the period from 1945 to 1964, see Francisco Weffort, "Sindicato e Política" (tese de livre-docência, USP, 1971). See also Rodrigues, *Sindicato e Desenvolvimento no Brasil*; Erickson, *The Brazilian Corporative State*; and Aziz Simão, *Sindicato e Estado* (São Paulo: Dominus, 1966). In addition to these general works, several case studies are also useful in understanding these relations in the pre-1964 period: see Annez Andraus Troyano, *Estado e Sindicalismo* (São Paulo: Edições Símbolo, 1978) on the São Paulo Chemical Workers; Maria Andréa Loyola, *Os Sindicatos e o PTB* (Petrópolis: Vozes/CEBRAP, 1980) on unions in Juiz de Fora, MG; and Lucília de Almeida Neves, *O CGT no Brasil 1961–1964* (Belo Horizonte: Editora Vega, 1981).

9. In the union federations and confederations in Brazil, the law requires that each union have only one vote regardless of its membership; thus the vote of São

Paulo Metalworkers' Union, for example, with a membership base of around 300,000, carries the same weight in the Metalworkers' Federation as that of a small union in the interior which might cover only 200 workers.

10. Timothy Fox Harding, "The Political History of Organized Labor in Brazil" (Ph.D. dissertation, Stanford University, 1973), 627–28.

11. Argelina Cheibub Figueiredo, "Intervenções Sindicais e o 'Novo Sindicalismo'," *Dados* 17 (1978): 136–45.

12. See Table 2. Data on the wage squeeze are available in countless numbers of the *Boletim do DIEESE.* Those used in the table come from the Sept. 1983 issue. See also Bolivar Lamounier and Amaury de Souza, "Governo e Sindicatos no Brasil," 144.

13. Under the previous system, workers with 10 years' seniority on the job had job stability. Under the FGTS system, stability was abolished, and both workers and employers made compulsory contributions to a fund which was to be used as severance pay if a worker was fired without just cause. The fund could also be drawn upon for particular purposes, such as purchase of a house or expenses for marriage. With seniority provisions abolished, union activists whose jobs had previously been secure could be fired at will.

14. James M. Malloy, *The Politics of Social Security in Brazil* (Pittsburgh: Univ. of Pittsburgh Press, 1979), 126.

15. For a discussion of the "renovação sindical" policy, see Maria Helena Moreira Alves, *Estado e Oposição no Brasil* (*1964–1984*) (Petrópolis: Vozes, 1984), 119–26. On the Osasco and Contagem strikes, see Francisco Weffort, "Participação e Conflito Industrial: Contagem e Osasco, 1968," *Estudos Cebrap*, Caderno 5 (São Paulo, CEBRAP, 1972). An interesting comparative article which looks at the Contagem and Osasco strikes in relation to labor uprisings in Argentina and Mexico is Elisabeth Jelin, "Spontanéité et organisation dans le mouvement ouvrier: le cas de l'Argentine, du Brésil, et du Mexique," *Sociologie du Travail* (April–June, 1976).

16. A set of legislation beginning with Law No. 4.725 of July 13, 1965, and culminating with Decree Law No. 15 of Aug. 1, 1966, established the formula whereby annual wage adjustments would be calculated. Decree Law No. 15 eliminated the unintentionally beneficial consequences of previous cost of living index calculations for workers, and stipulated that the indices for readjustment of the average real wage per category over the last 24 months would be determined monthly by presidential decree. This last measure became known as the "wage squeeze policy." For a detailed discussion of wage policy, the projects presented and the congressional debates involved, see Heloisa Helena Teixeira de Souza Martins, *O Estado e a Burocratização do Sindicato no Brasil* (São Paulo: Ed. Hucitec, 1979), 135–55.

17. For a detailed discussion of the strike law and its formulation, see Martins, *O Estado e a Burocratização*, 117–20.

18. For a discussion of social security reform under the military regime, see James M. Malloy, "Politics, Fiscal Crisis and Social Security Reform in Brazil," paper prepared for delivery at the 1984 conference of the American Political Science Association, Washington, D.C., Sept. 1984, and his earlier book, *The Politics of Social Security in Brazil.*

19. The Metalworkers' Union of São Bernardo and Diadema made its first major effort to win a collective bargaining contract—rejected by the employers—in 1975. This followed upon a gradual process beginning in 1970, during which, in attempting to separate the local union's demands from those of the federation, the union was

attempting to address particular problems present in the automobile industry (especially the huge gap between the highest and lowest paid workers in the auto companies). Assemblies began to be held at contract time beginning in 1973 to put forth, in addition to demands intended to augment the cost of living adjustment, a series of demands particular to the industries in the auto sector. See Luis Flávio Rainho and Osvaldo Martines Barges, *As Lutas Operárias e Sindicais em São Bernardo 1977/1979* (São Bernardo do Campo: Associação Beneficiente e Cultural dos Metalúrgicos de São Bernardo do Campo e Diadema, 1983), 29–33.

20. On the role of lawyers and action in the labor courts, see the contribution of Almir Pazzianotto Pinto in the round table discussion "Para onde vai o sindicalismo brasileiro?," *Escrita Ensaio*, São Paulo 2: 4 (1978), pp. 29–30. Union leaders who participated in the round table discussed the question of collective bargaining and the right to strike. See particularly the contribution of João José Albuquerque (Secretary of the Metalworkers Union of Santo André), p. 28.

21. Martins, *O Estado e a Burocratização*, 157–61.

22. Ibid., 127–33.

23. Law No. 6.386 of Dec. 9, 1976, rewrote Articles 580 to 592 of the Labor Code, which deal with the application of the union tax. See Campanhole and Campanhole, *Consolidação dos Leis do Trabalho*, 141.

24. Vilmar Faria, "Desenvolvimento, Urbanização e Mudanças na Estrutura de Emprego: A Experiência Brasileira dos Últimos Trinta Anos," in Bernard Sorj and Maria Herminia Tavares de Almeida (eds.), *Sociedade e Política no Brasil Pós-64* (São Paulo: Editora Brasiliense, 1984), 146–47.

25. Ibid., 155.

26. Ibid., 140.

27. Ibid., 152.

28. Duarte Pereira, "Um perfil da classe operária," *Movimento* (24 April–4 May 1980), 13.

29. IBGE, *Censo Industrial*, 1980, p. 4.

30. *Relação Anual de Informações Sociais* (RAIS), 1976, cited in Duarte Pereira, "Um perfil da classe operária," 13.

31. A fascinating case where a strong local labor tradition was maintained is that of Niterói, RJ, where the young metalworkers' activists in the 1970s and 1980s were the children of the dock and shipyard workers active in the 1960s. See Abdias José dos Santos and Ercy Rocha Chaves, *Consciência Operária e Luta Sindical: Metalúrgicos de Niterói no Movimento Sindical Brasileiro* (Petrópolis: Vozes, 1980); and José Sérgio Leite Lopes and Maria Rosilene Barbosa Alvim, "Metalúrgicos do Rio e Niterói: Ligações entre os conflitos de 1980 e as Lutas do Passado," *Aconteceu* (Rio de Janeiro, CEDI, Especial 7, June 1981), 20–23.

32. "São Bernardo: Uma Experiência de Sindicalismo 'Autêntico'," interview with Luís Inácio da Silva, *Cara a Cara* I, 2 (July–Dec. 1978): 62.

33. *Anuário Estatístico do Brasil, (1975–1980)* cited in Alves, *Estado e Oposição no Brasil*, 243.

34. See Jorge Guimarães, "Trabalhador do Campo," *abcd Jornal* (Dec. 1979), 16, containing mainly an interview with Vinícius Caldeira Brant.

35. Shepard Forman points out that the radicalism of peasant land struggles gives an impetus to the movement as a whole; these struggles automatically challenge the system in a way which the demands of wage laborers, which could presumably be met by legislative action, do not. See Shepard Forman, *The Brazilian Peasantry* (New York: Columbia Univ. Press, 1975), 190. For the development of a combative

position on the part of rural unionism, see CONTAG, "Anais do III Congresso de Trabalhadores Rurais" (Brasília, 1980); and Leonilde Sérvolo de Medeiros, "CONTAG: um balanço," *Reforma Agrária* (Boletim da Associação Brasileira da Reforma Agrária, Campinas, SP) 11, 6 (Nov.-Dec. 1981): 9–16.

36. See Maria Hermínia Tavares de Almeida, "Tendências Recentes da Negociação Coletiva no Brasil," *Dados* 24, 2 (1981): 164–65 fn. 5.

37. The text of the "Statement of Principles" is summarized in detail in Alves, *Estado e Oposição no Brasil*, 247.

38. The timing of the appearance of such organizations differed widely: in Rio Grande do Sul, for example, inter-sectoral discussions among unions began in 1975 with the "Independent Union Weeks," a series of seminars on general topics of union reform, and an Intersindical was already established in 1977. See Abílio Afonso Baeta Neves, Enno Dagoberto Liedke Filho, and Lorena Holzmann da Silva, "Rio Grande do Sul: Organização, lutas e debates atuais no movimento sindical," in CEDEC (ed.), *Sindicatos em uma Época de Crise* (Petrópolis: Vozes/CEDEC, 1984), 74–88. In most states such organizations were not established until after the strikes of 1978.

39. IBGE, *Pesquisa Industrial*, 1973, cited in Duarte Pereira, "Um perfil da classe operária," 13.

40. IBGE, *Censo Industrial*, 1980, calculations from figures pp. 8–19, using category "pessoal ligado à produção" as basis for number of manufacturing workers.

41. For an account of the Wage Recovery Campaign, see José Alvaro Moisés, "Problemas Atuais do Movimento Operário no Brasil," *Revista de Cultura Contemporânea* I, 1 (July 1978): 49; and Rainho and Bargas, *As Lutas Operárias e Sindicais dos Metalúrgicos em São Bernardo*, 39.

42. Lula discussed his meeting with Portela in an interview which appeared in *Pasquim* (24–31 March 1978), reprinted in Luís Inácio da Silva, *Lula: Entrevistas e Discursos* (São Bernardo do Campo: ABCD-Sociedade Cultural, 1980), 31–32.

43. Rainho and Bargas may have exaggerated the lack of attention to shop floor organizing in order to show that it greatly increased after the campaign. Humphrey emphasizes the fact that the São Bernardo union had up to 17 union officers working in plants at any one time, with stability of employment, and that at one of the plants he studied this had had a significant effect on unionization between 1975 and 1978. See John Humphrey, *Capitalist Control and Workers' Struggle in the Brazilian Auto Industry* (Princeton: Princeton Univ. Press, 1982), 140–45.

44. Rainho and Bargas, *As Lutas Operárias e Sindicais*, 42–43.

45. Oboré (ed.), "A Greve na Voz dos Trabalhadores da Scania à Itu," *História Imediata* (São Paulo: Alfa-Omega, 1979), 8–10.

46. Data on the number of categories on strike and on the strikers involved come from Alves, *Estado e Oposição no Brasil*, 251, 254, who based her estimates on detailed research in newspapers and union documents of the time.

47. Maria Hermínia Tavares de Almeida, "Novo Sindicalismo e Política (Análise de uma Trajetória)," 1983: (mimeo), 12.

48. *Anuário Estatístico do Brasil* (1983), p. 153.

49. While a skilled metalworker in an automobile plant might earn up to 8 or 9 times the minimum wage, the majority were concentrated in the range of 3 to 6 times the minimum. Thus, for example, at Toyota in São Bernardo in March 1985, 27% of the work force earned under 4 minimum wages; 75% earned under 6 minimum wages, and only 18% (which included foremen) earned over 7 minimums. Given the fact that most statistical breakdowns do not separate semi-managerial employees

and production workers, it is difficult to reach an exact estimate. In March 1985, the hourly wage for a skilled ironworker at Toyota ranged from 647 to 8,148 per hour, or approximately U.S.$1.00–$1.65 per hour. Some other automobile companies paid higher wages; at the high end of the scale, at Mercedes Benz, for example, a worker in the same position earned between 8,905 and 13,700 cruzeiros per hour. Data taken from a report by the Sub-Seção do DIEESE—Metalúrgicos, São Bernardo do Campo e Diadema, "Toyota do Brasil—SA" (July 1985).

50. Lula's description of his dream of becoming an auto worker is illustrative in this respect. See Mário Morel, *Lula o Metalúrgico: Anatomia de uma Liderança* (Rio de Janeiro: Editora Nova Fronteira, 1981), 33.

51. Humphrey, *Capitalist Control and Workers' Struggle*, 87–100.

52. Ibid., 55–105.

53. Maria Hermínia Tavares de Almeida, "O Sindicato no Brasil: Novos Problemas: Velhas Estruturas," *Debate e Crítica* 6 (July 1975), pp. 49–74. See also her articles "Tendências Recentes da Negociação Coletiva no Brasil," *Dados* 24: 2 (1981): 161–90; and "O Sindicalismo Brasileiro entre a Conservação e a Mudança," in Bernardo Sorj and Maria Hermínia Tavares de Almeida (eds.), *Sociedade e Política no Brasil Pós-1964*, (Rio de Janeiro: Brasiliense, 1983).

54. Humphrey, *Capitalist Control and Workers' Struggle*, chap. 9.

55. Ibid., p. 242. We should note that in her later work, Maria Hermínia Tavares de Almeida takes a more nuanced view of the political role of the new unionism as well as of the Workers' Party, both of which take anti-statist positions and stress class organization. She recognizes the importance of this as a contribution to the democratization of social relations, but denies its ability to effectively pose the question of the democratization of the State. See Maria Hermínia Tavares de Almeida, "Novo Sindicalismo e Política," São Paulo, 1982, mimeo.

56. *Brazil Labour Report* (Oct.–Dec. 1984), 3.

57. Cited in Luis Roberto Serrano, "Em busca de definições," *Istoé*, 26 Aug. 1981. The IBGE estimates the total Economically Active Population in 1980 at 43,235,712. For an economist's view of unemployment, see Roberto Macedo, "A Dimensão Social da Crise," in Adroaldo Mouro da Silva et al., *FMI X Brasil: a armadilha da recessão* (São Paulo: Forum Gazeta Mercantil, 1983), 217–49.

58. *Boletim do DIEESE* 1, 1 (1982): 13.

59. Inflation figures from *Almanaque Abril* (São Paulo: Editora Abril, 1983, 1985).

60. *Boletim do DIEESE* (Dec. 1982), 14–16.

61. The CLT provides for the creation of CIPAs composed of representatives of employers and employees, but unless the election is announced workers have no way of participating effectively.

62. For a detailed study of demand making by the different metalworkers' unions in 1981–82, see Márcia de Paula Leite, "Revindicações Sociais dos Metalúrgicos," *Cadernos Cedec* 3 (1984).

63. Clarice Melamed Menezes and Ingrid Sarti, *CONCLAT 1981: a melhor expressão do movimento sindical brasileiro* (Rio de Janeiro: ILDES, 1982).

64. For a discussion of the development of the different tendencies within the labor movement during this period, see Menezes and Sarti, *CONCLAT 1981*.

65. For discussions of the 1980 metalworkers' strike, see M. Keck, "Brazil: Metalworkers' Strike," *NACLA Report on the Americas* (July–Aug. 1980), 42–44, and José Alvaro Moisés, *Lições de Liberdade e de Opressão* (Rio de Janeiro: Paz e Terra, 1982), 161–96.

66. For an account of the struggle at Ford, see José Carlos Aguiar Brito, *A Tomada da Ford: O nascimento de um sindicato livre* (Petrópolis: Vozes, 1983). The text of the agreement between Ford and the São Bernardo Metalworkers' Union with regard to the factory commission can be found in the *Boletim do DIEESE* (Feb. 1982), 14–24.

67. See Menezes and Sarti, *CONCLAT 1981*, 43–57. For good discussions from the time of the conference, see Luiz Roberto Serrano, "Em busca de definições," *Istoé*, 26 Aug. 1981, pp. 70–73; T. Canuto et al., "Falam os Trabalhadores," *Movimento* (31 Aug.–6 Sept. 1981), 11–14. After the CONCLAT, the National Pró-CUT Commission published a booklet called *Tudo sobre a CONCLAT* (São Paulo: CIDAS, 1981), which contains the conference resolutions and short interviews with leading figures.

68. CONTAG, "Porque Decidimos não Participar do Congresso da Classe Trabalhadora e Somos Pelo seu Adiamento para 1983," Document signed by the president of CONTAG and presidents of 20 Agricultural Union Federations.

69. For an example of the position of the *autêntico* tendency, see the pamphlet "CUT Pela Base" produced by ANAMPOS, June 1982. Minutes of the dissident meeting of a section of the Pró-CUT Commission held in São Bernardo do Campo on Aug. 28–29, 1982, taken by Maria Helena Moreira Alves, describe the debate which followed the non-attendance of those who favored postponement of the conference. At this meeting it was decided to participate in the Sept. 11–12 meeting of the Pró-CUT to be held at CONTAG headquarters in Brasília, at which a final decision would be made about calling the next CONCLAT.

70. Labor leaders estimated the number of strikers at 3,000,000 in all of Brazil. In the state of São Paulo, aside from the capital, there were significant stoppages in eighteen other cities; in Rio Grande do Sul, aside from Porto Alegre and Canoas the strike was significant in nine cities in the interior. There were also strikes in Pernambuco, Espírito Santo, Rio de Janeiro, Goiás, and Paraná. For more details, see *Boletim do DIEESE* (July 1983), 17–18.

71. Joaquimzão, as he is called, became president of the union when appointed by the military in 1964 to replace the purged union president. He has managed to win union elections since that time, though in the face of a growing opposition. The strength of the opposition and the changing political situation have forced him in recent years to work hard to shed his *pelego* image.

72. As already noted, Federation and Confederation elections, under the CLT system, take place on the basis of one union–one vote. Thus the vote of the São Paulo Metalworkers' Union, the largest union in Latin America, which represents over 300,000 workers, carries the same weight as that of a metalworkers' union with only a few hundred members.

73. Roque Aparecido da Silva, "Sindicato e Sociedade na Palavra dos Metalúrgicos," paper prepared for a seminar of the Comisión de Movimientos Laborales of the Consejo Latinoamericano de Ciencias Sociales (CLACSO), Santiago, Chile, May 20–23, 1985. Cited with author's permission.

74. On this new "unionism of results," see "O Inimigo é o governo (Entrevista: Luíz Antônio de Medeiros)," *Veja*, July 8, 1987; "O Inimigo é o Estado (Entrevista: Antônio Rogério Magri)," *Veja*, Aug. 5, 1987; and "Uma Greve-Lição," *Veja*, Aug. 26, 1987.

75. The degree of dependence on the union tax varied significantly. In their 1985 budgets, the tax represented 22% of income for the São Bernardo Metalworkers and 26% for the São Paulo Metalworkers. For the Construction Workers of Curitiba, on

the other hand, it represented 85% of income. In addition, the 20% of the tax which remains with the Ministry of Labor accounts for well over half of the budget of that ministry. See *Folha de São Paulo*, June 2, 1985, pp. 35–37.

76. For a more detailed discussion of the changes in labor legislation made in the new Constition, see Mârcia de Paula Leite and Roque Aparecido da Silva, "Os Trabalhadores na Constituinte," Instituto Latino Americano de Desenvolvimento Econômico e Social (ILDES), *Documento de Trabalho No. 1*, April 1988.

77. These are Ministry of Labor Data (based on preliminary estimates for November and December 1987) taken from IBASE, *Políticas Governamentais: Uma Análise Crítica*, Dec. 1987–Jan. 1988, p. 12 (erratum).

78. On the denial of conflict, see Roberto da Matta's essay "Você sabe com quem está falando," in his *Carnavais, Malandros e Herois: Para uma Sociologia do Dilema Brasileiro* (Rio de Janeiro: Zahar, 1981), esp. p. 141.

79. The impact of this stratification for democratic transition is discussed by Guillermo O'Donnell, "Y a mi, que me importa? Notas sobre sociabilidad y politica en Argentina y Brasil," *Estudios CEDES* (Nov. 1984), esp. pp. 25–45.

80. This alliance is one of the ironies of Brazilian politics, in that Joaquim was originally appointed to replace Alfonso Delelis, purged from the presidency of the São Paulo Metalworkers because of his PCB membership and his leadership position in the CGT in the 1960s.

81. The usage of "re-present" is taken from O'Donnell, "Y a mí, qué me importa?," 33.

PART IV

Democratic Disclosure and Praxis: Evolution and Future

9

Associated-Dependent Development and Democratic Theory*

FERNANDO HENRIQUE CARDOSO

This return to the theme of democracy—after so much discussion about authoritarianism—is neither a defense nor a revision. In 1971 when I wrote my essay for *Authoritarian Brazil*, "Associated-Dependent Development: Theoretical and Practical Implications," I did not see any possibility of the Brazilian regime's metamorphosis.[1] Indeed, many considered the regime's persistence an irreversible tendency due either to the structural situation of dependency or to an authoritarian vocation inherent in the historical formulation of Brazilian society. On a central point, however, my essays in *Authoritarian Brazil* and *Democratizing Brazil* are consistent: in neither volume is my emphasis on reaffirming dependency. Rather, my task in both essays has been to explore new economic patterns, new social formations and, especially in this volume, new political contradictions and possibilities that emerge within this late twentieth-century structural-historical novelty—associated-dependent development.

So the consequences, as the old Counselor Acácio says in the novels of Eça de Queiroz, always arrive later, but they do arrive. And, in the Brazilian case, they have arrived. In the last two decades, we have seen an unprecedented transformation. Today it is useless to debate the nature of the development which took place in order to speculate about up to what point there has been a transfer of the productive system from the Center to the Periphery. It happened, and so quickly and significantly that dependency

*Editor's Note: This essay was specifically written for inclusion in this volume in March 1983. Because the author is not only a major social scientist but was a major party leader of the democratic opposition I asked him not to update the article, but rather to leave the text as a document of the period. A few footnotes have been updated.

theory began to be subjected to a strong critique from the position that instead of dependency there was interdependence.

Naturally, for authors who confused dependency with stagnation, and development of the Periphery with the renewal of traditional imperialist links, the example of Brazilian industrialization is enough to knock down their poorly constructed theoretical house of cards. However, in my judgment nothing has shaken the foundational observations of Raúl Prebisch and the United Nations Economic Commission for Latin America (ECLA) about the "deterioration of the terms of trade," about the differential speed of the fall in the price of primary-export products relative to industrial products in a period of decline of the economic cycle, or on the inverse relation with regard to periods of expansion.[2]

Nor has the analysis of the new structure of the international productive system, which described the dynamic role of investment in the Periphery, been cast into doubt. Multinational firms have reproduced the asymmetrical link between the Center and the Periphery through control of technology (in the production of inventions) and of the financial system.

It is now obvious that economic development on the Periphery is real; it is not mere "economic growth" without redistribution of resources and without deep structural transformations. But the links of dependency are not broken, and we are not just dealing with a gigantic process of "interdependence." In other words, the process of *domination* among nation-states—mediated by renewed economic channels—persists in the international capitalist system, in spite of the internationalization of the productive process, and even though there has been significant transformation in the social structure of dependent countries and a significant increase in the internal productive capacity of some of these countries.

In light of these historical transformations and given the perspective of time, the old polemic in which my essays on dependency were critiqued on the claim that they substituted the primacy of the *nation* for the primacy of *social class* is reduced to what it always was: a fundamental theoretical and political error. My whole theoretical thrust was to demonstrate that dependency produces a specific class situation in which on the political plane the question of classes and their struggle is inseparable from the question of the nation and its political expression, the State. A theoretical perspective which emphasizes the autonomy of class is of little help in understanding the dynamics of associated-dependent societies. Conceptually, "autonomy of classes" only makes sense if it incorporates a prior understanding of the "double determination" of classes. The first determination of class is the productive system, which in this case is internationalized. The second determination of class are the forms of domination: internal, consubstantiated in the state; and external, exercised by the central countries and their international regimes, such as the IMF.

Given this analytic perspective, how do we explain historically the fact that an associated-dependent development process in Brazil has opened up a range of political possibilities which resulted in the weakening of the authoritarian order? How do we theoretically and ideologically advocate a democratic position in countries marked by their heterogeneity between and within classes (which many call structural), by the persistence of pockets of poverty, and by inequality?

As a theorist of dependency and as an active democratic politician I am often asked (by foreigners more than by Brazilians) how I reconcile dependency theory and democratic theory. Since I consider this one of the most urgent questions of our day I will put aside issues relating to the form, the nature, and the economic limits of the associated-dependent development process (while maintaining the concept as a necessary tool for describing Brazilian industrialization) in order to concentrate on this political *problématique*.

In the pages that follow I will confront this challenge by examining four questions. First: What is the structure of the new society that the economic transformations of associated-dependent development have wrought? Second: Given that I do indeed think there is a certain "elective affinity" between the structures produced by associated-dependent development and the centralization of power, how and why was there a departure from the expected in Brazil—the emergence of democratizing spheres within a new and strengthened civil society? Third: (I do not believe there can be a modern democracy without democractic political parties. This absolutely should not be an issue, although in practice it is occasionally forgotten. The difficult question for the democratic parties in an associated-dependent setting such as Brazil is): What should be the correct relationship between social movements and parties so that both can carry out their democratizing tasks? Fourth: (Different types of societies have mobilized different sorts of interests and ideological defenses for democracy in different structural-historical settings; thus): What are the political arguments and concomitant actions within our context of associated-dependent development that are both appropriate and possible as vehicles for mobilizing and consolidating support for democratic institutions?

I. The New Society

The widespread belief that Brazilian society, with rapid industrialization through multinational investment, would end up generating a "new duality" disappeared before the historical results. We have seen neither the crystallization of a Belgium of prosperity nor the ocean of misery and marginality of an India. We do, however, have many new structures in our society. A close examination of the data from 1950 to 1980 reveals that at least eight major new material realities related to our style of industrialization have emerged.

1. There was a sharp increase in the number of workers employed in the secondary sector of the economy. There were 2.9 million such workers in 1960, 5.3 million in 1970, and 10.7 million in 1980 (see Table 1).

Table 1. Brazil: Distribution of Economically Active Population (10 years or older), by Sector and Sub-Sector, 1950–1980

	1950		*1960*		*1970*		*1980*	
Sector	*Number*	*%*	*Number*	*%*	*Number*	*%*	*Number*	*%*
Primary	*10,252,839*	*59.9*	*12,276,908*	*54.0*	*3,087,521*	*44.3*	*13,109,415*	*29.9*
Secondary	*2,427,364*	*14.2*	*2,940,242*	*12.9*	*5,295,417*	*17.9*	*10,674,977*	*24.4*
Manufacturing Industry	1,608,309	9.4	1,954,187	8.6	3,241,861	11.0	6,858,598	15.7
Construction Industry	584,644	3.4	781,247	3.4	1,719,714	5.8	3,151,094	7.2
Other Industrial Activities	234,411	1.4	204,808	0.9	333,852	1.1	665,285	1.5
Tertiary	*4,437,159*	*25.9*	*7,532,878*	*33.1*	*11,174,276*	*37.8*	*20,012,371*	*45.7*
Distribution (Commerce and Transportation)	1,581,233	9.2	2,455,615	10.8	3,415,359	11.6	5,926,848	13.5
Services	1,781,041	10.4	3,028,933	13.3	3,925,001	13.3	7,089,709	16.2
Social Activities and Public Administration	911,317	5.3	1,467,947	6.4	2,683,904	9.0	4,857,061	11.1
Other Activities	163,568	1.0	580,383	2.6	1,150,012	2.9	2,138,753	4.9
TOTAL	*17,117,362*	(100.0)	*22,750,028*	(100.0)	*29,557,224*	(100.0)	*43,796,763*	(100.0)

Source: Data drawn from advanced tabulations of the Censo Demográfico de 1980, Fundação IBGE and presented in Vilmar Faria, "Desenvolvimento, urbanização e mudanças na estrutura do emprego: a experiência brasileira dos últimos trinta anos," in *Sociedade e Política no Brasil Pós-64*, ed. Bernardo Sorj and Maria Hermínia Tavares de Almeida (São Paulo: Editora Brasiliense, 1983), 146.

2. Industrialization was spatially concentrated in the Center-South, but this process did not take place without the appearance of "industrial spots" in the Northeast, such as the massive petro-chemical pole in Salvador, and some metal-working in Recife and Salvador. Even in the North there are major plans for extractive and transformative industries by such varied interests as state enterprises, Japanese capital and Alcoa.
3. The form which accumulation and investment took *did not reproduce* decades later either the "Prussian model" (of concentration of investment in basic industry) or the "American model" of incremental industrialism. What we actually have is a kind of development based on the combination of the form of "inverted industrialization" (which initially featured finished products of the "mass consumption" type, such as automobiles, televisions, and refrigerators, and was established through technology transfer in black boxes) plus a form of industrialization which involves domestic entrepreneurs in technical progress and opens up investments in capital goods sectors (which grew at an annual rate of 16% in the 1967–76 period), and, on a smaller scale, in the key sectors of the second wave of the technological frontier, such as computers and military exports.[3]
4. This type of industrialization set off a strong chain reaction between strictly industrial investments and investments in services, to such a point that the hypotheses regarding a malignantly swelling tertiary sector (considered in this case as "not modern" and as a simple expedient to disguise unemployment) become unsustainable faced with the productivity of the "modern tertiary" sector directly linked to the expansion of industrial products. What had led many analysts to make extrapolations predicting the exhaustion of the industrialization process was that from 1950–60 tertiary services grew at 5.4% while the secondary sector grew at only 1.9%. But 1960–80 saw a strikingly different pattern: the secondary sector grew at a rate of 6.7% while the tertiary sector grew at 5.0% (see Tabie 2).

Table 2. Brazil: Comparative Rates of Growth of the Economically Active Population, by Industrial Sector, 1940-1980

	Years				
Sector	*1940–1950*	*1950–1960*	*1960–1970*	*1970–1980*	*1940–1980*
Total	1.5	2.9	2.7	4.0	2.8
Primary	0.5	1.8	0.6	0.0	0.7
Secondary	4.7	1.9	6.1	7.3	5.0
Tertiary	2.4	5.4	4.0	6.0	4.4

Sources: José Serra, "Notas sobre o processo de industrialização no Brasil" (paper delivered before the Seminar on Industrialization and Development, Inter-American Development Bank, Washington, D.C., Dec. 1982); and Vilmar Faria, cited in Table 1.

5. There was also the rapid capitalization of agriculture (from 1960 to 1980 the number of farms with tractors increased by a factor of 9 while the number of farms that utilized chemical fertilizers increased by a factor of 17). This process was accompanied by a steep rise in the rural proletariat and the emergence of the "bóias-frias" phenomenon where many rural wage earners in fact live in squatter settlements on the periphery of urban cities and are trucked out to their work sites. This capitalization of agriculture has three axes of dynamism: a. investments in pioneer areas; b. capitalization of family-based property or productive units; and, c. the emergence of large agro-business firms in areas of traditional agriculture.[4]
6. The dynamism unleashed by Brazil's particular version of rapid associated-dependent development produced brutal income concentration with impressive trickle-down. In 1960 Brazil had one of the worst Gini indexes of inequality in the world (50% of the population received 17.7% of the income, and 1% of the population received 11.9% of the income). By 1980 the situation deteriorated further, such that the bottom half of the population received only 14.2% while the top 1% garnered 16.9%. At the same time, trickle-down reached such flood proportions that whereas in 1970 only a quarter of Brazil's households had refrigerators and televisions, by 1980 more than half did. The apparent paradox is resolved when we take into account that in 1970–80 real average income per family increased almost 89% for the country as a whole. Contradictions abound, of course. In 1980, 73% of urban households were linked to the world by television but only 58% had sewage link-ups (see Tables 3 and 4).
7. This whole process took place in the context of growing participation—from the beginning—of foreign investment, and later required growing international financial support. With soaring interest rates, the world recession, and the consequent decrease in exports and their value, the expansion of the "external debt" was once again a necessary condition to sustain local economic activity.[5]
8. In spite of growing links with the exterior, the *internal* market absorbed most of the productive expansion and the "coefficient of openness" of the Brazilian gross domestic product to world trade did not increase. In fact, if we exclude petroleum-based imports Brazil has one of the lowest ratios of imports to total Gross National Product of any major market economy in the world (see Table 5).

These then are some of the complex phenomena we must think about. Hirschman, with his heterodox talents, called attention to the unexpected in economic development: an airline, for example, could be better managed and more efficient in an underdeveloped country than a railroad.[6] Something like this happens in a generalized way with associated-dependent development. When we wait for the "inevitable" to happen (in general

Table 3. Brazil—Social Structural Indicators, 1950-1980

	1950	*1960*	*1970*	*1980*
General Indicators				
Population	51,944,000	70,992,000	94,509,000	121,080,000
Gross Domestic Product[a]	12,309,000	23,774,000	42,885,000	97,838,000
Income Concentration Indicators:				
% of GNP to Poorest 50% of Population		17.7	15.0	14.2
% of GNP to Richest 1% of Population		11.9	14.7	16.9
Gini Index of Inequality		0.499	0.565	0.580
Trickle-Down Indicators:				
Infant Mortality Rate[b]	130	119	98	86
Literacy Rate[c]	52.7	66.6	77.0	84.3
% of Households with Sewage:[d] Total			26.6	43.2
(Urban)			(44.2)	(58.1)
% of Households with Television: Total			24.1	56.1
(Urban)			(40.2)	(73.0)
% of Households with Refrigerators			26.1	50.4
% of Households with Automobiles			9.0	22.7
Total Number of Secondary Students Enrolled			1,007,600	2,812,416
Total Number of Undergraduate Students Enrolled			456,134	1,345,000

[a]Expressed in thousands of 1970 constant U.S. dollars. The 1980 figure presented is based on the application of a 6% growth rate to the 1979 GDP (of $92.3 billion). Source: *Statistical Abstract of Latin America* (Los Angeles: OCLA Center for Latin American Studies, 1982), 280.

[b]Expressed as the number of deaths in the first year of life, per 1000 live births.

[c]Refers to percentage of population 15 to 19 years of age; *Anuário Estatistica Brasileiro, 1982*, 246.

[d]Public network or septic tank.

Sources: Adapted from the Demographic Census of Brazil, 1960, 1970, and 1980. Also from papers presented by José Pastore, Thomas W. Merrick, Helga Hoffman, and Vilmar Faria at the conference on "Social Change in Brazil Since 1945," Columbia University, New York, Dec. 3–5, 1984.

conceived as a tendency extrapolated from the history of the early developer), the "unexpected" happens.

This "unexpected" is basically a specific effect of the combination of structural forces which fuse the "old" with the "new," often in a contradictory manner and without a guarantee that the contradiction will result in a new synthesis. Brazilian society at the moment is an incomplete synthesis, between one dynamic set in motion by the internationalization of the productive system, and another involving successive and not always successful accommodations by economic and social interests which preceded this process. The command of social transformation lies unequivocally with the internationalized sector of the economy, to the point that the word "sector"

Table 4. Personal Income Shares of the Poorest 20% of the Population and the Richest 10% of the Population, for Brazil and Nine Other Market Economies

Countries	*Poorest 20%*	*Richest 10%*
Brazil (1960)	3.9	39.6
Brazil (1970)	3.4	46.7
Brazil (1980)	2.8	50.9
India (1964–1965)	6.7	35.2
Honduras (1967)	2.3	50.0
Peru (1972)	1.9	42.9
Mexico (1977)	2.9	36.7
Chile (1968)	4.4	34.8
England (1973)	6.3	23.5
Japan (1969)	7.9	27.2
United States (1972)	4.5	26.6
Norway (1970)	6.3	22.2

Sources: *World Development Report, 1979.* Data for Brazil for 1960 and 1980 is from the Fundação IBGE, cited in the José Serra report referred to in Table 2.

is inappropriate, since the "economic whole" moves within the context of internationalization. The social effects of this, however, do not correspond to what we would expect on the basis of a mere "functional sociology of convergence," nor is it merely an adaptation to the resistance of the "old structure." Both terms of the contradiction behave, in part, like an unresolved tension, as something which is created in the Periphery, as an original trait.

In less abstract terms, associated-dependent industrialization is creating a specific kind of society. It is a copy, but to paraphrase myself, an original copy.[7] And being a copy, it is also a "desired and programmed" copy. We

Table 5. Indication of Openness of Economies: Non-Petroleum Imports as a Percentage of Gross National Product, 1975-1981

	Years						
Country	*1975*	*1976*	*1977*	*1978*	*1979*	*1980*	*1981*
Brazil	8.1	6.2	5.2	5.2	5.7	6.1	4.5
Argentina	5.5	5.2	7.7	5.8			
United States	5.1	5.7	6.1	6.7	6.8	6.9	6.6
Japan	7.4	7.4	6.6	5.6	7.3	8.0	7.5
Spain	12.1	12.1	11.0	9.6	9.6		
Mexico	7.6	6.8	7.4	7.5	9.2	10.7	
Australia	11.7	11.9	13.0	13.2	13.9	14.3	
France	13.0	15.0	15.2	14.6	15.6	16.6	
West Germany	15.0	16.6	16.7	16.4	17.5	18.6	19.2
Canada	20.3	19.1	20.0	21.6	23.7	22.9	23.2

Source: International Monetary Fund, in José Serra report cited in Table 2.

are not witnessing a phenomenon of "irradiation" of a "cultural circle," à la Kroeber. Nor are we in the presence of a social and cultural dynamism which is a given because of "transfer of technology." On the contrary, there is a domestic debate on "the good society," a domestic strategy for reaching it (seen differently by the competing forces), and a will to plan, through the choice of policies, the steps to be taken.

It is this mixture of the "inevitable effect" of industrialism, the choice of forms of insertion into the "new world," and the dead weight of the past which cannot be thrown away, which gives vitality, presents difficulties, and, at the same time, opens intriguing perspectives for audaciousness in the interpretation of Brazilian society.

Just as an example: It is impossible to understand the patterns of social and geographical mobility, the aspirations for living and the form of social control in force, without considering that TV and airplanes are a fundamental part of the "new society." But it is also precarious to understand this society without perceiving that in the full dynamic of *Gesellschaft*—in the industrial heart of class society, the ABC region of São Paulo—the "ethical-political" moment, as Gramsci would say, of assertion of choice by workers and the insertion of the new São Paulo working class, took place through the revitalization—momentary but significant—of *Gemeinschaft*. The solidarity of the commune—the transcendence of the everyday and the corporative confrontation—before going through the party, came through the Church and through fraternity in neighborhoods.[8]

To take one of the poles of the dichotomy and bet on it as if it were the expression of the *essence* of peripheral industrial society is to dissolve the dialectic which built it in a mechanistic manner. To maintain the relation between the two parts in permanent tension is to give up on understanding the next moment, of eventual synthesis. To believe that we know beforehand which of the two contrary poles will win out without seeing that there might just as well be an unexpected fusion or a momentary solution through the more "traditional" pole, is to introduce into the theoretical framework a philosophy of progress like that of the nineteenth century, which could leave the observer perplexed before an unexpected turn in history.

It is in this spirit, which at the same time seeks regularities and rejects models, but is willing to accept "structural fractures" which break the regularities, that we have to understand contemporary Brazilian society.

One more example must be given, and this is crucial: the presence of the State in the economy (to which I will return later). Superficial observers have decreed that the multinational firm, with its international dynamic, will replace the presence of the State in production. The state has been condemned, by these same criteria, to the role of figurehead for external interests: from the political arm of the oligarchy, it would be transformed into the militarized arm of foreign capital.

Fortunate mistake: State investment in industrial firms and services grew and the regulation of the economy by the State intensified, to the point where it produced a (false) reaction on the part of the local bourgeoisie itself, which came to see "Statism" as the root of its problems (which, in reality, when they existed, were due to the competition of multinationals and to the crisis).[9] By 1979, if we rank the size of non-financial enterprises in Brazil by net worth, twenty-eight of the thirty largest were located in the public sector. Moreover, in the mid-1970s, comparative data demonstrate that Brazil had one of the most profitable State enterprise sectors in the world. (See Tables 6 and 7.)

The presence of the State in the economy became so strong that a sort of structural inversion of the old distinction between State and civil society took place. This distinction, which underwent a radical reformulation in the Marxist tradition through the work of Gramsci, needs to be reformulated theoretically to explain the type of relation which exists in Brazilian society. In effect, Marx replaced, in the terms of Hegel but going beyond, the natural law distinction which opposed natural society to civil society (the State) with the dichotomy between political society (the State) and civil society (the social order, classes, producers); Gramsci returned to Hegel, opposing political society to the private order, embodying it, however, as something beyond mere economic relations. In placing the moment of hegemony at the level of civil society, Gramsci broke with the traditional framework of relations between infrastructure-superstructure (closer to Marx's thought), in which the State is a part of the second, while the primacy of the contradictions which lead to overcoming class domination is given by the first, that is, by the social relations of production. Gramsci began to dissolve this rigid distinction and to a certain extent reabsorbed the State into Society, through Hegel's "Ethical State."[10]

So in the situation of countries like Brazil, in which the State is an important part of the productive order, it is also necessary to break with the notion that links between the political and private order are lacking, suggested by the old opposition between State and civil society. But the rupture takes place with regard to the opposite pole. As suggested by Gramsci's analysis we must not only stress that hegemony is developed at the level of classes as a struggle *in society* but also show that the State, in becoming a "Producer State," becomes part of the economic order and *ipso facto* of civil society.

It is clear that all this cries out for new theoretical frameworks able to understand both the new society and the new politics. It also cries out for

Table 6. Brazil: Ownership Distribution of the Thirty Largest Non-Financial Firms, Selected Years, 1962-1979

Ownership	*1962*	*1967*	*1971*	*1974*	*1979*
Public	12	13	17	23	28
Private	18	17	13	7	2

Note: Firms are ranked according to net worth (*patrimônio liquido*).

Source: Thomas J. Trebat, *Brazil's State-Owned Enterprises: A Case Study of the State as Entrepreneur* (Cambridge: Cambridge Univ. Press, 1983), 59.

Table 7. Profitability Performance of Public Enterprises in Selected Countries

Country	*Year*	*Profitability*
Turkey	1968	2.6[a]–4.8[b]
	1972	6.2[b]
India	1972–73	3.8[c]
	1974–75	11.5[d]
	1976	2.0[e]
Korea	1972	High
Egypt	1972–77	10–15[f]
Brazil	1974	11.4[e]
	1978	7.9[e]
Argentina	1972–78	Low
Mexico	1972–74	−3.8[e]
Chile	1978	−3.12[e]
	1978[g]	−1.04[e]
Peru	1973	1.4[e]

[a]Profits as percent of capital stock.
[b]Pre-tax earnings + interest in percent of net worth.
[c]Overall rate of return.
[d]Gross profits in percent of capital employed.
[e]Net profits in percent of net worth.
[f]Current surplus, net of interest, depreciation, and rent, in percent of fixed assets plus inventories.
[g]Excluding CODELCO, the state-owned copper concern.
Source: Thomas J. Trebat, *Brazil's State-Owned Enterprises: A Case Study of the State as Entrepreneur* (Cambridge: Cambridge Univ. Press, 1983), 177.

another phenomenology of classes, beginning from what I call the specific *blend* of dependent-industrial societies, and thus freed from the schematism of analogies with the early developers. It would recognize that while industrialization quickly creates a proletariat, however much this proletariat grows in absolute numbers it will not be one more class position which becomes generalized by its size, given that the new society is both industrial—in the old way—and also a service and "programmed" society, as described by Touraine. Likewise, in the agrarian structure the "typically capitalist" sector produces rural workers, who, in relation to the peasant economy on the pioneer fringe and with the family economy which is becoming capitalized, are not fast becoming the dominant form of rural occupation. And in spite of the vigor of the capitalization process, the "informal sectors" of employment (and the "marginals") are not disappearing. They reproduce themselves, at the old rate and at the new rate: for example, the "informal sector" of luxury crafts, which arises from an eventual combination of more sophisticated methods with a "liberal" employment of labor (as in the case of computer programmers, for example).

Moreover, in the real sense of the term, broad petit-bourgeois layers are becoming wage earners in services and in factories; former "liberal professions" are becoming wage earners; and the modern "putting-out system" is bringing about the rebirth of false wage earners—independents—who constitute a regular labor force for big industry, dispersed in familial productive units.

Returning the analysis to the top of the social pyramid, other surprises interrupt the placid vision of a society which reproduces the advanced economic order. A few examples: the managers of State firms become a significant layer of the dominant class; the old national bourgeoisie does not disappear, but forms a layer between the State and multinational enterprise, trying to reserve for itself relatively important economic spaces; big foreign capital appears socially as an international bureaucracy, made up of professional administrators, with a strong "structural presence," but with enormous difficulty in becoming a class for itself at the level of local politics.

Thus we can see that the society which associated-dependent development wrought broke in significant aspects with the images which the sociological literature elaborated to describe the "effects of industrialization" and of capitalization of the Periphery.

In the face of this new situation, how could we imagine a politics which conforms to old paradigms?

II. Breaking with the Expected: Democratization

Once again, this part does not intend to describe the political process which took place; other authors and I myself, in other works, have already done so. It is primarily intended to point out paradoxes, and then to try to explain them.

The discussion about whether or not there was a "democratic opening" in Brazil is today a non-question. There was an opening, with various consequences. The "theorists"—to whom I never subscribed—of the inevitability of fascism (whose only alternative would be revolutionary socialism) lost prestige even in the most radical intellectual circles; the force of events buried badly formulated interpretive caprices.

Nonetheless, some problems need to be clarified, given that even those who did not support the point of view that social revolution was inevitable as the antithesis of military authoritarianism, affirm and reaffirm (as I do myself) that there is a certain "elective affinity" between the structures produced by associated-dependent development and the centralization of power. The concentration of income, oligopolistic investment, the breakup of trade union structures, the generalization of mass apathy induced by the central power, the control of information, and the most repugnant aspects of the authoritarian order (such as torture and the deprivation of citizenship) were understood as symptoms of the relation between this pattern of development and a particular form of bureaucratic-authoritarian regime.[11]

It is true, and worth pointing out, that some authors, most notably Juan Linz, restricted the extent of their characterization of Brazilian authoritarianism; this would be more an authoritarian situation than an authoritarian regime.[12] It is also true that I tried to show (when there were already signs of redemocratization) that the "political form" (the regime) is distinct from the "pact of domination" (an alliance of hegemonic classes) which gives the State its social base. And I argued that the same style of associated-dependent development was consistent with democratic regimes such as Venezuela and Mexico.[13]

Nonetheless, for many Brazilians one of the most puzzling aspects of our recent history is that without any breakdown of the State apparatus, without any loss of military coercive capacity, civil society grew stronger, and, most important, in a society with many layers of authoritarianism there nonetheless was a deepening, a quickening of democratic aspirations.

Why is our society breaking with authoritarianism? Broadly speaking, we can identify three explanatory/prescriptive schools of thought about Brazil's authoritarian-democratic dynamic. All three schools are heterogeneous but one has a *functionalist* core, one a *Statist* core, one a *grassroots* (*basismo*) core.

The first ad hoc "functionalist" attempt comes in the form of a paradox: beginning from two theories which note the anti-democratizing effects of the historical process of change in Peripheral societies—dependency and bureaucratic authoritarianism—they end up stressing the "unexpected" democratizing aspects of those processes. As a result, given that there was *development* and given that bureaucratic control expanded the State machine and absorbed the military into it, an unexpected "space of freedom" was produced in civil society, at the same time as there was an authoritarian condensation within the State. Between the two, State and civil society, there was a *gap*, a sort of buffer vacuum. Democracy therefore was sown on the virgin soil of society, leaving the State wrapped in its splendid authoritarian isolation.

There were various versions of this hypothesis. Some were inspired by "modernization theory": the democratic subproduct was the consequence of social differentiation provoked by economic development, by the growing specificity of social roles required by growing secularization and rationalization of society, and by the need for standardization of norms appropriate for a modern industrial society.[14]

To a certain extent, almost all the authors in this school have shaped their work around this broad hypothesis, because there were in fact convergent processes which encouraged the re-elaboration of aspirations, conduct, and regular patterns of behavior commensurate with the universal aspects of industrialization. In an attempt to explain the return to democratic practices on the electoral level, for example, more than a few stressed that the democratic form of regulating the distribution of power would necessarily re-establish the competitive system of parties and representative mechanisms characteristic of "any democracy." Allied to the modernization theory, liberal political theory again resounded not only in the major press but also in academic texts and even in party programs.

The "theory of the gap" (between State and civil society) leaves us without a solution, however, to the central question: if it is true that the subproduct of socio-economic development is the demand for the autonomy of the social, of political representation of classes, and of a liberal-democratic creed, political change is moving towards an impasse: either the citadel of the State is conquered by the furor of classes demanding democratic power and becomes democratized, or, in a counter-offensive, the State advances still more in the direction of authoritarian processes, getting close to a situation of authoritarian control of society (in a version of the political process which is close to "socialism or fascism" vision, but rewritten as "democracy or dictatorship"). But, by postulating this kind of dynamic, what was clear in Marxist theory about civil society is hidden—that is, that it is ruled by *domination*—and that—taken by themselves alone—civil society and democracy have nothing to do with each other as such, given that the democratization of society requires struggles among competing classes and the overcoming of the contradictions between the exploited and the exploiters.

The hypotheses about the development of an opposition between society and the State outlined above are not the only ones sustainable from the point of view of a functionalist and liberal theory of democratic politics. Nothing prevents the conflict from remaining unresolved, with advances and retreats by the two opposed poles, or with a dialectic where conciliation of interests in specific areas and open conflicts in others are both possible. Such a hypothesis presumes slow and gradual transformation towards democratization based on the assumption of the effect of the spread of universalizing mechanisms of industrial society.

Within the Statist school—especially strong among the regime's ideologues—a version of the same process was developed from another angle. Rejecting the pervasive effects of the renaissance of civil society, these ideologues returned to Oliveira Vianna, Azevedo Amaral, and, against the express thinking of the author, Raymundo Faoro. They went back to seeing the modernization of the State as the main guarantee for a process of political opening and democratization which would escape the "pitfalls of liberalism." In this case, the emphasis was not so much on the gap between society and the State, but on the "emptiness" of society. They continued to interpret the country as if the only possible framework for a stable political order was one built upon the efficiency of the State machine, coupled with developmentalist policies and guided by an enlightened will which should moderate and conciliate, whenever pressure arose from "reasonable" private groups. The legitimacy of demands "from below" (and everything with roots in civil society, be it business, the press, or the Church, and not only the popular sectors, was " from below") would always pass through the filter of State will, the ethical-political collector of an unorganized people on the road to the constitution of the nation.

Obviously, the modernization of society by the State and the making of the people into a nation thought of in these terms is not part of "democratic theory." Nonetheless, as ambiguous and confusing as this Statist version of authoritarian ideology is, it is also "liberal-conservative." It postulates a series of transformations and a gradualism which do not in principle negate the legitimacy of a demand for a State of Law: they only postpone and attempt to conduct it on the "good road" to a "democracy without conflicts." But the logic of such an argument obliges its advocates to accept, as a thesis and as a principle, that a "good government," to be democratic, must break with authoritarianism.

The grassroots version of why Brazil is breaking with authoritarianism combines a radical vision of autonomy of civil society with a socialist critique of social domination. This version stresses the same process of constitution and autonomization of classes (an emphasis it has in common with the liberal-functionalist theory of political change). But it also views as essential the way in which the new capitalist-industrialist order, on the one hand, maintains class differences and, on the other, disaggregates those without class, the "poor," the "marginals," the "inhabitants of the periphery," or whatever name is given to the disinherited of the capitalist order. *Real* democratization will arrive (and is arriving, according to those who hold this perspective) as it is crystallized in the spontaneous solidarity of the disinherited. It lives as *comunitas*, experiences of common hardship which form a collective *we* based on the same life experience that is transformed only when, through molecular changes, the simultaneous isolation of the State and the exploiters—which will perish at the same time—comes about.

To those accustomed to Marxist literature these might sound like the old themes of Revolution and fusion between solidarity, equality, and democratic participation. But this is true only up to a point. In their most radical version (common among social movement activists) this reformist version which proposes the New Utopia is accompanied by a rejection of the State so strong (in theory) that it also excludes the party, which is seen as an institutionalizing force, and thus a cog of the State. In the radical formulation of this type of democratic theory there is a fusion of lay anarchism and Catholic solidarity thought.

Underlying these three schools (which simultaneously prescribe ideologies favorable to different types of democratization) there are explicit differences not only as to what "real" democracy is but also, and more important sociologically speaking, as to who are the "historical subjects" of the desired democratization. The distinction at the level of democratic ideas is simple:

— For functionalist liberal democrats (as anywhere else in the world) it involves establishing a competitive regime, which accepts differences of wealth and property (of classes), but which claims to identify the possibility of a Common Good (the Public Spirit) which would be exercised and controlled in particular spheres of the State (Legislative, Executive, and Judiciary) through explicit mechanisms of *representation* and *legitimacy* (delegation of powers, elections, etc.) which ensure the sovereignty of the citizen as "political being" *par excellence*, the individual as the subject of history.

— For Statist liberal-conservatives, the problem is to rebuild a political order founded upon the idea of a Public Good located in the Executive, whose excesses must be controlled on the one hand by the will of the Nation and by its "permanent aspirations" as perceived, expressed, and renewed by a privileged sector of the bureaucracy (the security apparatus), and on the other hand, by the existence of certain channels of representation (parties and assemblies) with the right to speak but not to act on the major decisions of the State. (In addition to this—as a concession—some of them add freedom of the press and, up to a point, freedom of organization in civil society (parties, trade unions, the Church) which, though they are controlled, exercise pressure and indirectly allow the guardians of order to correct the course of their policies.) The true subject of the political process is the State as enlightened bureaucracy, considered the incarnation of the metaphysical will of the people.

— For the grassroots democrats, the fundamental question is the autonomous organization of the population around concrete demands—almost always within the reach of and with direct consequences for the well-being of deprived groups of people. These demands should be made on the Public Authority without the ostensible mediation of parties and, if possible, without delegation of responsibility to elected representatives. The general will, in this case, is presented as the incarnation of a partiality which in its totality expresses a goal or a desire. The subject of the political process becomes a "living community": neighbors, workers in the *same* factory, landless tenants who measure their aspirations around the control of an area, etc. More than the rather abstract solidarity of a "class," what is needed is the solidarity of a professional "branch" or a specific segment of the people to give substance to the demand for democracy.

This typology of the kind of democratization desired and the actors expected to bring it about demonstrates that, with the exception of the liberal-democratic school (similar to what predominated in the democratization processes of early developing countries), the other schools which are most evident in the contemporary Brazilian process are, at the very least, "heterodox." One side comes from reformed authoritarianism; the other from Christian solidarity thought to be penetrated by anti-statist anarchism.

It would be wrong to imagine that the spectrum of democratizing pressures has been limited to these. The classic socialist vision, with all of its considerations about the relationship between social revolution and true democracy, is just below the surface among followers of parties which are small, but capable, at times, of strategic action: the (pro-Soviet) Communist Party, the (pro-Albania) Communist Party of Brazil, and various Trotskyist groups, all of which maintain the classical set of ideas and emphasize the "historical role" of the proletariat in the advance towards democracy. But either they attached themselves to the liberal-democratic view, adjusting their revolutionary aspirations to a "stage" which would follow a full State of Law, or they joined (in the case of the Trotskyists and the Albanians) with the grassroots pressures for a more direct democracy.

In addition, in the concrete process of political action, when these tendencies came together in the creation of parties (especially after 1979) and in the action of specific social movements (as in the unions and the movements in favor of a Central Workers' Organization (*Central Única dos Trabalhadores*), *basismo* was rarely immune to positions which supported the decisive political role of the oppressed under the leadership of the unions and the "working class" organized into a party; likewise, in the parties most influenced by an ideal of Western competitive democracy (like the PMDB and the PDT), the "Marxist-Leninist" ideological segments were sufficiently influential to give value to the idea that the "active presence of workers" was necessary for an effective redemocratization, thus compensating for the limitations of the liberal-democratic vision, which is more lenient about living with social inequality.

The reader versed in the history of political ideas will see in these distinct positions the debate between Locke, Hegel, and Rousseau, as (unwitting) inspiration of the existing polarizations. But the reader will also see that Brazilian liberals are willing to live with the presence of the State. The authoritarians with a liberal-conservative thrust are sprinkled with ideas not only from Montesquieu (each regime conforms to the nature of society, and in an industrial society there is a certain division and balance among powers), but from contradictory influences which go from the acceptance of a certain kind of planning to distorted Hobbesian formulations to authoritarianism. And the radically democratic, however much they make *basista* statements in favor of "direct democracy," incorporate a pinch of Gramsci, even mixing in a certain amount of Leninism, and cannot easily avoid the concern with the Party and the State.

But the society in which the political process is evolving has very little in common with the societies and the problems confronted by the classical theorists. This is the point where my perspective on "dependency" crosses with the debate about democracy.

New Ideas?

It would be strange if authors linked to a tradition of structural-historical analysis, faced with authoritarianism and the process of its transformation into a more liberalized order, critiqued authoritarianism and defended democracy from a Lockean vision of the two freedoms—the economic and the political—both foundations of philosophical individualism. Likewise, it would be inconsistent if the justification came from the side of philosophical utilitarianism, from Bentham and the idea of optimization of opportunities in the "political market."

However, there are other branches in the history of ideas and, moreover, in the history of socio-political practice, which can help to ground and justify theoretically and ideologically the defense of the "sphere of democracy." I called attention to the real (not epistemological) inversion of the relation between State and society, which characterizes the countries whose development takes place in the associated-dependent form (and which also occurs in many "advanced democratic" societies). Making use of other arguments, various sociologists and political scientists have proceeded in a similar way. For example, the notion of "regulated citizenship" developed by Wanderley Guilherme dos Santos tries to show the *similarity* and the *difference* when we compare the process of formation of citizenship à la Marshall and the process which takes place in Brazil. A citizen, among us, would be the worker whose right is *recognized* by the State.[15]

The weight of the entrepreneurial bureaucracy (what I have elsewhere called the "state bourgeoisie" in order to provoke reactions) is abundantly recognized in political analyses of contemporary Brazil.[16] And the key role of the National Intelligence Service (SNI) as a "party" of the armed forces and the upper bureaucracy was pointed out by Alfred Stepan.[17]

In other works I called attention both to the political form of the relation between the bureaucracy and the entrepreneurs (the bureaucratic rings) and to the fact that, unlike what happens in classical political philosophy, among us the subject to be constituted and justified is not the State, but the citizen and the class. All this means that implicitly and sometimes explicitly there is another paradigm of political analysis in process of elaboration.[18]

As before, I will cite a few examples to illustrate the argument. One example will relate to the formation of parties and the party system; another will relate to practical-epistemological aspects of the legitimation of democratic world view in process.

III. Parties Today

At the most abstract level, the discussion of the specificity of parties was already posed by dependency theory: the birds which sing here do not sing as they do there. But they are birds, and they also sing. They seem the same, but there are differences. Through this prism, the polarization between Conservative party and Liberal party in the Empire would seem at the same time like a fundamental dichotomizing core of political ideologies, a shadow cast by a stronger and more real radiating center—from the central countries—and almost mystified. An educated and intellectually nationless elite seemed to be using the struggle over *ideas* rather than the struggle over *interests* as its banners. In fact, under the clamor of parliamentary debates was the solid presence of plantations and slavery.

Meanwhile, at the same time as an engaged demystification should be able to show the extent to which the ideological prism was produced by the refraction of interests, it would have to show that the conservative and liberal visions also shaped decisions and adjustments which had an effect on the real. Thus we are not dealing with "mere alienation."

I think that from this angle today's party system suffers from similar vicissitudes. The ideological aspect, which was clearer in the polarization between ARENA and MDB, was real, however much on another level it deformed differences of interest, diluted distinctions, and encompassed concealed social accords. The current spectrum, which goes from the PDS to the PT, passing through the PTB, the PDT, and the PMDB, makes the differences more visible. It almost satisfies those who are always looking for an ethical Cartesianism which links the notion of "clear and distinct" with a categorical imperative: if someone is a worker he/she has to belong to such and such a party, and this will bring about the good of the universe. The party which represents this side will be able to establish a State of Virtue, which will be dissolved into the Community of the Future; if someone is a plantation owner, or a boss, he must purge himself of original sin, accepting responsibility for all the evils of society, and will become a prop for the founding violence of any State, because it is rooted in class society.

Moreover, as in Old Europe, each party would correspond, more or less, to a class situation, and if there be internal contradictions, they would correspond to an ideological nuance which expresses a class fraction. On the formal level, a center, a center right, and a right would confront a center left and a left. Insofar as the real parties do not fit the model, they are not authentic, are weak as organizers of social interests, are mere instruments of manipulation by the forces of order, and are fortified in the State, which rules by dispensing representation and democratic legitimacy.

Let us leave aside the discussion of whether the model worked in the paradigmatic historical situation, Europe. Let us not get lost in side discussions on the changes which took place in mass industrial society with the influx of internationalized monopoly capitalism and the action of the interventionist State. Let us assume that none of this affects the argument of the prospectors for party authenticity. Still, why should a dependent society, penetrated by international capitalism from head to foot, born from a colonial-capitalist-modern situation, based on slavery, organized around a State bureaucracy, end up with a class and political situation similar to the one which prevailed in societies organized by bourgeois-liberal and sometimes bourgeois-State dynamics, struggling on the one hand against the *ancien régime* and on the other against the plebes and the emerging working class? Only if the argument were anchored in a finalistic philosophy of history and overlooked the differences among historical situations could one expect that in Brazil the parties, the classes, and their struggles would take place in the image of the Single Mold of History.

Clearly there are differences. And it is not so easy to discard even the ideological arguments, à la Raymond Aron, about the equalizing effects of modern society. But the conception of the modern political party as "an organizing machine and a structured and articulated political program," like the one proposed by Umberto Cerroni, describes a historical situation and not History.[19] Cerroni opposes the notion of a party as "faction," born out of the electoral committee or club, which is discharged in Parliament. He takes as a model the European socialist (and Communist) parties, in which there is mass participation by members, and in which the organizational structure and a certain conception of the world (a political philosophy) constitute the vital core which animates them. However, as Cerroni says himself, the Social Democratic (Labor and SPD) and North American parties depart from this model insofar as they stress economic-corporative interests and separate these from the struggle for a conception of the world; "political operations" predominate in them over the overall political conception.

With this argument I am naturally reducing the extent of the classical (Marxist-Leninist) definition of the party: it is not always the expression of a class interest together with a transformational idea, a world view. But I do not want to eliminate the idea, which also comes from Cerroni, or from Gramsci, that whatever its form, the party is the place of *mediation between idea and interest*, the Gordian knot of all politics. Rather, I want to say that this mediation takes historically variable forms, of which the idea of the party as "embryo of a state structure" and therefore revolutionary because in conflict with the prevailing state structure, is the result of a historically specific situation and not the matrix of all *essential* definitions of the party.

In Brazil the parties were not born only in Parliament, nor were they the expression of an organizational machine which was set up to allow the mass membership of militants. Still less were their struggles and differences based on a global conception of the world.

Literally and paradoxically, the arbitrary legislation of an authoritarian military regime had a crucial role (in spite of its intentions) in the formation of Brazilian parties. It was to obey legal dictates that the PMDB and the PDS, the PTB and the PDT, and even the PT had to emerge from their shells made of agreements among Congressmen and of party leaderships which did not even have a bureaucratic structure because there was not a real party machine to be controlled.

Under the authoritarian regime, the MDB and ARENA were limited to expressing the will of the electorate in the choice of Congressmen and Senators—a will distorted by apathy, lack of information, and violence. The Congressmen, restricted by authoritarian law, opted between two parties whose leaders were also Congressmen. While there was not a "world view," there was a concrete opposition of ideas as to the form of government: democracy versus authoritarianism. Interests followed behind this dichotomy, without necessarily aligning themselves unequivocally with one "side" or another. But there was no articulated correspondence between society and the State, in which, if we consider the legislative branch to be a part, and a weak part, lay the roots of ARENA and the MDB.

The job of *articulation* within society (with its interests and its culture) was being carried out with difficulty during the years of authoritarianism. In Brazilian political language, everything which was an organized fragment which escaped the immediate control of the authoritarian order was being designated *civil society*. Not rigorously but effectively, the whole opposition—from the Church, the press, the university, the professional corporations, to the unions, business, and the parties—was being described as if it were the movement of civil society. And it was discovered, without anyone having called it scandalous, that what was happening here was the reverse of what Gramsci described with relation to Italy: we were returning to a Latin conception of civil society.[20]

Since this is not primarily a theoretical discussion, I will leave aside the (necessary) polemic on the relevance and limits of the Gramscian paradigm for the analysis of Brazilian politics. In his polemic against economism (and against the literal reading of the primacy of structure), Gramsci reintroduced the primacy of the party, just as in his struggle against the *dictatorship* he reintroduced the moment of the idea, within the discussion of hegemony. In this respect he was innovative and left a living heritage; but his re-elaboration does not help to describe the historical situation we are facing. The notions of the moment of hegemony and of the germination of liberty in civil society (no longer thought of as "natural society") are necessary and useful; but we must bear in mind that the boundaries of the old natural law distinction and also the Hegelian opposition between producers and the State have been blurred. The State produces, regulates economic relations, and is a key part of manufacturing and service society. There is a new amalgam, in which hegemony, the moment of freedom (ideas, intellectuals, the major regulative institutions) cannot be considered separately from political society. There is a pan-politization of the social and a socialization of the State, as Pietro Ingrao has noted.[21]

Thus when a new moment of politization erupted in Brazil, the parties were born *at the same time as* a state form; and, as an instrument or the organization of struggle, of classes, and of ideas, including anti-Statist ideas.

In this situation, some imagined that the party-form, to be *authentic*, had to incorporate the "social movements" and could be the incarnation of the liberating idea. And many, in their evaluations of the question of representatives, refused outright to consider that parties which were not born directly from "social movements" and which perhaps did not even aspire to include them and represent them were in fact parties.

But why not? If the new industrial society interpenetrates State and society, the real question about parties and their representatives does not lie in the polarization between the "society of producers" and the society of administrators and collaborators. It lies in the capacity (or not) of the parties to build movable bridges on both sides of the antimony, like the famous "forward and backward linkages" of economic development.

As spaces of mediation between interests and ideas, between cooptation, compulsion, and hegemony, between institutionalization and becoming, between administration, domination, and rebellion, contemporary parties are necessarily internally contradictory. Their capacity for articulation is always tenuous: the big corporative organizations (the trade union, business, the Church) do not dissolve into parties; on the contrary, the parties run the risk of being absorbed by them, as in the example of the Labor party. Nor are movements within society—the strike, the occupation of urban land, riots, the trade union and student movements, the Press itself—anchored in parties. The trajectory of parties and these movements may coincide; sometimes they will even be wedded, but soon after the wedding night will come nausea if not divorce. And they will have to propose unceasingly and routinely new adventures, whose result will be close to that of Bernarda Alba. Weak substitutes for the old parties, the nostalgic will say. *Too bad.* These are the parties we have, not those we want. Within this limitation, they carry on relevant political functions.

Returning to the factual: this was how, without many illusions, the parties we have today were formed, almost all of them. And in their eagerness to win a bit of power, they formed some relation with society.

I will not give many examples. Beginning with the PT (which in this respect is more like the old parties) it is undeniable that it built bridges in determined social sectors: the factory workers of ABCD, middle-class intellectuals, segments of the common people influenced by the liberationist Church. It did not capture the labor movement; it did not become nationally organized; it is penetrated by sectors of the "organized left." None of this in my view diminishes the fact that if it will not become the big mass party under the hegemony of workers, it will be, nonetheless, a party of sectors of workers and intellectual sectors able to propose an alternative society. There will be imperfections and contradictions in the proposal. But who does not have these?

If the thesis is clearest when we use the PT as an example, it remains weak: Wasn't the PT the party with the poorest electoral performance? There are two alternatives from which to choose: either this reference shows that the party closest (in its proposals) to the paradigm of authentic representation, by not doing well electorally, condemns the others to be the counter-proof of inauthenticity, or, on the contrary, one can argue that the PT *in spite of* its enthusiasm for direct representation, managed to become organized in the Brazilian political system.

I do not want to go deeper into these hypotheses. But I would say that it was more *in spite of*: the PT cut deep into those sectors (real and important, especially in terms of political renovation) sensitive to ideology. And it remained restricted, in terms of a more refracted, but diffused, style of representation, which was organized in view of the simultaneously *eventual* and *central* polarities of mass society.

In the ways that a party like the PT is weak as an expression of the collective will, the PMDB is strong. Observations that the party seems more like a "front" than a party miss the point: in mass societies democratic parties which are open to social variation are, in a certain sense, fronts. But they are nonetheless parties, on the condition that they take positions on the major questions, are diffuse but capable of producing a political cleavage which presents the voter with an *option.* And also on the condition that they have leadership able to promote simultaneously internal bargaining among the wings of the party (sometimes coopting, sometimes effectively opening participatory and expressive space) and demonstrating a symbolic consistency with regard to the major national questions.

Obviously, neither is the PT limited to the functions mentioned, nor is the PMDB organically disconnected from its base. I exaggerate the argument to stress differences. Suffice it to remember that there has *never* been a party in São Paulo with so many spontaneous members and so completely structured in *diretórios* (local party organizations) as the PMDB.[22] It is enough to look to the professions of the people in leadership positions within the party to verify that the PMDB is the political outlet of the middle class and of leaders of the popular sector: doctors, union lawyers, bank workers, teachers, social workers share power in the *diretórios* (grabbing the hegemonic positions) with the union leader from the interior, the president of a rural union, and the neighborhood leader.

The PDS itself, party born out of the clientelism of ARENA, political arm of the authoritarian bureaucracy, renewed itself partially and became somewhat more autonomous. This should not be denied. Today there are new leaders who are conservative but not immobilist, and who are having a certain amount of impact in the PDS. The 1982 victory of the party in the gubernatorial elections in certain states would not have been possible had it not been for the combination of official pressure, financial resources available, *and* the modernization of leadership.

With more trouble, the labor parties try to establish a profile of their own. Not so much the PTB, which, jolted by the personalism and inconsistent leadership of Jânio Quadros and Sandra Cavalcanti, is prevented from being more than a "front of individuals" and risks becoming only a screen for governmental interest, like the PDT. The latter, trying to emerge as the force representing a "social democratic" party, ran into a major obstacle: its electoral strength derives from a personal leadership (Leonel Brizola) ballasted in a poly-class movement which rebuffs an alliance without a future in a state (Rio de Janeiro) where the working-class base is small. In São Paulo, where the working class counts, the PT and PMDB occupy the space which a social democratic party would want to occupy.

There is not room in this essay to do more than refer to these facts. I do not want to analyze each party, but rather to counter the widespread idea that the parties are "inauthentic" and incapable of serving as a filter for the aspirations of the electorate.[23]

Withal, I do not negate that party institutionalization is far from complete and that there are serious problems of representativeness not only at the level of each party but at the level of the party system. Duverger called attention to the importance of electoral and party legislation in the institutional crystallization of political regimes. It is striking, in the Brazilian case, the degree to which the current legislation—arbitrary and conducive to the maintenance of interests in power since 1964—is an obstacle to democratization. The PP skidded and disappeared because of this legislation. The PT runs a similar risk if the district vote is established without a system of two turns. And the labor parties, if they do not unite, have little prospect of survival.

Even worse, the electoral system distorts the popular will in an alienating manner. The opposition won an eight-million-vote advantage and it made very little difference in the House of Representatives. Not to mention the Electoral College which will choose the President of the Republic, if things remain the way they are. But we should not confuse the disturbing action of authoritarianism perpetuated in this legislation with the incapacity of civil society to organize itself into parties.

If parties are not what they used to be, it is because they are more attuned to another type of society. Even if the electoral legislation is changed to allow democratization to advance, the parties will continue to be only partial instruments of the popular and national will, will contain very different wings (not to mention the regional differences within each party), will be in permanent and insoluble tension with the social movements and with the renewing eruptions of mass society, and will experience creatively (or not, depending on the leadership and the circumstances) the dialectic between front and party, between the function of interest aggregation and the ideological function.

One final comment regarding parties: if I tried to show that there is no reason to demand that the parties adjust to a classical European paradigm (on its deathbed there since the 1950s), it would also be naive to imagine that the future of the Brazilian party system lies in the direction of a United States-type division between Republican and Democratic parties. In Brazil, after the populist siege in which the masses erupted in the State (and stagnated there), social and regional contradictions and the traditions of absorption of intellectuals into the parties would fortify the "ideological nuclei" in each party. This is to say nothing of the presence of Communists and, to a lesser extent, the socialists, which would be a constant.

It is therefore better to keep an open mind about the future of the party system. We are not condemned by any structural law to a two-party system (in spite of the bipolarizing tendency characteristic of contemporary societies), nor to politics without ideology. We will construct a peculiar blend, in which the meeting between a European historical tradition and a society which is remaking itself, beginning from a colonial-slaveowning-exporting base and moving towards an industrialized and service society (but located at the periphery of the capitalist system), will deflect any tendency to convergence with Western societies. We will go in unexpected, but not inexplicable, directions.

IV. The Legitimation of the Democratic Idea

There remain a few comments to make on the utopian-theoretical-ideological foundations of the idea of democracy in a mass society in a country with an associated-dependent economy.

It is evident that "possessive individualism" and the idea of citizen-property owner as the basis for democracy is a weak basis for justifying the democratic struggle in an associated-dependent society such as Brazil.[24] What is at stake today is not the "freedom of the individual" versus the totalitarianism of the State. The subject of individual freedom (psychological, physical, political) is naturally an integral part of a political process which follows upon a struggle against a military dictatorship which oppressed and tortured. But the social inequality and the fragility of the individual before business and the bureaucracy calls for the legitimation of a "collective" historical subject—that is, the union, the community, the movement, and even the party—which appear as actors in the making to oppose themselves to arbitrariness and exploitation.

This non-individualist foundation of embryonic democratic doctrine is difficult to create and pays a price for its legitimation. Its emergence, however, is detectable even on the level of vocabulary and semantics: for the old *I* of the leader, today is substituted *a gente*—an expression that was rarely used in past decades and which expresses something similar to the *on* of French—the indeterminate subject—expanded with the concrete sense of "those present." To the extent that the democratizing demand today comes drenched with this character, destabilizing politico-institutional consequences are nourished.

As a result, the classical theory of delegation and representation, closely linked to the conception of citizen-elector (individual and rational being), is put in check, often through the strength of the "collective we," the only thing capable of legitimizing a general will which is becoming concrete. The result of this attitude is transparent: difficulty with or even horror at the delegation of power and the designation of leadership. This process was visible, for example, in the strikes, especially in those job categories, such as public employees and teachers, where "assembly-ism" put in check the process of "bargaining through representatives."

This radically democratic and collectivist attitude produces mistrust of members of Congress by the masses, not to mention the already discussed "gap" between mass society and the State.

Nonetheless, it would be wrong not to recognize that in spite of the problems that such values pose for institutionalization of democratic life, they have a positive side. They demonstrate the emergence of a will to renewal on the part of civil society which rejects the notion that the "political opening" remains at the level of a *re*-democratization, based on liberal-individualist principles which in the past safeguarded social injustice, class inequality, and traditional bourgeois domination.

There is also renewal on the opposite axis of thought about redemocratization. I am referring to the assimilation, in the sphere of participation and control of the State apparatus, of tendencies in the countries of European "advanced democracy."

If, on the one hand, the *basista* thrust and the constitution of a collective-popular subject so as to support a new historical subject of democracy breaks the confining bonds of past institutional forms, on the other hand the reform-democratic thrust which accepts the contemporary reality of the pervasiveness of the State breaks the illusions about the possibility of a democracy "of civil society."

There has not been enough progress in this regard, in terms of political movement and reform-democratic ideas, but we are going forward. There is an embryonic democratic thought which is not restricted to accepting the party-parliamentary game (although it remains a fundamental part, just as the defense of the dignity of the person and his or her rights remains fundamental to democratic collectivism) as a form of justifying the democratic world view. Without greater transparency of information and of the decision-making process in the firm (whether private or State) and in the bureaucracy (*idem*, *ibidem*), and without evolving mechanisms for participation and control both through parties and directly by the interested publics, the democratization process will be crippled and meet with little reception in a society in which the "private," in the strict sense of the word, is weak in relation to the organized, corporate, and State interest.

I do not believe that these ideas are rigorously "new." But their combination and especially their diffusion in Brazil are in fact new. I would not say that redemocratization, with the characteristics which it begins to show, can occur without there being at the same time a clash between an industrializing and urbanizing Brazil and the archaic set of practices and notions associated with the Authoritarian State. However, I would not say either that the current process of redemocratization was the "expected effect" of the general processes of social change to which I referred. At the intersection between unprogrammed "structural changes" and authoritarian practices sustained by groups in power, there were specific social struggles, universal currents of opinion which converged, leaders and political-organizational forms which became active, and "unexpected effects" all mixed together.[25]

The political process under way is the result—sometimes planned, sometimes imposed, sometimes remade by social and political struggles—of all this. Thus, certainly, it could not have been "any other"; but on the other hand, it did not have to be "this one." And in the future, perhaps it will be "another."

NOTES

1. "Associated-Dependent Development: Theoretical and Practical Implications," in *Authoritarian Brazil: Origins, Policies, and Future*, ed. Alfred Stepan (New Haven and London: Yale Univ. Press, 1973), 142–78.

2. See Fernando Henrique Cardoso and José Serra, "As desventuras da dialética da dependência," *Estudos CEBRAP* 23 (1979): 33–80; and Raúl Prebisch, "El desarrollo económico de la América Latina y algunos de sus principales problemas," *Boletín Económico de América Latina* 7 (1962): 1–24.

3. For the capital goods growth rate see page 145 of the article by Vilmar Faria cited in Table 1.

4. The data of tractors and fertilizers are from the interesting article by Bernardo Sorj and John Wilkinson, "Processos sociais e formas de produção na agricultura

brasileira," in Bernardo Sorj and Maria Hermínia Tavares de Almeida (eds.), *Sociedade e Política No Brasil Pós-64* (São Paulo: Editora Brasiliense, 1983), 188–89. Also see J.R.B. Lopes, *Do latifundio à empresa: unidade e diversidade do capitalismo no campo* (Petrópolis: Editora Vozes, 1981). Série Cadernos CEBRAP no. 26; Vinícius Caldeira Brant, "Do colono ao bóia-fria: transformações na agricultura e constituição do mercado de trabalho na alta Sorocabana de Assis," *Estudos CEBRAP* 19 (1977): 37–91; and G. Muller, "Estrutura e dinâmica do complexo agroindustrial brasileiro" (Ph.D. dissertation, Univ. of São Paulo, 1980).

5. See the essays by Albert Fishlow and by Pedro Malan and Edmar Bacha in this volume.

6. Albert O. Hirschman, *The Strategy of Economic Development* (New Haven: Yale Univ. Press, 1958).

7. Fernando Henrique Cardoso, "The Originality of the Copy: ECLA and the Idea of Development," University of Cambridge, Center of Latin American Studies, Working Paper 27 (Cambridge, June 1977).

8. Fernando Henrique Cardoso, "A crismá de São Bernardo," in *Album memoria de São Bernardo* (São Bernardo do Campo: Prefeitura Municipal, Secretaria da Educação, Cultura e Esportes, 1981), 27–93.

9. For the extensive debate about the role of the State in the economy, see my *Autoritarismo e Democratização* (Rio de Janeiro: Paz e Terra, 1975); Sergio H. Abranches, "Empresa estatal e capitalismo: uma análise comparada"; and Luciano Coutinho and Henri-Philippe Reichstul, "O setor produtivo estatal e o ciclo," both found in *Estado e Capitalismo no Brasil*, ed. Carlos Estevam Martins (São Paulo: Editora HUCITEC, 1977); Luciano Martins, "'Estatização' da Economia ou 'Privatização' do Estado?" *Ensaios de Opinião*, no. 9 (1978): 30–37; Werner Baer, Richard Newfarmer, and Thomas Trebat, "On State Capitalism in Brazil: Some New Issues and Questions," "*Inter-American Economic Affairs* 30 (Winter 1977): 69–91; and Thomas J. Trebat, *Brazil's State-Owned Enterprises: A Case Study of the State as Entrepreneur* (New York and Cambridge: Cambridge Univ. Press, 1983). Trebat's book also contains a valuable bibliography.

10. See Norberto Bobbio, *Gramsci e la concezione della Societá civile* (Milan: Feltrinelli, 1976).

11. Guillermo O'Donnell, "Corporatism and the Question of the State," in James M. Malloy (ed.), *Authoritarianism and Corporatism in Latin America* (Pittsburgh: Univ. of Pittsburgh Press, 1977), 47–87. See also his "Reflexiones sobre las tendencias generales de cambio en el Estado burocrático-autoritario,"*Documentos CEDES/G.E. CLASCO*, no. 1 (Buenos Aires: CEDES, 1975).

12. Juan J. Linz, "The Future of an Authoritarian Situation or the Institutionalization of an Authoritarian Regime: The Case of Brazil," in Stepan (ed.), *Authoritarian Brazil*, 233–54.

13. Fernando Henrique Cardoso, "On the Characterization of Authoritarian Regimes in Latin America," in David Collier (ed.), *The New Authoritarianism in Latin America* (Princeton: Princeton Univ. Press, 1979), 33–57.

14. See the summary of these positions in Fernando Henrique Cardoso, "Regime Político e Mudança Social," *Revista de Cultura e Política* 3 (1980–81): 7–27.

15. Wanderley Guilherme dos Santos, *Cidadania e Justiça* (Rio de Janeiro: Editora Campus, 1981).

16. Cardoso, *Autoritarismo e Democratização*, chap. 5.

17. Alfred Stepan, "O que estão pensando os militares," *Novos Estudos CEBRAP* 2, no. 2 (July 1983): 2–8.

18. See, for example, Cardoso, *Autoritarismo e Democratização.*

19. Umberto Cerroni, *Teoria del Partido Politico* (Rome: Editori Reuniti, 1979), 13.

20. Umberto Cerroni, in his *O Conceito da Sociedade Civil*, argues that "in the whole natural law tradition, the expression *societas civilis*, instead of designating a pre-state society, as would occur within the Hegelian-Marxist tradition, is synonymous—in the Latin usage—with political society, or the State: Locke uses one or the other term indifferently. In Rousseau, *état civil* meant State. Kant also, who—besides Fichte—is the author closest to Hegel, when he speaks of the irresistible tendency which nature imposes on man to construct a State (*nas Idee zu einer all gemeinen Geschichte in weltbuergerlicher Asicht*) calls this supreme goal of nature in relation to the human species *burgerliche Gesellschaft*" (São Paulo: Graal, 1982), 26.

21. Pietro Ingrao, *Massa e Poder* (São Paulo: Livraria Ciências Sociais Editora, 1982).

22. In 1982 in the state of São Paulo the PMDB had 400,000 inscribed members, and somewhere between 20,000 to 30,000 party members participated in the elections that produced the 600 local party directorates.

23 Fábio Wanderley Reis, "O Eleitorado, os Partidos e o Regime Autoritário Brasileiro," in Sorj and Tavares de Almeida (eds.), *Sociedade e Política no Brasil pós-64*, 62–86.

24 C.B. Macpherson, *The Political Theory of Possessive Individualism* (Oxford: Oxford Univ. Press, 1962).

25 I gave one example of "authoritarian electoral legislation" which animated more expressive parties; I could give another example: the rather surprising lack of interest on the part of the military regime in persistently manipulating (as opposed to periodically repressing) the unions; this situation allowed for a more effective use of that space by unions and their lawyers so as to have workers benefit from labor guarantees, as well as to allow the union space to be occupied in some categories by authentic union leaders.

10

Why Democracy?

FRANCISCO WEFFORT

"Why do you all talk so much about democracy in Brazil?," the political attaché from the American embassy asked me. "Why not revolution?" It was a difficult question to answer. He knew that, and tried his best to get over my discomfort by calling upon his erudition, discussing aspects of the French and the Russian revolutions. But the uncomfortable question was still in the air.*

A question like this coming from an American diplomat is likely to produce a spark of paranoia in any Brazilian intellectual. But paranoia aside, the fact is that the question is not unreasonable. The American diplomat touched a point which many Brazilians prefer to avoid. Only a short while ago, some Brazilian politicians used to talk about the risk of a "social convulsion," and others pointed to the possibility of a "rupture of the social fabric." But no one used the dangerous word "revolution."

There is in fact something very hard to explain in the political situation we are living through in Brazil. We insist on talking about democracy in the midst of the greatest economic crisis in the country's memory. The economists say that the country is drowning. They almost do not have to say it, because any citizen knows it already.** What we call the "political transition" began in 1974 with General Geisel's so-called "policy of relaxation,"

*The author would like to thank Margaret Keck for her translation and for her most helpful editorial assistance.

Editor's Note: This essay was specifically written for inclusion in this volume in March 1983. Because the author is not only a major social scientist but was also a major party leader of the democratic opposition, I asked him not to update the article, but rather to leave the text as a document of the period. A few footnotes have been updated.

**This article was written in March 1983 in an atmosphere of harsh economic crisis. In spite of numerous changes since then, including the brief interlude of the 1986 Cruzado Plan, the crisis remains.

just as the "economic miracle" was ending. It followed upon the most violent and criminal dictatorship in our history. By 1979, when General Figueiredo began his "*abertura* policy," the economic situation had worsened considerably, and the economy was already in a depression. The *abertura* has continued, even as the depression has become the gravest crisis in memory.

And yet in spite of everything, we Brazilians continue to believe that we are in a *transition to democracy*. "We have hot soup on the table," Air Force Minister Délio Jardim said once. "And what is interesting is that no one wants to upset the table." He was clearly thinking about the diverse groups which compose the regime of which he is a part; some of them support the government's *abertura* very unwillingly and, like everyone else, see the economic crisis getting worse.

But the idea that no one wants to upset the table is also true of the regime. We seem to recognize that the table Jardim was referring to stands on very shaky legs. After all, 1968, when parts of the left took up armed struggle in the hope of realizing the dream of a social revolution, was not so long ago. One did not speak of "rupture of the social fabric" then, but of revolution. At the same time, sectors of the right carried out another coup d'état with Institutional Act No. 5, which transformed the military regime into a bloody dictatorship.

At that time, the left believed that capitalism was breaking down in Brazil, and that a revolutionary move towards socialism was historically inevitable. On the right, the power-holders said that the country was undergoing a situation of "revolutionary war." But beginning in 1968, instead of breakdown we had a half-decade of extraordinary growth rates in Brazil—the "economic miracle." If revolution—or at least a belief in revolutionary action—were dependent upon economic collapse, how much more prevalent should talk of revolution be today!

No word has been more misused in post-1964 Brazil than the word "revolution." Even the coup which produced Institutional Act No. 5 was carried out in the name of "continuity of the revolution." Both sides spoke what they thought was the language of revolution, but in fact was the language of violence. Whatever meaning the word "revolution" had within the political cosmologies of the two sides, the fact remains that between 1968 and 1974 we lived under the sign of violence. How can we explain that we emerged from this phase talking—or at least trying to talk—the language of democracy?

What the American aide was really asking was whether 1984 could not not become another 1968. In doing so, he reminds us of a recent and painful past (a past moreover that is so recent that it almost cannot be considered the past). And we who have been struggling against the regime since its establishment are forced to recognize that in the little time that has transpired since 1968, the country has changed a great deal. There was no revolution, but Brazil has changed. And those who were defeated in 1964, and then crushed again in 1968, have also changed, sometimes more than they like to admit.

The past and the present are intertwined in our vision of this period. Are we making the changes or is it we who are changing? Evidently, many of those defeated in 1964 and 1968 maintain the same fundamental values, freedom and equality, the ideal of socialism. But can we say the same about our old conceptions of State and society? What can we say, especially, about the way we have traditionally viewed the meaning of politics and of power?

This essay is an attempt to understand how these changes occurred and are occurring. We will begin by looking at the way traditional political relations have shaped the ways in which both right and left have approached politics in Brazil—even to the point of shaping the very meanings of the words we use in political discourse. The superimposition of democratic forms on authoritarian relations, the prevalence of statist ideology even among those who called themselves anti-statist, and the resulting acceptance of coup-making as an everyday form of political action, are all characteristics which cut across left/right divisions. We are capable of calling authoritarianism democracy, and an act of usurpation is called a revolution.

And yet for many Brazilians, the dictatorship which began in 1964, and especially the violence which accompanied the Médici regime, rendered useless the traditional categories we used to analyze and act, forcing a re-examination of our own histories. If the State had formerly been the solution, now it was the problem. If before it had been possible to call "democracy" what were merely juridico-institutional forms of democracy, it was possible no longer. Out of an ambiguous historical legacy new meanings had to be developed, and, slowly and fearfully, democracy began to be seen not as a means to power but as an end in itself. Yet if politics were to have a new meaning, a new sphere of freedom for political action had to be developed. For political Brazil, civil society, previously either ignored or seen as an inert mass, began to signify that sphere of freedom.

A Legacy of Equivocation

> (. . .) We imported from abroad a complex system, full of presuppositions, without knowing to what extent it was adaptable to Brazilian conditions, and without thinking about the changes which those conditions would impose on the system. In fact, the impersonal ideology of democratic liberalism never took root among us. We only effectively assimilated these principles insofar as they coincided with the pure and simple negation of an uncomfortable authority, confirming our instinctive dislike of hierarchies and allowing us to deal with rulers on grounds of familiarity. Democracy in Brazil was always a lamentable misunderstanding.[1]

It is not surprising that Brazilians have difficulty in addressing the questions of democracy and revolution. There are countries which, at some point in their history, made a revolution and built a democracy, for example, England and France. Others, like Italy, built a democracy without having gone through a revolution. And still others, like the Soviet Union, made a revolution and did not succeed in reaching democracy. Brazil is an unhappy case of a country which never made a revolution nor had democracy.

These are words not to be used lightly. Today, some say, we are moving towards a democracy because under the circumstances we have nowhere else to go. I remember that in 1968 as well there were those who said that we were moving towards revolution by imposition of a historical necessity. Are the implications of the words "democracy" and "revolution" so devoid of human agency?

In the mountain of equivocations which we inherited from our eminently conservative political tradition, the first is about the meaning of politics itself. Even today, many people treat major political decisions as though they were derived from nature, unavoidable dictates of necessity. Our habitual equivocations on the meaning of politics are the result of a history in which politics has always been a privilege of the few; a history in which, until now, there was barely a public space where political activity, almost always limited to the dominant classes, could be differentiated from the activities of private life. Moreover, it is a history in which the conservatives are the eternal winners.

Marx said once that the "tradition of all the dead generations weighs like a nightmare on the brain of the living."[2] He was referring, as we know, to people who were living in an epoch of "revolutionary crisis." In addition, he was referring to the past "of all the dead generations." I leave aside for the moment the question of whether or not we are in a revolutionary epoch. In any case, I have the clear impression that in Brazil today our conservative past, rather than oppressing, is entering the brains of many politicians with great facility.

I know a left politician, for example, who spent a good part of his years of exile reflecting on the difficulties of the workers' movement and on democracy during the populist period. In 1964 the Goulart government was crumbling like a sandcastle. While he meditated on the causes of the fall, he followed attentively the saga of guerrilla movements in Latin America. He never had a weapon in his hand, it is true. But he never distanced himself so much that he could not now and then express his sympathy. It would be an exaggeration to say that he fell in love with the idea of revolution. But he certainly flirted with it. And, more than once, he talked to me about his own past like someone talking about lost illusions.

After his return to Brazil, for several years he warmed to the idea of contributing to the formation of a new political party. It would be a form of helping to strengthen the workers' movement and reorganize democracy in the country. It would have to be a new kind of party, at least by Brazilian standards: open, democratic, rooted in popular struggles. "It would have socialism on the horizon," he would say. But when the moment for action arrived, he desisted. "There are two paths," he explained to me, "one from below, rocky and difficult; and another, easier one, from above." Below is the popular movement; above is the State, or at least the vicinity of the State. The differences between the two appeared so obvious that he did not even think it necessary to justify the choice.

There are many cases like this, and they make me think about the real significance of Marx's famous phrase. I have the feeling that in Brazil today, the dead weight of tradition is getting mixed up with the weight of the pasts of the living. Sometimes it is the past of the oldest dictating the path for the youngest. Sometimes it is the past of a generation dictating their paths in the present. Or even the past of an individual defining his path now. Do our own pasts weigh upon us as much as those of the "dead generations"?

In the Brazilian tradition, this mixture of the past, of conservative sensibility with good intentions, is what passes for political realism. Politicians claim that the difference between their public views and their more civilized private views is a matter of real politics. This is the source of the bad faith, so common among our politicians, as they make promises that they never fulfill. Their so-called realism is really mere conservatism mixed with simplistic Machiavellianism—politics is what you do to win (or keep) personal power and State power. Machiavelli would be covered with shame.

The left as well as the right is susceptible to this traditional mentality. Left conservatism, more subtle than the real conservativism, is nonetheless real. Our left politician, for example, had been sure that things were going to change. The coup d'état, exile, the distant experience of guerrilla movements—circumstances, finally, had imposed a change. But in the end he returned to the usual path, the same "Statist" practice whose limitations had appeared so clear when he himself was expelled from the game. Everything happened so "naturally" that he did not even feel the need to explain his decisions. He did not seem to recognize that the return to his old road would lead to the same place that he had criticized so much when he was in exile. The presence of tradition in his decision was so strong that it was unrecognizable as such.

The conception of politics which the conservative tradition left us is a perfect alibi. If the paths that you take are not understood as a choice, you are absolved *ex ante* from any political responsibility for your actions. You had no alternative. You only responded, like a good functionary of history, to the demands of the situation. When this conception of politics is taken seriously, democracy will always be crippled and revolution will always be impossible. Any historical situation has its requirements, and part of political wisdom obviously consists in recognizing these. But the truth is that those who insisted in 1968—the left and the right—on the inevitability of revolution make the same mistake of those today who insist on the inevitability of the democratic path. They grasp one aspect of the political reality but lose another. They miss the essential point if what they want is to understand democracy and revolution.

If we are going back to Machiavelli, let us remember then that true political realism consists of seeing events as "*cose a fare*" (thing to be done). This means that is is *always* possible to take different paths of action, and that it is not the conditions under which this takes place that define its meaning. Much as the conditions weigh upon the situation, much as the past imposes itself, there are always choices to make. A political actions is, *par excellence*, an act of freedom. It takes place only in the present and in the face of a future which is always open and uncertain.

Brazilian conservatism bequeathed to us an authoritarian conception of democracy. When Figueiredo said in 1978, "I have to make this country into a democracy," he was summarizing our whole tradition. It is an unsustainable contradiction, obviously, on the logical plane, but one whose pointedness is blunted by the long conservative tradition, which reaches into the depths of Brazilian political history. It is composed of all the ideas we inherited from the past about society and the State, about power and freedom.

The Brazilian tradition has included both real authoritarians and true democrats. For example, there were the fascists who expressed themselves in the 1930s and '40s in the integralist movement. And there were also the democrats who appeared at the moments of attempts at historical renewal, and who, often, ended up renouncing politics. But the most important part of the conservative tradition is the equivocation by which many can be, or pretend to be, both authoritarians and democrats at the same time. The French sociologist Alain Touraine captured the historical meaning of Brazilian development from the '30s to the '60s with the expression "democratization by authoritarian means."[3] This means that Brazilians have done the miracle of discovering some democratic effectiveness inside authoritarianism. More difficult for us to understand has been the real meaning of democracy.

An authoritarian conception of democracy requires a good dose of cynicism besides a certain taste for ambiguity. Suffice it to look at the innumerable proverbs which are part of Brazilian political language. "Voting doesn't fill your stomach"; "In politics it's the version that counts, not the fact"; "The law, oh sure, the law"; and so on. They are proverbs sometimes shared by the people. Although they appear to be just everyday expressions of cynicism, in fact they translate into sophisticated criteria for political action.

Their cynicism is mirrored in the ironic smiles of the powerful, for whom the lack of popular expectations serves as a license to act arbitrarily. These "popular sayings," whose implications are profoundly anti-popular, are the complex result of a long and difficult job of ideological training, in which, over time, all the oligarchies and all the dictatorships of this country have collaborated.

In the Brazilian political tradition, the idea that democracy is only one possible instrument of power among many, only a means, is so deeply rooted that it is difficult for us to conceive of democracy as an end in itself. There are exceptions, for example, the brilliant essay by Carlos Nelson Coutinho, the first Brazilian intellectual to treat democracy seriously as a universal value.[4]

Some people think that the origins of these ideas about democracy as an instrument are in the left. This is not true. These ideas have their roots in the conservative privatism, *soi disant* liberalism, of the Old Republic oligarchies. What was the Brazilian State before 1930 other than a sort of annex to the big plantations and the lordly domains? What kind of notion of *res publica* could large landowners have who were proud of the fact that the law never entered the gates of their land? What kind of social sense could oligarchs have for whom the "social question" was defined as a "police question"?

The liberalism of the lords of the land was essentially privatism. It was identified only with *economic* freedom, and, even so, was limited exclusively to property rights. For this reason, the political liberty that these people were capable of understanding did not go beyond the justification of their own privileges.

Instead of being an invention of the left, as is often asserted, the idea of democracy as an instrument is a conservative trap into which the Brazilian left fell, in part because it itself is, in the end, a product of the same tradition it is fighting. While it has struggled for a long time against tradition, it has been forced to do so with weapons provided by that tradition. This is the meaning of conservative hegemony in Brazilian politics: it imposes its discourse, or at least its logic, even on its adversaries.

On the other hand, it is true that the left allowed itself to be seduced very easily. Some of the ideas it derived from a badly digested Marx pointed in the same direction. Among those, the principal one was the idea of democracy as the form, *par excellence*, of bourgeois domination. The problem is that a lot of people preferred to forget that what was true for Marx's Europe was already ceasing to be true for the Europe of the old Engels. The constant of history, finally, is change. Whoever doubts this should read Engels's famous preface to Marx's essay *The Civil War in France*.

In any case, the idea of democracy as a privileged form of bourgeois domination has little or nothing to do with Brazilian history. The forms of political domination adopted by the dominant classes in Brazil have been very different. The Empire, from Independence to 1889, was a badly disguised dictatorship by the Emperor. And even if we consider the Regency period—*intermezzo* between the first and second reigns—how could one fail to see there the dictatorship of an oligarchy of landed proprietors? What was the Old Republic, from 1889 to 1930, other than the undisguised rule of landed oligarchs? There are those who prefer to call the landowners of that epoch an agrarian bourgeoisie, in order to emphasize the capitalist nature of the Brazilian social formation. But this changes nothing about the patrimonialist form—as violent as it was paternalist, and liberal only on the surface—of the regime which they established.

Since the 1930 revolution, we have had two long dictatorial periods: the first from 1930 to 1945, and the second from 1964 to today. Between 1945 and 1964, there was a fragile democracy, which depended less on the enthusiasm of the bourgeoisie for democratic forms than on the social pressure created by a popular urban mass which had recently entered the political scene.

The truth is that in 160 years of our history as an independent country, Brazil has never had the opportunity to test the hypothesis of democracy as a form of bourgeois domination. With its origins in the aristocracy and oligarchy, the instrumental conception of democracy runs through our history like a curse. If Marx had been Brazilian, he would have said that dictatorship was the form, *par excellence*, of bourgeois domination. And perhaps he would also have said that democracy is the form, *par excellence*, of popular rebellion.

The political tradition molded by the oligarchs and the dictatorships is still with us. A good example is the typically oligarchical confusion between right and privilege. If democracy is only an instrument, then the law is nothing more than a direct and immediate reflection of interests or of force. Those who run the post-1964 military regime have abused the notion of law so much that a general designation of its supposedly legal artifacts was created: they are called *casuismos*. [Note: The traditional meaning of the word *casuísmo* in juridical parlance was obedience to the letter of the law or formalistic appeal to the jurisprudence of the courts. In current usage it has come to mean laws—generally issued by decree-law—containing a strong measure of self-interestedness and arbitrariness.—Ed.] But the legal aberrations which it designates go way back.

In the 1940s, fascist jurist and thinker Francisco Campos coined a particularly cynical phrase. "There's law that takes and law which does not take." "Take," when it responds to force or to important interests; "does not take," when it neglects to pay attention to them. It is the same thing as saying that the law has little or nothing to do with the public, in the democratic sense of the word. It is not shaped by a public opinion nor does it respond to a public interest. It is anti-law, a simple pretext for the exercise of particularism or of force. Whatever our discomfort with such cynicism, this is *de facto* a description of some aspects of the reality of power in our country until today.

In the first years of the Old Republic, in a desperate effort to stop President Floriano Peixoto's systematic violation of the new Constitution, Rui Barbosa brought suit before the Supreme Court asking for a writ of *habeas corpus* for political prisoners. Asked to comment about the possibility of a Court decision against his interests, the President is reported to have commented with irony: "If the judges of the Court grant the politicians *habeas corpus*, I wonder who will grant them the *habeas corpus* they in their turn will need tomorrow?"

In 1955, in spite of the fact that Congress had declared a state of siege to forbid him access to the courts, President Café Filho appealed to the Supreme Court alleging that military constraints were preventing him from returning to the Presidency after a short period of absence for health reasons. It was a difficult moment, with coups and countercoups in the air, and the President's attitude was considered by his adversaries to be just another maneuver. But what is important to note here is that the Supreme Court refused the suit, and one of its members, the jurist Nelson Hungria, justified the measure in almost the same words used by Floriano a half-century before. That is to say that the Supreme Court responded to a situation of force. Restraining Café Filho might have been *de facto* a counter-coup, a necessary though painful measure for the re-establishment of a threatened Constitutional normality. But there is no doubt that the procedure used, involving the Supreme Court in a juridical farce, had a debilitating effect on the meaning of law and democracy in the country. In a certain sense, it served to prepare for today's philosophy of *casuísmos*.

In any case, it is certain that the 1946 democratic Constitution was violated a number of times before it was definitively abolished in 1964. And when this happened, the same Francisco Campos, once again called upon to give shape to arbitrary rule, introduced his old cynicism into the first Institutional Act of the military regime. "The Revolution"—once again confusion over words, he meant violence—"is the creator of law." Once again, the meaning of law for those in power does not go much further than a pretext for arbitrary rule.

This is particularly clear when one examines the trade union laws. The decree-law which regulated the right to strike from 1946 to 1965 was so vague and gave so much discretionary ability to the government that the latter could decide according to its own interests which strikes should be considered legal or illegal. In fact, it suppressed the right which it claimed to regulate. Everything came to depend on the tolerance (or intolerance) of the governments. And in fact, over time, governments manipulated the law according to their own criteria, whether they repressed strikes or allowed them to occur.

At certain moments after 1954, when some governments came to see the workers' movement as a possible ally, one had the impression that the anti-strike law fell into disuse, became a dead letter. Some ingenuous trade unionists even came to think that it had been repealed in practice. But the law remained on the books, and after 1964 it was used extensively by the first military government. It lasted until 1965, when it was replaced by another law, which was equally harsh and arbitrary. In the end, if democracy was only a means, how could law have been other than a parameter by which interests or relations of force could be measured?

If democracy is only a means, the end of politics is power. Thus usurpation becomes a normal political procedure. Usurpation—the coup—has always played an important role in Brazilian history, even when democratic practices have been adopted. In the two decades from 1945 to 1964, there were coups or attempts at coups d'état in 1945, 1950, 1954, 1955, 1961, and 1964. In the period from 1930 to 1945 we had coups in 1935, 1937, and 1938, to say nothing of the beginnings of a civil war in 1932. From 1964 to the present, three of the four successions involved elements of a coup d'état. Costa e Silva imposed himself on Castello as his successor by force, and when Costa e Silva died, the Vice President was removed from power by force as well. The replacement of Geisel by Figueiredo was only possible after having averted at least one attempted coup by General Frota in 1977.

The coup mentality is generalized to all kinds of political activity, even those which do not involve the use of violence. A PDT leader talked to me about his happiness over the electoral victory of his party in the state of Rio. "We will have a base there for building the party in the country." But when he talked about a base, he was not referring to a popular base or to the large quantity of support which the PDT had won in civil society. He was referring explicitly to the spoils of power: state money, state jobs, political positions in the state. Another example of this is called "*aparelhismo*" (instrumentalization, making use of an apparatus or organization for particularistic ends). Common in the left parties, in which individuals or groups take over and occupy party positions with no legitimate right to do so, it is not limited to the left. While the word is not used of the center or the right, one can find *aparelhismo* in the numerous cases of corruption, manipulation of employment practices, nepotism, and so on.

When the coup mentality becomes generalized practice, it is evident that political procedures become perverted. Lying and manipulation take the place of debate and persuasion. The levels of tolerance required to build democracy fall frighteningly. And verbal violence prepares the way for physical violence. The prevalence of the coup as a form of normal, everyday political practice thus prepares for its big moment: the coup d'état, the violent rupture of pre-existing legality, be it democratic in origin or not. And this, the use of violence to break the law through a coup d'état, is what political tradition in this country calls a Revolution.

The confusion between democracy and authoritarianism thus belongs to a long tradition, in which a word begins to mean its opposite, in which meanings are themselves cynical functions of power relations. To call authoritarianism democracy is no more strange than to call a coup d'état a revolution. To give yet another example: Brazil today is ruled by an amendment imposed in 1969 in the most dictatorial manner possible by a military junta. And this amendment, which no one except a half-dozen military officers discussed or voted upon, is called a Constitution. And the greatest irony of all is that this typically dictatorial measure opens with an affirmation of popular sovereignty: "all power comes from the people and will be exercised in its name."[5]

The explanation of this apparent contradiction is simply that after the legitmation of power through the divine right of kings was exhausted in modern history, it was no longer possible to talk about power without mentioning the idea of popular sovereignty. It is common knowledge that virtually all variants of authoritarianism, or, even totalitarianism, like to pay homage to the idea of democracy. Is this the whole explanation for the "misunderstanding of democracy" that Sérgio Buarque de Holanda was talking about in his *Raízes do Brasil*? Or might it be possible that we have always sought some form of democracy, although we have always been incapable of defining it?

Since 1974, the conquest of democracy has been the *leitmotif* of Brazilian politics. Look at the party programs and the declarations of the politicians.

Look at the concepts which flourish in the enormous political record of the period. In spite of differences, there have been more arguments in favor of democracy than at any other time in our history.

And so we have to ask ourselves whether the political transition we have been undergoing since 1974 has finally incorporated democratic values into Brazilian political culture. As always, I am referring to political Brazil, to leaders and militants of political parties, to citizens in general, and also to those who, without having yet attained full citizenship, are struggling for it. After all that has been done to make the terrain where democratic values should flourish sterile, can we believe that democracy is becoming a *core value* in political Brazil?

To make clearer what I mean by core value, let me cite an example. The idea of economic development has been, as I see it, a core value in Brazilian politics since the 1950s. It seems so natural to us now that this is a political parameter that few of us remember that this was not always the case. Politicians, parties, and citizens in general might differ—and in fact do—about the best path towards development. But the great majority of them are in favor of development as such.

And it is not just a matter of seeing development as a historical necessity, derived from the economic laws of capitalism or from the sociological laws of modernization. If this were all, we would be talking about a technical or scientific consensus, not a consensus on a political value. But apart from whether or not a historical or sociological law is involved, development in Brazil is conceived as a goal *which is worthwhile for its own sake.* More investments, more production, more jobs, better wages, greater opportunities for consumption—all this goes way beyond economics or a preference for one economic system or another. We have the right to prefer socialism or capitalism as a road to development. But in any case, we consider development to be a condition for gaining a more dignified life.

If the 1950s were the years of the constitution of development as a core value, perhaps we will be able to say that the 1970s and 1980s were those of the constitution of democracy as a core value. But it is still an ongoing process, and therefore one which is difficult to analyze or even to recognize. There are evidently those who remain attached to a rigorously authoritarian vision of politics. And as for the others, who I think are the majority, the way in which traditional patterns of behavior mix with newer ones as they express their preferences for democracy make it difficult to characterize the implications of these choices with certainty. But the question is still worth asking. Are we on the way to constituting democracy as a value in itself?

The word "democracy" has been used in so many ways to characterize the Brazilian political transition that sometimes we have to wonder whether it has any meaning left. General Geisel, for example, tried to characterize his period of government as "relative democracy." The very mention of that expression, which was famous from 1974–78 is enough to launch a storm of polemic.

The General, who spoke very little, liked ambiguous phrases. In a period when the struggle for human rights was above all a struggle against torture, illegal imprisonments, and press censorship, Geisel made the statement that the extension of water and sewage networks ought also to be considered one of the human rights. While this might even be true, given the extremely precarious sanitation conditions in Brazil, to the opposition the General's statement sounded like intolerable cynicism.

The opposition could perhaps accept that Figueiredo talked of "relative democracy" as a description of his "*abertura* policy." But during the Geisel government, the most tolerant members of the opposition rejected even the term "relative dictatorship" for fear that the adjective would soften the harsh image of autocratic power. Senator Paulo Brossard compared the Geisel government to the absolute power of the Roman emperors, and Deputy Ulysses Guimarães went even further, when he made a comparison to Idi Amin Dada. The historical precision of comparisons like these does not matter much. Their intention was to describe the greatest concentration of autocratic power in Brazilian history.

But at the same time, neither Brossard nor Ulysses, important liberal leaders, nor even the most radical sectors of the left, would deny that it was precisely under Geisel that the process of transition in which we find ourselves began. And therein lies part of our problem about the meaning of democracy in Brazil today.

The regime tried to apply adjectives to the word democracy. Geisel talked about "relative democracy"; others talked about "social democracy"; still others spoke of "strong democracy," "Brazilian democracy," and so on. The opposition rejected the use of any adjective whatever. But what the regime tried to define in an apologetic spirit was perhaps not so very different from what the opposition was talking about. Both sides sought to understand the difficult reality of a transition that was full of ambiguities and contradictions.

There are any number of examples. There is no question, for instance, that the National Security Law is a totalitarian artifact. It is an anti-law, and to call it a law at all is to make an extreme linguistic concession. But does the totalitarian—or, if you will, authoritarian—character of laws like this suffice to characterize the regime which elaborated them? The same regime which made such laws allowed them to be denounced in the newspapers, on the radio, and on television. Another example: at their inaugurations in 1983, various opposition Congressmen refused to declare allegiance to the dictatorial amendment which serves as our Constitution. It would be absurd to claim that a dictatorship which allows such occurences has become a democracy. But how can one deny that it is becoming less of a dictatorship?

It seems undeniable that both the regime and the opposition are seeking in some way or another to relate to a common reality. Or might it be to a common value? As strange as it seems, between the apologetics on one side and the criticism on the other, from 1974 on something like a reciprocity of perspectives has begun to be designed. While from both sides come harsh words, between the lines one can perceive the emergence of a subtext in which similar meanings are being elaborated. The oppositions emphasize how much is left to do; the regime, from Geisel to today, make much of what has already been done. But the idea of an arrival point appears the same.

Since Geisel, the regime has said that one day there would be "full re-establishment" of democracy in the country. This implies from his point of view we were going through only a partial and difficult re-establishment. And none of the critics of the regime would refuse to recognize that, from 1973 to 1983, we have gone down a piece of the road. It is this consensus around the idea that we are moving towards democracy that we have to examine. First: What makes us all agree that the change which we are undergoing has a direction? Isn't that the same thing as agreeing that we all see something on the horizon to reach? Second: If there is something on the horizon to reach, why must we all agree that it is democracy?

Our traditional "political realism" is of little use when we confront questions like this one. "Political realism" would tell us we are moving towards a new form of dictatorship. In fact, if this were a more widespread belief, that alone would be enough to make it a highly probable hypothesis. I have said that there is something astonishing in the common vision that we are moving towards democracy. Could this not be a signal that behind our vision of democracy there is a belief? We all have different reasons for believing that we are moving towards democracy. One of these, though, is basic. It is the reason of will: that is where we want to go. It is the reason of belief. How can we explain how it has arisen in the context of a tradition as authoritarian as ours?

The political transition has been going on since 1974 and promises to continue for some years more. Alfred Stepan, one of the most knowledgeable students of Brazilian military politics, notes wonderingly in a *Jornal do Brasil* interview that if the regime manages to maintain control of the succession to Figueiredo, the transition could go on until 1991. Seventeen years is a long transition, a lot longer than the ten years of military regime. How can we explain the continuity of the transition?

The most obvious explanation is that continuity is ensured by the capacity of the regime's highest leaders to command in the military area. The military commanders of the regime are also the political commanders of the transition, as with Geisel and Figueiredo. This attributes an eminently conservative character to the transition, something which Geisel had also defined in his famous metaphor: "slow, gradual, and secure decompression."[6]

To this first explanation we can add another. The transition has been limited to the elites, or at least has affected the elites much more than it has the popular masses. Suspension of prior censorship of the press, re-establishment of *habeas corpus*, amnesty, party reorganization, direct elections for state governors—all this, important as it is, has taken place at the juridico-institutional level. Other institutional measures which would directly affect the people, such as changes in the trade union laws or guarantees of the right to strike, have not been enacted. The economy is ruled by an obviously anti-popular "austerity policy." All this means that the popular masses continue to be almost as marginalized as they have been since 1964. The continuity of this transition is therefore ensured by its class premises.

These two explanations are complementary. Geisel's, and later Figueiredo's capacity to command the political process as a whole cannot be attributed entirely to their undeniable control over the armed forces. Nor would it be sufficient if we added to this the obvious control by the president over the PDS. The truth is that if the President—whether Geisel or Figueiredo—has been capable of directing the transition, it was also because his initiatives were meeting with some response on the other side of the wall. The opposition wanted a State of Law and Geisel wanted the whole of the armed forces under his control. Under the circumstances, these two objectives traced a field, however precarious and narrow, where there was a coincidence of interests. The traumatic dismissals of General Ednardo in 1975 and General Frota in 1977 are only two examples where this came into play.[7]

Something similar has occurred more than once during the Figueiredo government. The best example took place before the 1982 elections, with the RioCentro episode, when a fortunately unsuccessful murderous action by right-wing groups put the whole political *abertura* process at risk. From one moment to the next, all the parties, from the PDS to the PT, met around the same table. Differences apart, all were moved by the same desire to say "no" to the terror which came from the pores of the State itself, via the actions of the so-called "intelligence community." It was a fleeting gesture, perhaps only a formal one. But it was enough to make it quite clear that the government and the oppositions had at least one point in common.

If terror had carried the day, some left leaders could have died, many unions would have been closed, some liberals would have been imprisoned, prior censorship of the press would certainly have returned, and the 1982 elections for state governments would probably have been suspended. But if this had happened, what would have been left of Figueiredo's "*abertura* policy"? The President himself would have become completely dispensable. Those who ordered the RioCentro action were sufficiently mad to create a situation which could have caused hundreds of innocent deaths. Had they been successful, why keep a President who would have been entirely demoralized?

Thus, since Geisel an area of consensus has been created around the elimination of terror; a common ground which, however, could not last if it were limited only to the relations between the government and the liberal sectors of the opposition. If the transition was to be from above, why should other political sectors not try unexpectedly to upset the table? If the continuity of the transition is assured by its class premises, why do the masses continue to wait for the soup to cool?

It is quite probable that the political sectors which remain outside the advantages of a managed transition simply do not have the force to change the direction of things. To upset the table is a question not only of desire but of power. What interests me, however, is the prior question of will. I believe that not only are they unable but they also do not want to. In 1968, university students and the workers at Contagem and Osasco also lacked the power to change the situation, but they made the attempt anyway. There must therefore be some over-riding reason why they do not want to do so now. Opportunities to upset the balance have certainly not been lacking.

In 1980 and 1981, when groups from the "security community" burned newsstands which sold publications of the alternative press, it was clearly a provocation. The problem for the left was how to respond. The great risk was obviously that some left group would decide to respond to violence with more violence. This was a risk which increased in proportion to the dispersion of the Brazilian left in a number of tendencies and groups with extremely precarious, almost nonexistent means of communication. It became all the more probable from a political point of view when one considers that a managed transition leads to a perspective of conciliation among the dominant groups which menace the left groups with political isolation. If this was to be a return to 1968, would that not have been a likely time for it to happen?

The courses of action not taken often have as much historical significance as those taken. The violent provocation by the right was answered by democratic attempts to mobilize public opinion. Perhaps one might say that in the case of burning newsstands, like in the case of RioCentro, the response of the left groups was irrelevant in terms of their size and in influence. This does not, however, detract from the significance of their involvement in a process of transition in which they thought to see a possibility of a commitment to democracy.

The Failure of Violence

Whoever wants to understand why Brazil in 1984 will not repeat the experience of 1968 must understand several other things as well. One of these is that the violence which brought down the political system in vigor before 1964 also dealt a serious blow, in 1968 and afterwards during the Médici government, to its ideological traditions. It produced ruptures in political relations and, especially, in the way in which people saw these relations. Our tradition of cynicism, in a situation where many people were afraid, gave way to a new appreciation of democratic values for their own sake during those years. It is hard to be cynical when you are afraid.

Juan Linz has noted that Spain's transition from dictatorship to democracy can be explained at least in part by the memory of the civil war with its million dead.[8] In Brazil, one cannot really talk about a civil war, neither in 1964, nor in 1968, nor during Médici's reign of terror. But it is nonetheless true, too, the continuity of the transition cannot be understood without examining the effects of violence.

There is violence, and there is *violence.* We say that in Brazil since 1964 and especially after 1968, right-wing violence took on industrial proportions, while the left never got beyond the craft level. The first was produced out of the State apparatuses, while the second came from small political groups. Nonetheless, the two had similar effects as factors of disorganization and rupture of the political system. Moreover, the use of violence proved itself a failure, both as a technique for winning power on the part of the left, and as a technique for maintaining power on the part of the ultra-right. It was sufficiently effective to destroy to political arena, but entirely ineffective in terms of the political goals which its agents proposed to carry out.

In the ten years from 1964 to 1974, the political system formed during the democratic period was entirely destroyed. Not only the party system, abolished in 1965 and replaced by a pretense of bipartisanism with ARENA and the MDB. In the years of terror under the Médici government, even the press was reduced to a caricature of itself. And finally, it was the State itself, agent of terror which, after destroying the political system, went on to destroy itself.

I think that situations like these hark back to ancient political laws. Paradoxical as it may seem, it was never so easy in Brazil as during the years of terror to explain the classical theories on the origins of the State to students. If the State is—as Weber says—a group of individuals who claim for themselves the monopoly of legitimate use of violence,[9] how does one consider those individuals who, in the Brazilian State, use violence illegitimately?

Whatever theory one adopted, the evidence was there. The State, said Engels, arises when society divides in an irreconcilable conflict.[10] The State is, therefore, an organ capable of presenting itself to society as its sovereign and ensuring its cohesion. That somewhere there was an irreconcilable conflict was obvious. As obvious as the lack of an entity capable of bringing it under control.

Madison said that two conditions are necessary for a democracy: the first is that there be a government capable of governing, and the second is that there be a society capable of controlling the government.[11] During the years of terror both were lacking. What we called a government at that time had a great deal in common with a group of gangsters. They only managed to give the country the illusion that they constituted a government because we were in the middle of the "economic miracle" period. What we called society was little more than a mass of frightened individuals. Nonetheless, out of these years of confusion and fear arose in the country a new attitude in relation to the State, society, and democracy. Let us look at the few examples of how this happened.

Perhaps the most constant theme in Brazilian politics since 1968 has been the re-establishment of a State of Law. In the beginning it was almost exclusively a demand of the liberal opposition. Subsequently it became a commonly used term among the opposition, including sectors of the military regime itself which were considered liberal.

As the use of the term was extended, the confusion as to its meaning grew. For the opposition liberals, it meant certainly the re-establishment of the *rule of law*, in the strict sense of the term. But I have no doubt that for many of them, and particularly for some liberals linked to the military regime, the demand for a State of Law was also the opportunity to demand the re-establishment of the State as such, in other words, of a political organ capable of assuring the cohesion of society and keeping violence under control. For other sectors of the opposition, particularly the left, the re-establishment of a State of Law would also have to mean the re-establishment of *democracy*. As time passed, the demands for a State as a cohesive force, for a State as the rule of law, and the State of Law as democracy, *become shades of the same general democratic aspiration.*

Ambivalence with regard to democracy has been endemic to Brazilian politics since long ago. My impression is that under the circumstances of violence of the Médici period, many of these ambiguities were resolved in favor of what might come to be, in contrast to tradition, a democratic conception of democracy. Even now there are evidently sectors of Brazilian politics, both on the right and on the left, which have refused to bend. There are some small groups on the left which still bow down to the Stalinist mythology, and on the right, there are fascists grouped in certain segments of the PDS and what remains of the so-called "security community."

However, I believe that the other sectors, fortunately the great majority of the political spectrum, have been impelled by the inadequacy of traditional approaches to politics to take a step forward. For many of them, both in the liberal camp and in the left, the appreciation of democracy could appear as a rediscovery. Many believe that they have always been democrats even though they did not have, in the past, the conditions for expressing themselves as such. While I do not think that this is true, I do think that the effort that many make to seek in the past the roots of their present convictions is the beginning of a process of review which will end up, sooner or later, in a break with tradition.

For the Communists of the PCB, for example, their current appreciation of democracy might mean, instead of something new, the deepening of positions which are found in their own tradition. In 1945, in the context of the defeat of Fascism in World War II and the fall of dictatorship in Brazil, the PCB had a moment of democratic enthusiasm which unfortunately lasted as short a time as its position as a legal party. It disappeared afterwards, little by little, in the climate of political repression which came with the Dutra government's alignment with the United States government's Cold War policy. It should be noted, however, that it was not only the Communists who lost their enthusiasm for democracy after 1947. A large number of liberals grouped in the UDN demonstrated almost from the beginning their disappointment with the democratic regime based on the 1946 Constitution, which they saw as a disguised continuation of the dictatorship of 1937. And a good part of the plotting characteristic of the democratic period can be explained by the disaffection of these liberals who rapidly evolved towards the right.

The Communists evidently cannot be considered exempt from practicing coups of their own during those years, nor in fact can any other political force. But after 1964, the party sought democratic routes for opposing the military regime. It is notable that the Communists who participated in the guerrilla struggles had to leave the ranks of the PCB. Whatever our opinions on PCB policy, there is no reason to be very surprised if the struggle for democracy is on their agenda today.

So what about the sectors of the left which threw themselves into armed struggle in the wake of 1968? How can we explain the fact that the majority of those who yesterday were committed to armed actions against the military regime today are participating in the struggle for democracy? Here too, it is necessary to go beyond the current prejudices and look at the facts as they are. In the first place, the majority of armed groups were acting on the Rio–São Paulo axis, the center of the country's economic and political system. Second, the majority of guerrilla groups of the period were made up of young people, in general, students, who began in politics after 1964 and found their road to citizenship barred by the military regime.

To understand the left's experience of armed struggle we also need to remember a whole set of circumstances which characterized national politics at the time. First: the sentiment of democratic enthusiasm which accompanied the government of Juscelino Kubitschek had been thoroughly undermined by the resignation of Jânio Quadros. The circumstances of quasi-civil war around the inauguration of Goulart convinced many people that the country was entering a crisis of the State for which there was no peaceful solution in sight. Second: Jânio Quadros' resignation coincided with the onset of a depression in the country which would last until 1968, and for which many sectors of the left did not see a possible way out within the limits of capitalism. And it is important to remember that it was not just the youth of 1968 who thought that way; there were also such important Brazilian economic thinkers as Celso Furtado with his "theory of exhaustion" of the Brazilian economy.[12]

Added to these first two elements was the fact that between 1964 and 1968, all Brazilian political forces—all, without exception—were undergoing a serious credibility crisis. After the failure of Castelo's political project, who could believe in the democratic claims of Costa e Silva, member of the "hardline" in 1965? Who could take seriously the Constitution which he granted to the country in 1967? There were also serious grounds for skepticism about the democratic intentions of the liberals, in or out of the regime, who had previously been enthusiastic about the 1964 coup d'état. It was difficult to put one's trust in the Church, when the "'marches with God," organized by Church leaders in support of the coup, were fresh in one's mind. Even the PCB, defeated in 1964, was now seeking an incongruous alliance with Lacerda, Goulart, Kubitschek, *et alia*, in the famous and unsuccessful Broad Front.

The 1968 left was born out of all these crises. By trying to confront the regime with arms, it broke with the political tradition of the country and with the tradition of the Brazilian left itself. I do not have the information to evaluate what importance they attributed to the question of democracy. In any case, their often overblown rhetoric apart, it seems to me that in the context of those years, the few armed actions that they managed to carry out had, within them, elements of a democratic resistance struggle. And I do not believe that under all circumstances can one call the use of arms incompatible with the democratic struggle. This would make incomprehensible the political meaning of phenomena as important for modern democracy as the Italian and French resistance to fascism.

The young people of 1968 proposed to "overthrow the dictatorship" as a means, a first step towards socialism and national liberation. We knew that they would not get beyond the first step, if that. After the defeat of their armed actions, their goal of ending the dictatorship remained. Once again in politics, a means became an end in itself. For many of the young people of 1968, their struggle against the military dictatorship was the point of departure from which they began to open up to the idea of democracy.

If General Geisel, with his extraordinary authoritarianism, did democracy any service, it was to re-establish discipline in the armed forces, preparing the conditions for a government with capacity to govern to exist in the country. General Geisel talked about "relaxation" to define his policy because he believed that his task was to restore peace and order to a country at war. And perhaps the best translation for this policy was the phrase invented by one of his aides, directed at the opposition: "you get your radicals under control and we will control ours."

The truth is, therefore, that in 1974, when Geisel became President, there were no more "radicals" in the opposition, at least not in the sense that the regime used the word. By 1974, the guerrillas were defeated and the whole of the left had turned to democratic struggles. The only "radicals" who survived at that movement were in the "cellars of the regime," in the famous "system," in the so-called security community. The peace and order that reigned in Brazil was the order of the garrote and the peace of cemeteries. Geisel's accomplishment was to re-establish peace in the armed forces and the State, which since Médici had been under the control of right-wing groups which used terror as an everyday political tool.

The discovery of the value of democracy is inseparable, within the opposition, from the discovery of civil society as a political space. More than the "economic miracle," the terror years produced a real "political miracle," by undermining traditional ideas on the relations between State and society. And the concept of politics was placed on its true foundations.

During the Médici period, many people entered into deep confusion over the meaning of the State which they saw before them. Those who had supported the 1964 coup had inherited the classical liberal suspicion regarding the State. They were against João Goulart in the name of a democracy, which was supposed to mean, for them, less State intervention in the economy. For many of them, Goulart, as heir of Getúlio Vargas, also had to be the heir of the 1937–45 dictatorship, which they thought of as the creation of the interventionist State. To overthrow Goulart was to break with this tradition.

But now, in the midst of the Médici period, marginalized from the power which they had helped to create, they saw from the outside, buried in confusion, the monstrous fruit of the seed they had planted. They were looking at a dictatorship much more violent and corrupt than the one from 1937 to 1945. And, besides, they were looking at a State as interventionist as the others. In the first half of the nineteenth century, when French expressions were common usage in Brazilian culture and politics, a liberal coined the classic phrase about liberal disappointment in face of struggles, which contributed more grist for the mills of the conservatives and authoritarians. They were the "*journées des dupes*," (the days of the dupes). In the 1968 shift, for many Brazilian liberals the movement of 1964 became yet another *journée des dupes*.

There was no less perplexity on the side of those who had defended Goulart. Against the classical liberals generally grouped in the UDN, many democratic liberals, generally in the PSD, the PTB, and the whole left, were convinced of the usefulness of the State as instrument of democratization of the economy and society. This belief dates back to the revolution of 1930, when the growth of the State was a move against the privatism of agrarian oligarchies. Even with the detour of dictatorship, in 1937, the growth of the State continued, during the war and after the war, in industrialization policies which stressed public investment to support economic growth.

However realistic this perspective may have been as historical analysis, it reinforced, during the whole democratic period, the belief in the State as a factor for greater economic development and greater social equality. The traumatic reality of politics in the Médici period, however, showed many that the dream of an interventionist and egalitarian State had turned into a nightmare. Instead of the democratic and egalitarian State which they had desired, they had before them a dictatorship which promoted simultaneously the growth of the economy and the misery of the masses. Médici himself, in one of those rare moments in which he broke his deafening silence, defined the results of his ungovernment: "The economy is doing well but the people are doing badly."

The perplexity increased still more because it was easy to see that, during the Médici period, if there were criminals and victims, no one could, strictly speaking, consider himself completely innocent. Everyone had a suspicion—if not a certainty—that in some sense all had helped to pave the way which the armed forces used in coming to power. Some more, others less, but everyone had prepared the way for violence. Whether or not they believed in the virtues of State intervention in the economy, everyone, without exception, was strictly Statist in their political conceptions. If we can speak of a "Brazilian ideology" for the epoch beginning with the revolution of 1930, it would be an "ideology of the State" in the sense in which Bolivar Lamounier uses the word.[13] As Dahrendorf says for Weimar Germany, and for in Brazil as well from 1930 to 1964, "Everything is opened for the state side of life."[14]

The fundamental assumption of political activity in Brazil during that period could be summarized by Gramsci's famous metaphor about Oriental countries. The State was everything, and society, inarticulate and gelatinous, was nothing. No Brazilian intellectual translated this "oriental," strictly authoritarian conception of relations between State and society better than the conservative and pro-fascist Oliveira Vianna.[15] His was the conception of a demiurgic State, from which none of the Brazilian political parties of that period entirely escaped. All of them conceived politics and the State itself with the same palacian narrowness of their horizons. In addition, absolutely nobody involved in politics after 1930 was exempt from the common sin of knocking at the door of the barracks. Some more than others, but all political parties had their favorite generals and brigadiers.

Even the Communist Party in this respect demonstrated a trait which differentiated it from the great majority of Communist parties in Latin America and perhaps in the whole Western world. Begun as a small group of anarchists, the party came under the leadership of Luis Carlos Prestes, the best-known name in *tenentismo* in the 1930s. Some of his comrades from *tenentismo* joined along with Prestes, giving the PC a fair amount of influence in the Brazilian military milieu until 1964.

Whatever the importance of the Communists at that time, the example serves to affirm that in the armed forces at the beginning of the 1960s, military politics was cultivated as something which would be decisive with regard to the directions of civilian politics. There were generals, admirals, and brigadiers belonging to the center, the left, and the right, the last being obviously the majority. If the State is all, isn't recourse to the hard core of the State, which is the military, inevitable?

Thus all parties thought of society not as a space for politics, but as something amorphous, something supposedly incapable of becoming organized. So it is understandable that when after 1968 liberals and leftists of all tendencies turned away from their habitual exaltation of the State, they began to exalt civil society in the same way. But the fact that the discovery of civil society resulted from deep ruptures in the ideological traditions of the country does not mean that it was primarily an intellectual discovery.

In fact, the discovery that there was something more to politics than the State began with the simplest facts of life of the persecuted. In the most difficult moments, they had to make use of what they found around them. There were no parties to go to, nor courts in which they could have confidence. At a difficult time, the primary recourse was the family, friends, and in some cases fellow workers. If there was any legal chance at defense, they had to look for a courageous lawyer. And, above all, someone who is persecuted can always, as an old Brazilian proverb says, "complain to the bishop."

What are we talking about if not civil society, though still at the molecular level of interpersonal relations? There were years during which phrases of black humor circulated, such as "we are all equal under torture." While the number of those who were imprisoned, tortured, and killed in Brazil is doubtless much smaller than in Spain, Chile, or Argentina, it was nonetheless enough for "State terror" to carry out its plan to frighten everyone who had any political inclinations. It was a large enough number for everyone to realize that the persecuted were not exceptions to the rule. On the contrary, they were the rule. And in the resulting climate of fear there was no essential difference, from a political point of view, between those who were free and those who were in prison. As Raymundo Faoro once said, the only difference between a free man and a prisoner is that the first was tolerated and the other was not.

"State terror" had reduced all its opponents—generally on the left, but also many liberals—to their common denominator as unprotected and frightened human beings. Civil society was born out of this experience of fear. It was born in the family, which although it was sheltering a son or a father who was being sought by the repressive apparatus, was not, as an institution, politically suspect. After the family, in ascending order of participation, came the Church, the bar association, and the press. Eventually you could also go to politicians—whether of the government or the opposition—under the condition that they had courage and "access." But, strictly speaking, you could not talk to the parties about these things. The mission of resistance to arbitrary rule belonged especially to civil society which, when Médici was out and Geisel in, came to include unions, business groups, cultural associations, and so on.

In a situation of enormous ideological perplexity, the discovery of civil society was much less a question of theory than of necessity. Perhaps this is why the expression is sometimes used in an ambiguous manner. The development of resistance within civil society to the arbitrary rule of Médici and later to the violence which still continued for a time under Geisel was fueled in part by the results of the "economic miracle."

It was no accident that it was in the most industrialized states of the center and south—first São Paulo and then Rio and Minas—that the democratic resistance was strongest from the beginning. Opposition to the regime organized for the gubernatorial elections of 1965, and actually won the governships of Guanabara and Minas Gerais. The giant student demonstrations of 1968 in Rio won support from other sectors of the population and from the Catholic Church, leading to attempts by the "Commission of the Hundred Thousand" to negotiate a liberalization with the government. Trade unions, supposedly quiescent after most militant leaders had been replaced by government appointees, organized the Inter-Union Anti-Wage Squeeze Movement, and the Catholic Workers' Pastoral stimulated organization in the factories. The failure of these movements to bring about a change in the situation contributed to making these the areas in 1968 in which the most important *focos* of the armed resistance were found, rather than in the peripheral states, as a certain revolutionary romanticism of those years would have us suppose.

This is not to attribute to Médici and Delfim Netto any undeserved success. The military regime's economic wisdom was as irrelevant in producing the "economic miracle" as it is now, in 1983, in lowering the inflation rates which are running around 250 percent per year. But one cannot deny the fact that following up on a process of structural change of industrial capitalism which comes from the Juscelino Kubitshek government in the middle of the 1950s, the military regime pushed the country a little further towards industrial growth. For example, between 1970 and 1980 the population employed in the industrial sector doubled, as it did between 1960 and 1970.

Whoever participated during the years of the resistance or even now in any activity within civil society knows its weaknesses and its real strengths. And for this reason it is always surprising how much respect the public in general—and to a certain degree the State itself—reserves for its actions. I know many intellectuals who trembled in their boots when in 1975 they joined the protest demonstrations against the death of Vladimir Herzog. But it is evident that the public viewed these demonstrations as expressions of the courage of civil society. In retrospect, the first actions of the Church during the terrible period from 1969 to 1970 appear timid. And withal, they were a moment of indisputable political relevance. It always struck students from other countries that Brazil in those years had more studies on the working class than the latter deserved given its long silence since 1964, broken only by two strikes in 1968. The point is that such studies, even though they dealt with the past, were ways of calling attention to the possibility that the workers' movement might begin to move again in the present. The reality of civil society stayed somewhere between the amorphous nullity of traditional conceptions and the image produced by its great public resonance.

In these considerations on civil society, what I called the unlikelihood of the Brazilian political transition comes up again. We want a civil society, we need it to defend ourselves from the monstrous State in front of us. This means that if it does not exist, we need to invent it. If it is small, we need to enlarge it. There is no place for excesses of skepticism in this question, because it would only serve to make the weak even weaker. It is evident that when I speak here of "invention" or "enlargement" I am not using these words in the sense of propaganda tricks. I use them as signs of values present in political action, which give it meaning precisely because the action intends to make them real. In a word, we need to build civil society because we want freedom.

NOTES

1. Sérgio Buarque de Holanda, *Raizes do Brasil* (Rio de Janeiro: Livraria José Olympio Editora, 1982), 119 (first published, 1936).

2. Karl Marx, "The Eighteenth Brumaire of Luis Bonaparte," in Robert C. Tucker, (ed.), *The Marx-Engels Reader* (New York: Norton, 1978), 595.

3. Alain Touraine, "Industrialisation et conscience ouvière à São Paulo, "*Sociologie due Travail* III, no. 4 (1961): 77–95.

4. Carlos Nelson Coutinho, "A Democracia Como Valor Universal," *Encontros com a Civilazação Brasileira*, no. 9 (March 1979): 33–47.

5. Constitution of the Federal Republic of Brazil (1969), Article 1, Section 1.

6. The phrase was first publicly used by Geisel in his August 1974 speech before ARENA party leaders opening the campaign for that November's congressional elections.

7. See Skidmore's article in this volume.

8. Juan Linz, "The Transition from an Authoritarian Regime to Democracy in Spain: Some Thoughts for Brazilians" (paper delivered at the "Conference on Democratizing Brazil," Yale University, March 2, 1983).

9. Max Weber, "The Fundamental Concepts of Sociology," in *The Theory of Social and Economic Organizations*, ed. Talcott Parsons (New York: Free Press, 1964), 156.

10. Frederick Engels, "The Origins of Family, Private Property and the State," in Marx and Engels, *Selected Works* (Moscow: Foreign Languages Publishing House, 1958), Vol. 2, pp. 155–296, esp. pp. 288–96.

11. James Madison, *The Federalist Papers, No. 51.*

12. See, for example, his *Dialética do Desenvolvimento* (Rio de Janeiro: Editora Fundo de Cultura, 1964).

13. See Bolivar Lamounier, "Ideology and Authoritarian Regimes: Theoretical Perspectives and a Study of the Brazilian Case" (Ph.D. dissertation, University of California at Los Angeles, 1974).

14. Ralf Dahrendorf, *Sociedad y Libertad* (Madrid: Editora Ternos, 1979), 249.

15. For an examination of the thought of Oliveira Vianna, see Evaldo Amaro Viera, *Oliveira Vianna e o Estado Corporativo* (São Paulo: Grijalbo, 1976).

11

The Brazilian "New Republic": Under the "Sword of Damocles"*

MARIA DO CARMO CAMPELLO DE SOUZA

Although some political scenarios are more likely to materialize than others, the Brazilian political transition is so muddled that predicting the route it might take is a risky exercise. Among political analysts, however, there is extensive agreement on two points: (1) the enormous political control of the armed forces, not paralleled by the Argentine, Spanish, or Uruguayan transitions; and (2) the incompleteness of the transition, since a new phase, with direct presidential elections and the promulgation of the new Constitution, is still to begin. Reviewing the political process of the "New Republic," I shall try to detect some of the limits and potential of the current Brazilian democratic undertaking. To embark on such an exercise means to work on uncertain terrain, given the usual problems in any attempt to comprehend events still poorly understood and to which we are too close to have a proper perspective.

If in this analysis special emphasis is given to variables of a basically political nature, this does not mean that a democracy can be constructed solely on the force of juridical principles or through sophisticated institutional arrangements. A democratic regime must demonstrate its ability to represent the interests of society. The problematic character of the Brazilian transition results essentially from the problems that come with both an inequitable socio-economic order, marked by the concentration of wealth, and the growing political awareness among social groups of the need for changes in the system. Nonetheless, if it is true that political democracy is fragile in the absence of social democracy, we cannot assume that a social

*The author would like to thank Margaret Keck, Maria Lúcia Montes, and Scott Martin for their assistance with this article.

democracy will be achieved automatically in capitalist countries without democratic political conditions.

In the sections that follow, I shall limit my exposition to some observations regarding (I) the modality of the Brazilian transition process, emphasizing opposition strategies and the ideological process I call "invertebrate centrism"; (II) recent political tendencies, focusing on the relationship between government performance in the socio-economic sphere and the credibility of the new democratic regime; (III) the capacity of democratic political institutions—particularly political parties and the Congress—to control the executive branch, the military, and the bureaucracy; and (IV) the mechanisms of inclusion and exclusion developed by the new civilian government to deal with the many new social movements and political organizations that emerged during the dictatorship.

I. The End of Authoritarian Rule

A. Opposition Strategies

The creation of democratic political and institutional conditions in Brazil depends largely on transformations in the patterns of participation and exclusion that characterize the political system; on changes in the roles some groups within the State—especially the military—believe they should play in the political system; and also on an extensive redesigning of the State apparatus.

The difficulties of such a wide-ranging process are increased by the absence (now clearly perceived) of a coordinated project which could orient the constitutional debate; by the negotiated form of the Brazilian transition, which established a framework of marked continuity from authoritarian rule to the present regime; and by the concentration of tasks placed on the political agenda.[1]

Towards the end of the authoritarian regime, the strategy of the opposition was conditioned by a consensus that a constitutional policy for the new regime was either too hazardous or hardly necessary. Between 1984 and 1985 some opposition groups made only timid evaluations of their own strength, understandable after twenty years of military dictatorship. This kind of diffidence, together with the fact that the negotiations involved in the Brazilian transition were conducted behind closed doors, ended up confining opposition demands. In contrast, other opposition sectors displayed growing optimism concerning the evolution of the political process, the result of an exaggerated trust in the impetus of civil society and in the impact on the military of the prevailing consensus on the illegitimacy of the authoritarian regime.

The political arena was clearly dominated by a generic agreement on the need for the new government to transform the agrarian structure, the tax and wage structures, and the distribution of wealth, issues which, in similarly broad terms, had already been raised by the opposition during the authoritarian regime. As to the new constitution, however, it must be said that agreement between the various opposition forces was limited to such demands as direct presidential elections, which have yet to be instituted, and the carting away of the "authoritarian refuse" (*entulho autoritário*), a good part of which remains in operation.

Albert Hirschman quite correctly points out the danger that projects with rigid prerequisites, inflexible, built-in positions, or too many interlocking elements pose for the consolidation of democracy.[2] In the Brazilian case, however, the problem was quite the opposite: a lack of minimally articulated positions prior to the inauguration of the new government that could serve as a basis for subsequent public debate.

Center-left and leftist groups held a misleading conception of political parties, causing them to evaluate the institutional force of the party system almost exclusively based on the vigor of social groups' participation. Although they were aware of the structural limits of the transition, left-wing sectors hoped that the mobilization of civil society (which had massively demonstrated against authoritarianism during the campaign for direct elections) and the activities of social and political organizations (which had fought against military rule) would continue to grow and foster democratization. Thus, in order to break through the last obstacles keeping an institutionalized party system from flourishing in Brazil, many on the left thought it would be sufficient to install civilian rule and to widen the scope of political party options.

The issue of reorganizing the State apparatus was possibly one of the least debated points among the opposition forces. Although they have managed to occupy positions at various levels in the State apparatus—thus beginning to erode the model characteristic of the former regime, in which decisions concerning economic and social issues were concentrated in the Finance and Planning Ministries—this has not as yet been sufficient to produce a rearrangement of the decision-making structure as a whole.

Moreover, the entrepreneurial sectors who supported the transition process also lacked a politically articulated project of their own, if by this we understand something more than the demand for a generic pledge from the new government to preserve private property and to maintain a less drastic version of the current policy of capital accumulation. Only recently has the action of these sectors begun to show some degree of articulation, in the form of public demonstrations and statements through the UBE (Brazilian Union of Business Leaders) and the FNLI (National Front for Free Enterprise) and by attempts to coordinate into a unified strategy their pressure on the Constituent Assembly.

The character of the Brazilian transition, managed as it was in coparticipation with the military,[3] also inhibited open discussion about a "policy for the military" to be endorsed by PMDB opposition sectors. These sectors seemed to pin their hopes on the power of the armed forces becoming attenuated or outweighed by the coming into operation of more solid and representative institutions, particularly political parties and the Congress.

Three years after the inauguration of the New Republic, it is obvious to any observer that, rather than moving in line with the expectations outlined above, the country is heading in the opposite direction, as we observe the following:

> Political parties have failed to carry out the functions of governing and representing the great majority of the population. The definition of ideological boundaries continues to revolve around populist discourse, and political party consolidation is dependent upon political networks that feed on State clientelism;[4]
>
> Aside from a considerable degree of union mobilization, there has been an ebb in the movements of so-called "civil society," which during the military regime reached an intensity unseen during the populist-democratic period, both in absolute and relative importance. It is almost as if whatever in the last years of the dictatorship civil society was able to mobilize, the first years of the New Republic have demobilized;[5]
>
> There continues to be a high degree of military control over the political process.[6] Through vetoes and explicit threats, the military have prevented innovative measures and reforms at all levels of Brazilian politics. Although invested with the mission to establish the breakthrough to a new democratic order, the New Republic has not promoted the hoped-for return of the military to the barracks. To many analysts, the New Republic has in fact brought the military into politics for good. Free of the onus of the direct exercise of power, military presence in politics has become more difficult to remove than ever because it appears to be a "normal" part of civilian rule.

B. "Invertebrate Centrism"

The sense of public relief and enthusiasm resulting from the political opening that in 1985 brought a civilian leader to the presidency of the Republic highlighted the positive aspects of the political process of opposition to the dictatorship, thus concealing, for the most part, the obstacles that the Brazilian modality of transition could pose for later democratic development.

It is well known that the Brazilian transition established the New Republic on the institutional foundations of the authoritarian regime rather than on its ruins, allowing most of the political elites and administrative personnel of the former regime to remain in control of the country's political course.

If a transfer or partial surrender of authoritarian power to a democratic opposition seems more propitious for the consolidation of democracy than an overthrow of such power by implacable antagonists, this version of democratic transition can involve extremely high costs that may even jeopardize the democratization process.[7]

When we compare the Brazilian transition with that of Spain—a case always mentioned in the debate about negotiated transitions—an important difference between the two must be stressed: the public character of the Spanish negotiations between opposition groups and sectors allied to the Franco regime. The democratic agreement articulated by Adolfo Suárez and King Juan Carlos prior to the Moncloa Pact was negotiated with representatives of the Spanish socialist and communist parties (PSOE and PCE); the radical right-wing parties (Alianza Popular and Fuerza Nueva) did not take part in it.[8] As stated in the 1984 "Pledge to the Nation," which resulted in the Democratic Alliance, the Brazilian version of negotiation did not involve the presence of left-wing parties or representatives of its allied sectors, nor was it conducted in a manner visible to society. The Brazilian agreement followed the logic of a pact among regional elites supported by military factions; it was not an accord reached by spokesmen of party institutions responsible to their voters and constituencies.

The present party-parliamentary process stems precisely from this type of agreement. I would call it "invertebrate centrism," since it is characterized by a vast center whose boundaries and backbone are unknown, an ideological and political arena in which everyone seems to be in alliance with everyone else, yet each is actually alone. It is not surprising, therefore, that the government faces tremendous difficulties in trying to build its parliamentary support or that society seems unable to make heads or tails of the political process.

Certainly, important continuities can always be observed in non-revolutionary transitions, with regard both to the former authoritarian regime and to the configuration of the dominant sectors who supported military rule. Even revolutions do not completely transform political systems, and in the study of such processes debates about change and continuity are of great relevance. The continuities of less drastic transitions are even more ambiguous and difficult to grasp.[9] In Brazil, this phenomenon is far more apparent than in other cases. The degree of continuity among administrators and politicians from the former regime is extraordinary. Even in Spain, where great continuity of political elites could be observed between the Franco regime and the post-Franco era, the transition was more rapid than in Brazil, and politicians linked with the authoritarian system lost elections with reasonable speed.[10]

Statistical data confirm the enormous presence of the old regime in the two parties that sustain the New Republic and at the head of many ministries, raising questions about the conservative profile of the present Constituent Assembly and whether solutions to pressing socio-economic problems will be found.

It is significant that the largest delegation in the Constituent Assembly is not formed strictly by longtime members of the PMDB but rather, seen from a 1979 perspective, by the old ARENA. No less than 217 of the 559 members of the Constituent Assembly were at some point before 1980 in that party, which supported the authoritarian regime. Of the 298 representatives in the PMDB delegation to the Constituent Assembly in 1987, 40 members belonged to the PDS in 1983 and another 42 were in ARENA in 1979. Only 137 representatives of the PMDB in 1987 came from the MDB of 1979 or from the PMDB of 1983. If we add those with no previous party affiliation—the 47 who entered the PMDB in 1982 and another 28 who were elected in 1986—the whole delegation of representatives coming from the "historical" PMDB in 1987 adds up to only 40 percent of the members of the Constituent Assembly and is not, therefore, the "hegemonic" delegation of 53.3 percent that the party has *de jure*.[11] In addition, three years after the reinstatement of civilian rule, the highest federal executive office is still occupied by the ex-president of the party upon which the parliamentary power of the military regime was based (ARENA/PDS). Furthermore, an offshoot of this party, the PFL, holds increasingly more control over important posts in the cabinet.[12] If to this we add the fact that the Constituent Assembly has the prerogative of both writing the new Constitution and performing the regular functions of the national Congress, and the continued existence of ultra-authoritarian legal provisions (such as the National Security Law, which has been used at various moments since the beginning of the new government), then some of the constraints imposed on the redemocratization process by the Brazilian modality of transition become clearly visible.[13]

It seems to me, however, that far too much weight has been placed on variables such as the social origins or former party affiliation of present members of the Constituent Assembly by several analysts in their evaluation of the prospects for the constitutional reform or in explaining the difficulties that the construction of party consensus must face.

The difficulties encountered in aggregating support and in coming to parliamentary agreements cannot be explained solely by "continuism" or the one-dimensional ideological character of political elites. The linear causality between social origin and political beliefs is clearly insufficient in terms of its predictive capacity.[14] This is also true of congressional representatives' party affiliations during the authoritarian regime; in the Brazilian party system, the central variable in the explanation of party membership is access to State clientelism.

The real weight of all these variables becomes clear when we examine the rather conflicting data we have on the attitudes of members of the Constituent Assembly. It is interesting to note that at the beginning of 1987, 58.4 percent of the members claimed to be in favor of employment security for workers. Also significant is the fact that, among the majority (81.9 %) who defended an increase in measures to improve working conditions, an important number came from the Northeastern states, which are economically less developed and regarded as strongholds of political conservatism. Though the positions they take in the actual constitutional deliberations undoubtedly reflect various other factors, a surprising 84 percent claimed to have become candidates to the Constituent Assembly because of social issues. Moreover, although given the prevailing political uncertainty it is unlikely that the Assembly will decide to alter the military's constitutional prerogatives, 50.7 percent of all parliamentary representatives stated they were against conceding the armed forces responsibility for internal security. Despite the overall conservative tone that the new Constitution certainly will have, to emphasize a generic conservative *continuismo* clearly limits our understanding of delegates' positions on different issues and is insufficient to explain parliamentary conflicts and outcomes.[15]

Those analysts who emphasize continuity perceive the tendency always to seek close ties with sitting governments, whatever their stripe, as an "atavistic" trait of the Brazilian political class, thus obscuring a basic structural constraint on the relationship between parties and the State: the necessity of *State clientelism* for the survival of parties. This is an historical characteristic of mass parties in Brazil, which has been present since their establishment in 1945.

In fact, the conditions surrounding the origins of the political parties are of particular importance in determining their character as organizations. Thus, whether or not a party comes to use the distribution of clientelistic benefits to gain support from social groups closely relates to the way in which the party leadership *initially* established ties with its popular bases of support. This initial connection marks the character of the organization in terms of the mechanisms it must activate in order to preserve its electoral base, while within the parties it serves as the basis of the great bargaining power that politicians who make use of clientelism hold vis-à-vis their eventual opponents.[16]

An approach that does not examine the relationship between political parties and the State, the structural characteristics of inter- and intra-party competition, and regional variables as central criteria in the establishment of parliamentary alliances will not illuminate the most basic conflicts and cleavages in the constitutional debate.

Exclusive insistence on continuity of political elites has obscured other important points that party strategies must take into account and which will be of great relevance in understanding the future political process. In fact, many politicians visibly identified with the military regime have been defeated, and the 1985 and 1986 elections favored a bloc of parties that, roughly speaking, could be said to represent the "center-left." The weak performance of the PFL, the most important of the parties that congregate the political forces of the former regime, is particularly notable.[17] The emphasis on *continuismo* makes it difficult to comprehend the character and evolution of the majoritarian party, the PMDB. By stressing conflicts within the party, viewed as a product of the swelling of its ranks with politicians connected with the former regime, the analysis fails to focus on questions such as the reasons for the timing of these internal splits, that is, conflicting positions regarding alternative models of economic policy, an issue that was already controversial before the New Republic.

If today we have a government whose legitimacy and ability to function depend on the electoral and parliamentary support of an amorphous and divided PMDB, this is not entirely a result of the continuing presence of politicians identified with the authoritarian regime.[18] It is primarily a consequence of the transition itself, which brought to light and exacerbated existing internal fissures. The need to maintain party unity—induced in particular by the constraints of an electoral competition centered on access to State clientelism—makes the party seem, from a popular point of view, quite distant from the image it had created during the military regime.[19] The PMDB was expected to lead the process of democratic transition. Yet today the party is not even the main instrument of communication either from the government to the people or vice versa.[20] If the party's internal differences are not resolved, moreover, a sizable group may soon leave to found a new party.

Together, all the factors we have noted so far—the modality of the transition, the absence of minimally articulated policies for the new regime, and disarray among political parties—have led to the difficulties that the political process faces today. They have also created a context in which little has been done to alter or minimize the weight of some factors which are frequently said to maintain and reproduce both the instability of political democracy in Brazil and the social injustice that ends up delegitimizing democracy:

(1) the "praetorian" characteristics of the political system, which have led to constant military intervention and—worse yet—an acceptance of such intervention as legitimate by the "producing classes" (as the economic elites call themselves) and the majority of the population;

(2) the reinforcement of corporatism and of the direct integration of social interests into the State, without the mediation by political parties; and

(3) the high costs of political information for the popular sectors, which either favor support for populism, scarcely controlled by the State or parties, or lead to apathy and indifference, which are no less disastrous.

II. The Politics of the New Republic, the Dimensions of the Brazilian Political System, and the Credibility of Democracy

Without going into the historical and sociological debate over the relationship between government efficacy, dissatisfaction with the economic system, and support for political democracy, it is possible to say that, insofar as the government is sovereign and democratically elected, authorities have in principle the power to transform the socio-economic structure. Popular support for a democratic regime may be accompanied by disappointment in its efficacy. As long as there is hope for change through democratic means, belief in the system can remain strong.

In an important study about the impact of the 1929 economic depression on European democracies, Ekkart Zimmermann demonstrates a relationship less direct than previously thought among economic crisis, its various indicators, and political crises in democratic regimes.[21] The conclusions of his study indicate that, in breakdowns of democracies, greater weight should be attributed to political factors, especially what he calls the presence or absence of "system blame." In Germany and Austria, "negative evaluation of the democratic system" was so pervasive that, when some sectors finally came to the defense of the regime, they already were such a minority that they were unable to prevent its fall. For many other European democracies (such as Sweden, Holland, and Belgium) which survived the impact of the depression—whose consequences were as strong for them as for Austria and Germany—the post-depression era was, on the other hand, a period of intense creative political engineering, when political coalitions and new policies were established which guaranteed the resilience of the democratic regime.

Research on the recent Spanish transition shows that, although the deterioration of the Spanish economic situation after Franco's death led to a decline in the public's evaluation of the democratic regime's performance, this has not prompted the Spanish population to lose faith in democracy as the best political system for the country.[22] Under the Franco regime, the unemployment rate in the early '70s was one of the lowest in Europe, averaging 3 percent. It has risen continually since then, and, under democratic rule, the present 20 percent unemployment rate is one of the highest in Western Europe. The GNP growth rate, which between 1960 and 1974 averaged 7 percent—one of the highest in the world—reached only 1.7 percent between 1975 and 1985. Nevertheless, the positive response in surveys to democracy as the best political system for Spain actually grew 6 percent between 1978 and 1983. Even more significant, despite disappointment with the socio-economic performance of democracy, practically the whole Spanish population rejects a return to military government.[23]

In a country like Brazil, with world record-high levels of economic and social inequality,[24] the total absence, over a period of almost three years, of comprehensive economic policies designed to alter minimally these conditions—except for the brief interlude of the Cruzado Plan[25]—could represent yet another complicating factor for the process of extending democratization. The economic inviability of, or the political refusal to undertake, attempts to solve at least the most extreme inequalities is exacerbated by the government's failure to persuade the populace that at least it is committed to finding a solution. The crucial intervening variable that explains the relationship between the socio-economic efficacy of the regime and the credibility of democracy is the public's perception of how much a democratic regime can actually achieve. This perception has proved to be increasingly negative in Brazil, because of peculiarities in its political system and the management of the recent political process.

Little can be said a priori about how democratic institutional mechanisms and practices could prove more resilient against negative evaluations of the regime's socio-economic performance. Ideally, there are reasons to believe that the legitimacy of democracy can somehow be insulated from negative public perceptions about its socio-economic efficacy through the following: acceptance of the democratic procedures that produce government officials;

faith in the electoral process as an instrument that leads to the turnover of those in power and to new economic policies; system identification engendered by party competition; and the collective memory of the abuses committed by the previous authoritarian regimes.

Due to several factors, this separation between democratic legitimacy and socio-economic performance is not being made in Brazil, or has been complicated by conflicts characteristic of a period in which political actors' evaluation of their own power and that of their antagonists is still uncertain. In general terms, these factors have to do with the central role of the State in the economic development of the country and with the limitations of a political system in which institutions are insufficiently autonomous to buffer shocks. The extent of State intervention in the Brazilian economy explains why, in the process of attributing responsibility for the economic crisis, the public looks towards the political rather than the social sphere to find "guilty parties." Entrepreneurs consider the federal government responsible for a crisis because they claim that State intervention distorts the normal course of the capitalist system. For the great majority of the people, the federal government *and* the President of the Republic (seen as the virtual personification of the regime) appear as responsible for the current turmoil.

Likewise, the constraints that the political regime still imposes on social life and the peculiar absence of a competitive and institutionalized opposition make it difficult either to link the government's low efficacy in the socio-economic sphere to different political actors or to locate it at different institutional levels. The multiplicity of institutional centers that can be held responsible in a more expanded democratic system frees the government from accountability in a crisis. In Brazil today, it is difficult to distinguish between the political system as a form of authority legitimation and the government, a distinction still harder to make when almost all the competitive political forces are included simultaneously under the government and the opposition labels. Thus, mistrust of the governing authorities of the day, particularly of the President, is more rapidly carried over to the entire political system. In November 1987 the majority of Brazilians (62.7 %) wanted to have a new President installed by March 1988.[26] If the current formula holds, however, President Sarney will not leave office until March 1990.

In the following section I will discuss six other factors that, with varying impacts on different social groups, have contributed to a negative evaluation of the democratic regime.

A. The Weak Electoral Claim to Rule

It is well known that President Sarney reached the office of chief executive due to special circumstances. As Vice President, he succeeded the deceased Tancredo Neves, who had just been indirectly elected President by Congress after negotiations among opposition leaders, military sectors, and pro-regime regional party groups. These procedures could hardly be viewed as democratic by Brazilian society, especially given the extensive mobilization of millions of people all over the country in favor of direct elections during the final months of the military regime.[27] The contrast with other countries is striking. In Spain and Argentina direct elections enhanced the legitimacy of Adolfo Suárez as Prime Minister and Raúl Alfonsin as President of the Republic. Both men rose to authority with the support of a well-defined party alliance.

In Brazil, the electoral issue is becoming ever more serious. President Sarney has rigidly refused to accept a reduction in his term of office by the Constituent Assembly, and subsequent disagreements over the length of the presidential term have been the focus of almost all the constitutional debate. These clashes have brought about erratic attempts to create a core of party-parliamentary support for the government, thus leading to a paralysis in decision making on crucial issues. Ultimately, the latter have come to depend upon an evaluation of the length of the term of the present chief executive, which is to be established by the Constituent Assembly. Moreover, Sarney's leadership is increasingly seen as illegitimate both by the national Congress and by society as a whole. For instance, a survey conducted by the *Jornal da Tarde* in the city of São Paulo and published on October 26, 1987, shows that only 20 percent of those interviewed approved of the President's performance.[28]

Federal intervention in the debate over direct elections and the term of the chief executive takes forms that are comparable to those which occurred during the military regime. A particularly revealing example is Radiobrás, the news and information center of the Ministry of Communications, which prohibited all its journalists and broadcasters from making any reference to the campaign for direct elections in 1988 or publishing interviews with politicians who demand that President Sarney step down after four years in favor of a directly elected President.[29]

With existing data it is difficult to assess the population's perspectives concerning the electoral process and its ability to find solutions for pressing socio-economic problems. According to the survey conducted by the *Jornal da Tarde*, most people living in the city of São Paulo still seem to have faith in direct elections' supposed capacity to bring about changes: 76 percent of those interviewed want *immediate direct elections* for *all* legislative and executive offices.[30] Other surveys, carried out by IBOPE among 800 respondents in the metropolitan area of Rio and São Paulo and published in the *Jornal do Brasil* on May 31, 1987, show that in answer to the question, "What type of government do you think is best for the country?," a strong majority of people responded, "Democratic with direct popular participation." But the relation of such answers to the current political atmosphere in Brazil is problematic, since the survey was limited geographically and also because the options the respondents had to choose from were obscure. How did they interpret terms such as "direct popular participation"? Would this indicate a delegitimization of the representative process? Preferences for answers like "democracy with direct popular participation" bear the mark of a more conjunctural negative evaluation of current political representatives and of dissatisfaction with the present version of Brazilian democracy brought about by the 1986 elections.

In any case, on this question there is consensus among political analysts: the role of political parties and of the government in economic policies has contributed to a process of delegitimization of the present parties. The same survey published by the *Jornal do Brasil* indicates that 57 percent of the population believes that political parties "are necessary, but perform their role badly." According to data presented by the *Jornal da Tarde* on October 26, 1987, 76 percent of the population believes that the candidates for the next elections should be independent, that is, not nominated by parties.[31]

Since it is possibly the most pedagogical example for analyzing the relationship between economic policy and the political party process, what does the entire process surrounding the Cruzado Plan reveal?[32] Two important issues have been forgotten in the chronicles of the New Republic, since they were supplanted by criticism on the negative economic effects of the Plan. The first is that the Cruzado Plan was more vulnerable to the electoral process, and therefore more democratic, than any exercise of power from the former period, when no less "pragmatic" and short-ranged policies responded almost exclusively to the demands of pressure groups that had privileged access to the bureaucracy. The second—it is worth repeating—is the tremendous power the implementation of redistributive measures, even if short-lived, proved to have as a means of aggregating party support for the government.[33]

The electoral administration of the Cruzado Plan led to vast popular mobilization—effected by the government through the mass media, especially the Globo television network. That mobilization was frustrated abruptly by the virtual abolition of the Plan immediately after the 1986 election ballots were counted. Popular support for the Plan was reflected in the vast number of votes received by the PMDB in November 1986. After it was discontinued, President Sarney's popularity fell to a level similar to that of April 1985, one of the lowest since the beginning of the New Republic.[34] Once the election was over, the PMDB—even if it had wanted to—could do little to promote similar measures, given its own structural fragility. The way this process was conducted damaged the credibility of the electoral process in the public's eyes, shaking its belief in the "illusion of choice" through the vote.

Following the pattern of "praetorian" systems, political parties and the government expanded mobilization to gain power, thus increasing expectations, so that eventually it became difficult to separate measures in the sphere of socio-economic policies and those regarding the current constitutional reform. More articulated political efforts vis-à-vis the population by the parties would not only have permitted the expansion of economic redistribution but also guaranteed society's support for necessary readjustments in the Plan during the months of July and August 1986.

An analysis of the Cruzado Plan also reveals the difficulties encountered at the governmental level in the establishment of party and intra-party agreements regarding economic policies, both prior to the decisions and at the time of their implementation. Such difficulties derive from the process I called "invertebrate centrism," which, although different in nature from the process of "polarized pluralism" described by Sartori,[35] leads to the same paralyzing consequences in the governmental decision-making sphere. This "invertebrate centrism," based on regional party politics and the "feudalization" of the State apparatus, fragments State authority in the implementation phase of economic measures.[36]

B. The Waning Power of Nationalism

The Brazilian economic crisis is, to a large extent, a consequence of the international situation and the foreign debt problem. The government constantly emphasizes this international connection, trying to make the public understand the limits it imposes on national economic sovereignty.

Although the government may consider appealing to nationalism if new restrictions imposed by the international economy or by American protectionism emerge, such appeals, though they were often used to a great extent in former periods, are less and less frequent today. They have been increasingly restricted to parliamentary debate, where they are used primarily by representatives of a few regional interests, especially from the Northeast.

At any rate, appeals to nationalism seem no longer to be able to temper the impact of negative evaluations of government performance, as two complementary processes—which are perceived mainly as the product of internal political conditions—take hold in the decision-making sphere: *decision-making paralysis* in some areas of economic policy, coupled with *precipitious decision making* in others. The former is induced by disagreements among parties and between parties and the government, as evidenced by the impasse on agrarian reform; the latter is derived from a centralized, bureaucratic decision-making model with insufficient articulation of internal and international political support, the most clear examples of which are the deactivation of the Cruzado Plan and the moratorium on debt service payments.[37]

C. Entrepreneurial Ambivalence

Entrepreneurs' perceptions of the democratization process are crucial to the public's evaluation of the democratic regime. The New Republic has assured owners of capital of direct and privileged access to the State apparatus, a privilege they have always enjoyed. Theorists of the power elite and of neocorporatism say that, among all the means at their disposal to protect their interests in a democracy, it would be sufficient for capitalists to maintain this access to have their power guaranteed. Yet uncertainty and growing decision-making paralysis, whose negative impact on this sector can sometimes be stronger than the content of some public policies, frequently have the effect of upsetting the balance of costs and benefits, leading it to withdraw its support for the democratic regime.[38]

Unlike Argentina, where the authoritarian regime's model of economic policy led to a dismantling of the economy of the country, in Brazil the economic policies of the military regime enabled many entrepreneurs to amass their present wealth. Consequently, the mild and ambiguous support given by these sectors to the democratization process can be more easily reversed. In June 1978, a few days before the publication of a document that came to be known as the democratic manifesto of the Brazilian bourgeoisie, a group of one hundred businessmen sent a letter to the President of the Republic in which, although expressing their dissatisfaction with the extent of State intervention in the economy, they demonstrated their apprehension regarding the debate then occurring over liberalism, explicitly questioning whether it would not bring about a return of the Communists to the Brazilian political scene. Despite its old-fashioned tone, this document cannot be ignored, given its similarity to the alarmist discourse of these sectors before the military coup of 1964.[39]

Yet although their actions may become harsher, given the way they perceive the current course of events in the Constituent Assembly—to which they attribute the label "anti-free enterprise"—this does not mean that as a whole business has adopted a position of open opposition to the regime, as it did in 1964. Moreover, it is interesting to note that the recent activities of entrepreneurial organizations reveal an unusual new development in Brazilian politics: for the first time, this sector is undertaking political action publicly and through class organizations,[40] instead of expressing its demands directly and exclusively at the administrative-bureaucratic ranks or contriving conspiracies within the barracks.

D. Weak Human Rights Claims

Recent military interventions in strikes and popular sector mobilizations or in areas concerned with issues of economic and social reform, together with the measures taken by the President in agreement with the leadership of the strongest parties to organize the 1986 election, could deepen the mistrust which many left and center-left forces feel regarding the new regime's willingness to protect their political rights and their chances to succeed in party competition.[41] With State clientelism as a basic condition for party survival, the regime restricts both their space for political action and their electoral chances.

The leading forces of the regime would thus seem to be about to waste one of their most important assets for building a political democracy—its legitimacy among left-wing sectors (as revealed by Weffort in this volume). This legitimacy has been expressed in the discourse and the action of the majority of the representatives of these sectors in the Constituent Assembly. The common critique of the democratic order has as its main spokespersons those in the conservative sectors, who frequently attribute to left-wing groups a merely rhetorical support for democracy.

The violation of human rights in Brazil did less to delegitimize the military regime than it did in other Latin American countries. Unlike the

Uruguayan, Chilean, or Argentine cases, the political crimes of the Brazilian authoritarian regime affected relatively small sectors, particularly in the middle class and the intellectual *milieu*, and they were less visible, given the size of the country. Moreover, the very length of the Brazilian transition, one of the longest known, helped to diminish the saliency of human rights abuses perpetrated during the military regime.

Thus, it is understandable—although painful to say—that the majority of the Brazilian people, whose existence has been constantly marked by the violation of almost all their rights, have not been affected to the extent one would expect by movements that criticize the abuses of the former regime, linking them exclusively to the violation of political rights in restricted sectors of society. This is even more understandable when we consider that the political liberalization, which led to the closing of prisons for political crimes and the end of arbitrary measures against freedom of opinion, was not extended to the sphere of police actions infringing on the rights of the popular sectors. On the contrary, arbitrary measures continue as always, in a country where being poor or black is frequently considered synonymous with being a criminal.[42]

Furthermore, organizations seeking the defense of human rights or amnesty for political crimes—which held great appeal for the middle class during authoritarian rule and which at the beginning of the new regime tried to organize their campaigns with more wide-ranging objectives—have undergone a process of continual erosion or repression. The leaders of several of these movements had imagined that, with the democratic transition, the demands which had earlier been used in defense of small groups of dissidents could be transferred to struggles for the protection of the impoverished majority. This somewhat innocent assumption clashed with the very mechanisms of social reproduction and domination in Brazil. The affirmation of these rights threatened—or was perceived to threaten—rules of obedience and the extremely rigid hierarchy within Brazilian society. Since one of the bases of this system is the threat of illegal violence, attempts made by human rights groups to control this violence—both in the institutions legally entitled to use it (prisons and mental hospitals) and in its illegal forms (torture of prisoners and suspects)—provoked a dramatic reaction by sectors in civil and political society, who came to reject principles they widely defended during the resistance to the dictatorship.

Some of the media have carried out a campaign clearly intended to deny groups connected with human rights movements access to the spheres where social policies were decided. The Church was a favorite target, continually denounced for its commitment to the causes of civil society movements and Liberation Theology. In some radio programs, the campaign in favor of human rights came to be identified as the "cause to defend outlaws." The Church and the commissions connected to it were viewed as accomplices in the "impunity of delinquency." These programs were never subjected to any official control by the new regime. The media campaign was heightened and reinvigorated when in 1985 Pope John Paul II imposed a vow of silence on the main theoretician of Liberation Theology, Friar Leonardo Boff.[43]

The strategy of the Church's high echelons to force a retreat by some of its more aggressive sectors—as Ralph Della Cava correctly points out in this volume—has been reinforced under the new regime, in light of a series of events purposely intended to delegitimize their activities.

The actions of the government, political elites, and the armed forces[44] (due to their direct involvement with these issues) have contributed to a loss of collective memory of the atrocities perpetrated by the former regime. Thus, a significant difference is to be noted here in relation to the centrality that the issue of human rights had in the Uruguayan democratization process or still has today in Argentina.

E. The Media and the Erosion of Legitimacy

The intervention of the Brazilian print media, radio, and television in the political process of the country demands systematic study of its own, focusing on the "adversarial discourse" present in the mass media with respect to democracy.[45] Nevertheless, given their importance in the formation of public opinion,[46] it seems possible to say that the media have played a major role in extending the aforementioned process of "system blame," not to mention the explicit fascist appeals made by some radio and television programs. The media also reinforce a traditional aspect of Brazil's political culture—a deep-seated mistrust of politics and politicians—thus reinforcing the lack of faith in the very structure of party-parliamentary representation.

This is not to deny the important role the media have played in denouncing blatant privileges, arbitrariness, and corruption, all of which are pervasive in the activities of current Brazilian government officials. In this respect, they guarantee one of the fundamental bases of democracy, the right of criticism. Yet "behind the scenes" coverage of the New Republic is frequently partial and selective, and on many occasions it clouds more than it clarifies.

Analysts have claimed that some sectors of the media appear to have mounted deliberate campaigns to delegitimize the present government or some party groups. For example, the press insists on attributing the responsibility for the mistakes of the Sarney administration almost exclusively to the PMDB and only to a lesser extent to the military or the PFL, the latter of which in fact controls a large number of Ministries that have a high degree of financial and political power.

There is nothing surprising about the partisan character of the media (although many present this as if it were a Brazilian invention). The most harmful aspects of existing practices among the Brazilian media are their "pseudo-partisanship" and irresponsibility—a consequence of the elitism and lack of transparency that continues to prevail in Brazilian political decision making—which frequently appear in the form of sourceless denunciations with slippery semantics, publicized through social columns as mundane news intended to reach the middle classes in the large capital cities.

I believe that here attention should be focused on the lenses through which the political process is interpreted by those who claim to be identified with the democratic project. Though with varying objectives, their criticism ends up adding to the attacks of the authoritarian right, insofar as both versions transmit the idea—so dear to the authoritarian thought of the '30s—that democracy is impossible in a country with Brazil's characteristics. For this, there are various reasons.

On the one hand, frequently the denunciations are not followed by the necessary comparative information about similar events in the former regime. After all, under authoritarian rule conditions were worse, given government repression and censorship, together with the self-censorship of newspaper, radio, and television editors, all of which made it difficult, if not impossible, to disclose events that were unfavorable to the regime.

On the other hand, the accusatory character of a large part of the information ends up establishing—particularly for those under twenty years of age, who have no political history and who represented nearly 50 percent of the population in 1980—a direct and harmful connection between the demoralization characteristic of the present conjuncture and the very substance of democratic regimes. This interdependence is reinforced when one notices that the favorite target of these denunciations is the sector that has come to be called the "political class," especially those who entered it via the electoral process, while "efficient" leaders more identified with the military regime's technocracy are constantly being rehabilitated. "Partisan political interests" has become an almost obligatory addendum to the exposition of denunciations of all sorts.

Despite the evident responsibility of the great majority of the political class for the somber development of the Brazilian political process, the mass media have blamed its members as if they were *homogeneous*, forming a sort of "kleptocracy." Yet at the same time their criticism has spared other commanding sectors of the New Republic. Often one gets the impression that corruption, cynicism, and arbitrariness are the monopoly of party politicians or Congress;[47] although subject to criticism, the military appear as an intransigent group, notorious for their interventionist spirit, deeply rooted attachment to hierarchy, and identification with order, characteristics which are not necessarily viewed as negative or rejected in Brazilian political culture. The selective character of information and the partiality of some campaigns bring to light yet another question arising in the current political system: Who are the leaders and institutions that command the greatest power? On the whole, they are those attacked the least by the media.

With regard to the smaller parties, viewed as radical by the establishment, the analyses of the conservative printed media and various radio programs emphasize the "destabilizing" aspects of their action, rather than their contributions to the work of the Constituent Assembly. From another perspective, it is also important to point out the consistent omission by the media of any information regarding attempts—feeble as they may be—to implant some reform measures in the country. On the contrary, the mass media, especially the television networks, are packed with information about redemptory presidential projects, as gigantic in size as in fantasy, that would turn the country into a land of abundance and equality. Thus, some characteristics typical of praetorian systems are once again reinforced—namely, belief in the all-powerful and magical capacity of the President and the growth of illusory social expectations.

F. The Private Appropriation of Public Goods

Nonetheless, the media reveal the lack of "authenticity" of the Brazilian public scene, which is a result of both the patrimonial aspects of the State and the actions of economic and political elites, for whom the boundaries between public and private goods are obscure. This trait, rooted in the political development of the country, was reinforced during the authoritarian regime, and became more visible with the political liberalization. If for Brazilian politicians, contented with their role as "State clientele," political action cannot or should not be geared towards the representation or the debate over the public good, for the people political action is oriented towards the discharge of deep-rooted frustrations or the imposition of ritualistic punishment on politicians through the electoral process.

The lack of authenticity and the demoralization of Brazilian public life malign party institutions and related processes. When the main lines of economic development or of modernization are not themselves objects of public debate, the ballot can prove to be, as has happened innumerous times in Brazil, one of the main instruments by which the bulk of society expresses in a plebiscitary manner its rejection of political life. When only private needs are projected into the public arena—or the process is perceived as such by society—and economic growth decreases and insecurity and uncertainty grow, voters tend to turn to those politicians who call themselves "pragmatic" or "efficient." It is much easier to talk about the need for stimulating economic growth and increasing the GNP than to discuss ways of doing it. The right is the first to ventilate the need for "immediate politics," declaring the superiority of "efficacy" over "politics" and establishing an incompatibility between these two terms, while its representative organizations pronounce themselves to be spokespersons for such efficacy and enemies of the "waste" brought about by politics. Authoritarian solutions follow both implicity and explicity from this discourse.

One of the most pernicious legacies of the Brazilian civilian and military elites—which was deepened in the twenty years of authoritarian rule and remains unaltered in the New Republic—is the insistence that politics be reserved almost exclusively for the satisfaction of private needs and well-being rather than for the debate about and achievement of the common good. For a country trying to build a democratic regime, this is a heavy legacy to bear.

The visible loss of confidence in political representatives in the country constitutes something more than the ever-present tension between representatives and the represented, inherent in all representative political systems. In the present context, it takes on an expanded and all-encompassing form, which increases the illegitimacy of the very structure of party and parliamentary representation. The lack of credibility of political representatives does not necessarily extend to the entire system, yet the rupture of institutional channels for the expression of demands, which often occurs along with the delegitimization of structures of representation, does weaken the legitimacy of the regime as a whole.

III. Political Institutions and System Governability

Up to this point we have examined the fragility of the democratic system built through the Brazilian transition. The question of the resilience of the democratic regime can be analyzed from another, more structural perspective, becoming an issue of which particular institutions designed for the supervision of government continue, and will continue, to be accepted by the citizenry in view of the "changing boundaries of the political."[48] This brings us to the analysis of some aspects of representative structures and of their effects on system governability.

As far as political integration and representation are concerned, Brazilian democratization is developing out of the great modernization and industrialization undergone by the country in recent decades and in the context of the increased ability of society to make demands. These demands can no longer be resolved within the institutional arrangements that regulate these processes, especially considering that some of these arrangements have prevailed since the 1930s. The marks that the great economic and social transformations of these last few years have left on political actors, their social and political organizations, and the political agenda became even more visible with the end of authoritarian military rule. Besides the traditional political actors—workers and entrepreneurs, whose present political "weight" is in a state of flux—Brazil now has the active voice and more organized actions of rural workers, as well as new forms of political participation and organization of the popular sectors, the middle classes, and urban and rural business sectors.[49] Coexisting with these new actors are traditionally institutionalized patterns of political integration and representation, such as the State corporatism of the unions, clientelism, and populism, which, although powerful, show signs of being insufficient to handle the conflicts generated by a modernized society.[50]

In examining these processes, I shall try to outline their development in the context of the New Republic.

A. Political Parties: Populism, Regionalism, and Clientelism

World history shows that strong parties with clearly defined programs and a body of integrated militants are a temporary phenomenon. Generally they are formed in periods of intense social change (in particular, those caused by migration and transformations in the occupational structure) and strong pressures on the part of new interest categories trying to enter the political system.

Does the present stage of Brazilian political development contain such conditions? Are there parties which have some of these characteristics? Quite correctly, Fernando Henrique Cardoso affirms in this volume that it is very unlikely that in the future the Brazilian party system will either move towards the North American model or follow the classical European paradigm of class parties.

In general terms, then, what is the profile of the parties and of their evolution under the New Republic? How has the new regime altered some of the attributes imprinted on the political party system by the authoritarian regime?

Present-day Brazilian parties are different from their predecessors in their names, the size of their constituencies, their programmatic positions, and even the militancy of their cadres. Yet they still form a fragile system with respect to the functions of governance and decision-making and the representation of the interests of the majority of the population.

Among political analysts there has been consensus that the organization of the Brazilian party system is one of the most serious obstacles to the functioning of a stable democratic system. No doubt this is still true.

In the majority of modern representative democracies the evolution from "class parties" to "mass parties" shows that the ideal of cohesive parties always geared towards programmatic action does not correspond to reality. Most modern mass parties are in fact criticized for behaving as clientelistic vote-getting machines. Indeed, it seems inevitable that even the most paradigmatically "modern" mass parties will continue to be only partially instruments of popular will, and will always experience the tensions existing between party fronts and parties, as well as those between interest aggregation and ideological functions. In any event, there is consensus regarding the fact that Brazilian party institutions fall far short of the classic modern mass party paradigm.

In the current situation, which is marked by skepticism, perplexity over the future of the country, and the gravity of the economic problems that afflict the population, a discussion about this theme can always run the risk of appearing to some observers as a formal or secondary exercise. Therefore, it is necessary to add to the aspects already pointed out some considerations about the present conjuncture, so as to emphasize the importance of reflecting upon party issues:

a) Although there seems to be no compelling reason for the armed forces to oppose a democratic type of capitalist modernization in Brazil, it is unlikely they will accept such a normalization through "non-institutionalized" processes, which occur mainly when there exists a party system whose ideological and integrative role vis-à-vis the popular sectors continues to rest on populist appeals;

b) At this moment, little can be said about the institutional, economic, and organizational conditions that make political and social pacts possible. There is some consensus that, if political pacts demand less "principled" and rigid political parties, they require even more a minimally representative party system capable of gaining the support of significant sectors in society. As far as social pacts are concerned, although they depend upon arrangements between corporative organizations, parties are fundamental to their implementation, acting as conveyor belts for the transmission of the legitimacy of these arrangements to society.[51]

c) Without denying the weight of economic factors in aggravating the Brazilian crisis, I believe that it is in the functioning of party life and related areas that we can locate some of the factors that make the current national political process problematic, especially in relation to the "lack of government" expressed in the previously mentioned decision-making paralysis and rashness. These phenomena expand the feeling of impotence of the elites and aggravate social conflicts.

1. *The Electoral Imperative and Populism.* For many authors, Brazil represents a unique case of the survival under authoritarian rule of structures and administrators inherited from the former democratic period (1945–64). The military, which came to govern the country after 1964, did little to create new institutions distinct from those of the former period and managed Brazilian politics by distorting pre-existing institutions rather than by destroying them.

Partly as a consequence of this institutional situation, the impact of authoritarianism on the party system was decisive. On the one hand, it legitimized the electoral process as the main form of resistance to the dictatorship and allowed social sectors somewhat greater familiarity with party life. On the other hand, however, it reinforced plebiscitarian support for parties and aggravated the dependence of parties upon the State, through the expansion of the bureaucratic, centralized decision-making process.

The New Republic faces the following legacy of the authoritarian regime: the consolidation of the electoral process as the institutional instrument par excellence for affirming citizenship—eventually leading to an almost mystical belief in the power of direct elections—coupled with a party system that is still a loose articulation of regional elites, distant from that image of a body of institutions which, through lasting structures permanently subjected to public exposure, guarantee the currency in which negotiations are established in the political market.[52]

While in other Latin American authoritarian regimes the electoral process was extinguished or relegated to a minor role, the singularity of Brazilian authoritarianism was the importance of elections for the legitimacy of the regime. This "electoral imperative" turned the party and electoral spheres into objects of continuous government intervention. The existence of three different party systems during military rule emphatically reveals the high degree of intervention of the regime and the negative effect this had on party consolidation.[53]

The maintenance of the electoral calendar and the presence of direct elections at different levels of the political system, along with government interventions to disorganize the system, led throughout 1964–85 to intermittent mobilizations, in which political action on the part of the population was oriented by a kaleidoscope of contradictory signals.

Consequently, as many surveys confirm, the military regime eradicated ideological and party references that were being created—even if only incipiently—among the great masses of voters toward the end of the '50s. This was unlike the situation in Uruguay, where, according to Karen L. Remmer, there are no indications that the authoritarian regime produced significant fractures in previous voting patterns.[54] Moreover, the Brazilian authoritarian regime reinforced adherence on the part of the population to the type of plebiscitary clash induced by the two-party system implanted at the time, resulting in massive party identification in favor of the MDB/PMDB, the party which stood in opposition to the authoritarian regime (as Bolivar Lamounier explains in this volume). Unfortunately, identification with the opposition party developed within a pattern of electoral competition that aggregated voters almost without differentiating among them, that is, confronting them with the simple options of approving or rejecting generic and abstract propositions. Uninformed and incapable of articulating a view of the socio-political universe in terms of specific problems (due to the failure of the parties in fulfilling this function), the average voter perceived at least the existence of camps in which "the people" and "the elite," or "the poor" and "the rich" confronted each other, and he or she consistently chose the side of "the people."

Massive identification with the PMDB—which in the 1986 elections was expressed in the casting of over half the total national votes for the party's candidates, after a significant PMDB loss in 1985[55]—underscores the volatile character of populism and the potential for radicalized contestation due to its dynamics.

Polls carried out by Bolivar Lamounier in 1985 in the city of São Paulo reveal the progressive dispersion of preferences formerly concentrated on the PMDB: that year only 45 percent of former party sympathizers remained identified with the party in São Paulo.[56] According to Lamounier, party identification, the main factor used in predictions of voting behavior during the two-party period, would thus seem to have been relegated to a secondary role by a significant part of the population.

The victory of Jânio Quadros as mayor of São Paulo in 1985 and the successes obtained in Rio in 1982 and 1985 by Leonel Brizola—both politicians of the 1945–64 period, clearly unconnected to parties or party organizations—are significant signs of the existence of sectors still not attracted or integrated by parties. In certain aspects, the favorable performance in recent elections in some areas of the PT would seem to fit into the populist model of political support. But it also means something more: the need to establish links between political party institutions and demands for social democracy.

The consequences of populist identification for the process of building democracy are well known: it represents a threat to system governability given the high degree of uncertainty it brings to the electoral process, particularly for elites, since it can result in a vast opposition mobilization of the popular sectors at the national level, as illustrated by the events in the years preceding the 1964 coup.

2. *Regional Interests and National Parties.* As in the United States, regional politics decisively mark the configuration of the Brazilian party system, thus making it difficult to characterize the political parties as national organizations. In the North American system, however, the institutional soundness of the federal Congress and the fact that local politics developed over a long period of time account for the different effects of federalism on the loose articulation of the party system at a national level. In Brazil, the national Congress is not an institution with the strength of its American counterpart, and Brazilian political history is marked by the discontinuity of party systems (there have been six since the proclamation of the Republic), of parties, and of political groups within each party. Not only do parties disappear with each new regime but the regional elites, in a dance of frightening speed, change from one party organization to another during a single, short-lived presidential term.

The creation and preservation of political allegiances should be fundamental functions of political parties in representative regimes, for citizens need stable structures to which they can refer in order to acquire political identities. The Brazilian political parties are far from fulfilling this function. Nonetheless, while the Brazilian party system is not institutionalized (in the sense in which Huntington uses the term) for the majority of the population, it is an important channel for the political expression of regional elites in Congress and for their articulation in the political system.

With the New Republic, such aspects have been aggravated by other factors:

a) the *abertura* itself, which, by allowing for growth in the number of parties, made the electoral process more confusing for less informed voters and which, by inducing wider-ranged party competition, may also have led to electoral fragmentation. The formidable aggregation of votes achieved by the PMDB in 1986 still needs more detailed analysis. It seems clear, however, that the initial success of the Cruzado Plan had an extraordinary influence on these results, and there is evidence—revealed daily by the press—of the illusory and temporary character of that electoral alignment;

b) the President's decision to pursue legislative and party politics based on state governors rather than parties, and the fact that, since the beginning of the new regime, the President has acted to stimulate internal disputes in the majoritarian party, the PMDB, that should provide his base of support; and

c) the course of action pursued by party leaders and the government, who, based on a "hyper-liberal" conception of party systems,[57] permitted the establishment of an excessively wide span of parties (29 registered for the 1986 elections), while at the same time they took measures to regulate the 1985 and 1986 "founding" elections of the New Republic, which had extremely negative effects on the expansion of political information and the promotion of a national party system.

The New Republic has seen a replay of the course of action that Presidents and the military have taken historically: while hoping for legislative majorities to support them, they have feared the existence of cohesive majorities that could question or govern them.

The federal Congress voted that elections for state executive and legislative offices in November 1986 would coincide with the election of a federal Congress which would also function as the National Constituent Assembly. This undermined the importance of the constitutional reform, turning it into a secondary issue in the elections.[58] Furthermore, with the delay of the constitutional reform until the beginning of 1987, all other constitutional issues were taken off the agenda, so that any initiative in this area came to be viewed as illegitimate.

As a result of presidential pressure, and with the support of the leadership of the majoritarian party front, the High Court of Electoral Justice (in decision #12,288 handed down on September 10, 1985) decided to proscribe two forms of campaign activity. First, it prohibited television campaigns of party leaders who were not candidates, while those running for office were forbidden from appearing on television in states where they were not campaigning. Following the pattern of authoritarian *casuísmos* (provisions for legal but arbitrary rule-making), this resolution, which explicitly sought to better the chances of parties already based on national machines and to prevent Leonel Brizola from campaigning throughout the country, had deleterious consequences: it hindered a national articulation of mass parties and confined the action of political leaders with national potential within the narrow limits of state boundaries.[59] Moreover, the court also forbade media interviews with the candidates during the three months prior to the elections, thus making it very difficult for debates on the constitutional reform to be held and favoring those candidates who generally benefit from the public's lack of information.

3. *Parties and State Clientelism.* The present disarray in the Brazilian party system has a structural cause. The historical relationship between parties and the State in Brazil produces a party system deprived of power in the decision-making process.[60]

This statement is not based on the assumption of an idealistic or normative view about parties in modern societies. There is a consensus among political scientists regarding the fact that in contemporary societies parties attend more to attracting electoral support than to designing coherent political alternatives. The transformation of class parties into mass integrative parties seems to have resulted in a growing inability of political parties to play the role of central actors in public policy decisions. The more diffuse and socially heterogeneous a party constituency is, and the harder the party tries to dilute its ideological identity so as to be accepted by the greatest possible number of groups, the less this party is able to make clear choices with regard to public policies. This erosion of party identity, coupled with the failure of parties to position themselves as agents for the definition of the "popular will," allows corporatist interest structures to surface and generates a need to find more predictable and reliable means for coming to agreements about public policies. Nonetheless, under modern democratic "neo-corporatist" arrangements, policy proposals on key issues frequently emerge from identifiable sectors within parties, particularly the governing party, and, once adopted, the latter takes responsibility for them.

What stands out to the observer of the present Brazilian situation is the schizophrenic structure of the party-government relationship. The "governing" parties (PMDB and PFL) do not govern, nor do they oppose the government, and they exempt themselves from any responsibility for government policies. The governing parties' strategies—which are designed primarily to prevent the internal implosion of their organizations—have been either to withdraw some issues from the political agenda or to hand them over to bureaucratic and administrative levels further removed from public opinion. Alternatively, they may allow their respective parliamentary delegations to act with complete autonomy.

The account by C. A. Sardenberg of "Cruzado behind the scenes" is extremely important in revealing how the historical Brazilian decision-making pattern has been maintained. In particular, he discloses the "well-intended elitism" displayed by the techno-bureaucratic groups during discussions of the Plan. Party spokesmen were conspicuously absent from these meetings. Particularly significant, given the enormous impact this economic policy was to have, was the total lack of interest on the part of PMDB leaders in challenging the techno-bureaucracy's monopoly of power or in altering the "political pragmatism" that defined the party line. According to Sardenberg, when informed of the government's plans, party president Ulysses Guimarães abstained. For him, "the best, at that point, would have been for the Plan to be officially presented as an act of the Sarney administration and not of the PMDB." Confronted with the observation that, among the authors of the reform, only members of the PMDB were to be found, Guimarães is said to have replied that those cadres were "acting as government officials, not as PMDB militants."[61]

The processes involved in the decisions behind the Cruzado Plan exemplify the ways in which the New Republic has reinforced deep-seated, negative characteristics of the institutional and political culture of the country, making party institutionalization extremely difficult.

As a consequence of the Brazilian State's central role in capital accumulation, which was accentuated by the authoritarian regime, the pattern of decision-making just described was reinforced. This pattern has traditionally led political parties to enhance their competitiveness through State clientelism. The reinforcement of clientelism is a result of more people becoming dependent on State benefits, a deepening of the bureaucratic, centralized character of the decision-making structure during the military regime, and also the declining importance of the federal Congress and parties in that sphere. Finally, in a context marked by continuous electoral mobilization and debates that have frequently been censored or repressed, the usefulness of clientelism has increased as electoral results become increasingly uncertain and State resources are expanded.

The importance of clientelistic resources continues to be if not a sufficient, at least a necessary condition for the survival of Brazilian parties, even those that are more "ideological."

As in many other countries, the Brazilian party system is being consolidated on the basis of party machines sustained by State patronage.[62] Unlike, however, the European and American systems, where party systems developed mostly around legislative assemblies, in Brazil the party system's consolidation is based on municipal and state executives.

Today the process of party competition features a virulent contest for control over State resources and the use of State patronage by the President to gain parliamentary support as a substitute for his almost non-existent political leadership. Deprived of clientelistic resources, the smaller parties either try to acquire them through "pragmatic" alliances or are forced to develop "outbidding" strategies based on populist mobilization.

The theme of clientelism has received special attention in social science research, either with respect to the traditional aspects of Brazilian political culture or because of its pernicious effects on the consolidation of more "ideological" parties. Such studies have stressed the effects of "pacifying social conflict," so to speak. Without entering into the theoretical debate between "ideological authenticity" and "electoral pragmatism," central to any discussion of parties, some little-studied issues raised by the analysis of Brazilian clientelism are:

A. the nature and volume of pressures exerted on the State apparatus from the ongoing "associative-corporatist" surge in Brazilian society and the vast electorate of almost 70 million voters mobilized by mass parties (in a system where voting is obligatory);
B. the failure of clientelism to integrate the majority of the so-called popular sectors, the more disorganized groups from large urban centers whose demands are only precariously, if ever, met through State clientelism;
C. the "ideological" groups linked to parties that compete for State resources and simultaneously direct their ideological claims toward the micro distributive arena of clientelism and the macro spheres where decisions on more general issues of income and property concentration are made;

D. the numerical expansion of middle-class groups, liberal professionals, and intellectuals who bring clientelistic pressure on both the regulatory and the distributive arenas, but for whom obtaining benefits does not translate into either political passivity or "loyal" behavior towards the party; and
E. the transformation of regional elites, who, in response to electoral results since 1974, have moved away from the label of "oligarchs" to become professional politicians, basing their power on access to State clientelism.

These observations seek—though in a preliminary way—to raise questions about the argument that "political complacency" and governability result from clientelistic politics. Research on these issues would benefit from a comparative approach, taking into account the different dynamics of party machine consolidation in other countries, such as the Mexican one-party system, the loosely centralized two-party system of the United States, or even the Italian system, where the Communist Party possesses considerable electoral strength from its control over a solid patronage machine.

The erosion of clientelism—if possible—would not necessarily guarantee the existence of ideological parties. On the large scale typical of Brazil, clientelism can produce varying effects on the democratization of the State and system governability, depending on the linkages made by political organizations between distributive and redistributive demands and on the pressures they bring on a State whose role in economic development is central.

Recent Brazilian elections reveal the relative importance of State patronage in consolidating popular support for parties. However, among the popular sectors, patronage is certainly not as strong as some political analysts have suggested. This is confirmed by a number of factors: the extraordinary electoral growth of the MDB during the military regime, even though the party's access to the State apparatus was reduced; the erosion of the electoral strength of the government party ARENA, which for many years monopolized State clientelism; and the collapse of the party machine controlled by Chagas Freitas in Rio de Janeiro, which, given its clientelistic power, was viewed as invincible in the 1982 elections.

In fact, the 1982 elections are of great significance, for they demonstrate the limits of clientelism even at the municipal level. Because the government party controlled the majority of elected municipal offices (mayors and city council members) in all the states, except Rio de Janeiro, the government decided to synchronize municipal with state and federal elections. The *voto vinculado*, a ballot forcing voters to choose candidates from the same party at all levels, was established in the hopes of "municipalizing" the elections, with municipal votes carrying over to the state and federal levels. This worked in the Northeast, but the opposition won in the North and Center-South, an area containing almost 70 percent of the population, GNP, and tax revenues of Brazil. In the Center-South, the public's identification with the opposition in the interior annihilated the government party at the municipal level, despite its powerful clientelistic network in those areas. Thus, while in 1980 the PMDB had 38 mayors in São Paulo state, by 1983 it had 307 (against the 253 affiliated with the government party). In the state of Goiás, the increased power of the PMDB allowed it to win in 185 cities and towns, as opposed to the sixty it formerly controlled, while in Paraná the number of PMDB candidates elected as mayors increased from 13 to 183.[63]

The pervasiveness of this phenomenon in the present political system must be pointed out. Just as in 1946 (as I have pointed out in earlier works[64]), today the centralization of power and the patrimonial character of the Brazilian State are hindering the development of a new party structure and to a great extent neutralizing the momentum of democracy. The democratic transition is taking place in a context where politicians neither contemplate or intend to undertake an alteration of the power monopoly maintained by State bureaucracies in important decision-making arenas.

As in 1946, parties now have power in the clientelistic arena, from which they derive their electoral strength while simultaneously weakening their institutional status. And intentionally or not, party leaders have been endorsing a process that results in their own delegitimization and in that of political parties.

A survey published by the *Jornal da Tarde* on October 26, 1987, shows that only 25 percent of the respondents approved of the performance of the members of the Constituent Assembly, 16 percent of that of state representatives, and 21 percent of that of political parties. The "crisis of representation" that these data point to is driving middle-class groups out of party activity. At the same time it reinforces the tendency of a large part of the population—whose situation in the labor market is insecure, who are not affiliated with unions, parties, or other associations, and who have no real contact with any structure of political representation—to vote or not according to temporary allegiances, to express themselves through populist voting, or, worse, not to participate at all in political life.

B. Parties, Social Movements, and Unions

As many have pointed out, during the authoritarian era mobilization of movements and organizations in civil society reached high levels by Brazilian historical standards, although, according to the analysis presented by Mainwaring in this volume, the prospects for the continuation of this mobilization were doubtful. There are innumerable examples of movements that have simply lost momentum. Movements united in the resistance to the dictatorship ended up breaking apart and have even begun to attack one another.[65] However, there has been a clear and steady increase in labor mobilization throughout the first years of the New Republic.[66]

Some of the factors that explain the ebb in mobilization or the decline in visibility experienced by these movements are: (1) the centrality of electoral party politics, a phenomenon which occurs naturally in democratic transitions; (2) the declining impact of certain demands in the context of a political opening which at least symbolically responds to them; and (3) the neutralization of social movements through the co-optation of their leadership.

The end of the dictatorship gave absolute priority to political parties and to some institutions that came to be identified with the new regime.[67] Soon after President Sarney took office, "political actors" clearly started to predominate over "social actors." The freedom of party organization and the convocation of a Constituent Assembly focused what were formerly unified opposition demands of civil society and political society on specifically political channels.

A review of the political process of the initial years of the New Republic reveals not only how parties have left vacant some "political space" that was opened by various social organizations during the authoritarian regime but also how, in failing to occupy that space, they have lost much of their legitimacy. In fact, parties have opted for or been forced into a form of consolidation based on the establishment of vast clientelistic machines throughout the country. In terms of governability, this consolidation succeeds in toning down social conflict and diluting political discourse. Yet it also leads to direct pressure on the State apparatus, as noted above. Moreover, by blocking the proliferation of political opportunities emerging within social movements at the end of the authoritarian regime,[68] this form of consolidation may make it more likely that unorganized popular sectors will turn to populism. In the populist model, various antagonisms come to be concentrated in a single conflict that divides the whole social domain into two opposing camps, the "dominant" and the "dominated," thus creating conditions in which the ideological resistance of elites shows a marked continuity with the dictatorship.

One of the most dramatic changes caused by the military regime was in the structure of unions through their expansion to various wage-earning sectors, especially in rural areas, where unionization reached astronomical proportions. (See note 49.) The chapter in this volume by Margaret Keck shows how government measures during the military regime eventually led unions to develop strategies to build strong local and national organizations.[69] This differs from the Spanish transition, in which the labor movement split into twelve different organizations, none of which was sufficiently strong to act as a valid interlocutor on behalf of the labor movement as a whole.

Over the last few years the most important feature of the evolution of unions has been the quest for greater autonomy vis-à-vis the State and political parties.[70] Although the literature on the developing relationship between parties and unions in recent Brazilian politics is still sparse, after the 1979 party reform—as Keck claims in her chapter—"organic" links were not established between the party and union spheres, despite an effort by leftist parties. The insignificant impact of unions on electoral results—even the performance of the party most closely associated with union organizations, the PT[71]—is clear in the 1985 and 1986 elections.

Thus, unlike political parties, the unions—or at least the largest ones—are seeking to break their links with the State, as well as with existing parties. The New Republic has therefore witnessed the fracturing of the party-State-union coalitions developed in the 1945–64 period and the search for new institutional forms by means of which the three can establish relationships with one another.

The immense transformations recently experienced by Brazil in terms of migratory flows and the occupational structure have led to shifting political identities. The labor movement must choose between two strategic options—either to expand in order to increase pressure on the State or to pursue a policy of pacts, in response to the current economic recession. Continuous mobilization in the rural areas,[72] despite the recessionary tendencies in the economy, indicates that labor unions may be expanding their base, a situation which might lead to a more inflexible posture. This rigidity will conflict with the current indecision among parties regarding rural reforms, hindering the aggregration of support necessary to government policies.

Despite the increase in labor activity and strikes, some important unions seem to be retreating from confrontation and moving toward greater flexibility in bargaining, forming "pacts" between business and workers (although these are limited to collective bargaining contracts involving single enterprises or specific professional categories). Though such a process could result in a less unified expansion of social and political democracy, it may also make the system more governable.

During the August 1986 Second National Congress of the CUT (*Central Única dos Trabalhadores*, or Central Workers' Organization), a peak labor association that represents some 1,000 unions, the focal point in the debates turned out to be whether or not the organization should present itself as a political party. Although the final outcome of the discussions was not exactly clear—the very term "party" aroused suspicion among workers—the conclusions reached at the meeting were summarized by the *Jornal do Brasil* in its August 3, 1986, headline, "CUT, the Effort Not to Look Like a Party."

Given the tenuous nature of a party system incapable of organizing the socio-political universe of the population and unable to articulate its interests, the peak labor associations (*centrais sindicais*) are called upon to fulfill the role of parties among urban and rural workers.[73] Of all possible forms of political parties, this would arouse the greatest resistance on the part of the other establishment "parties"—regional elites, the Church, and the military.

Whatever strategies they eventually adopt, it seems clear that the unions—and above all the peak associations—are the organizations least likely to be eliminated from the unfolding political process, regardless of their combativeness. If, in seeking the much-needed expansion of social democracy, they continue to sever their links with parties or try to replace them, their activities may serve only to complicate the construction of political democracy.

The great growth over the last few decades in peak associations of urban and rural workers, popular sector organizations, and associations of civil servants and liberal professionals raises the important question of neo-corporatism and its relationship to democracy.[74] Besides the classical problem of the functions of such corporate organizations in relation to the role of party-parliamentary systems, the Brazilian case draws attention to questions of political development and to the relationship between parties and unions. Could Brazil be skipping the pluralist stage, experienced by the European countries, based on class parties and the Parliament and be moving toward the neo-corporatist phase observed today in those nations? Although we have no systematic analyses of the 1986 elections yet, preliminary data reveal the importance of particular corporate actors in the electoral results.[75]

Further exploration of the above questions is beyond the scope of this essay. They do, however, raise issues important for the analysis of the present evolution of Brazilian democracy.

The "praetorian" characteristics of the political system, the loss of momentum by social movements, which has diminished the resistance of society to the "authoritarian virus," and the threat that the large peak labor associations represent for the ultra-conservative world views of traditional business elites, especially landowners, all clearly bring the armed forces to the center of the political scene. The military's increasingly explicit power does not come from their political and social characteristics alone, but is related also to the existence of institutional prerogatives in effect since the end of the Brazilian Empire. These prerogatives were extended during the authoritarian regime and remain untouched by the civilian government.

In his analysis of the role of the military in newly democratic regimes, Alfred Stepan presents a dismaying picture of the institutional prerogatives of the Brazilian military, as compared with those held by the armed forces in other countries that have recently embarked on processes of democratization.[76] During the first three years of the present civilian government, the Brazilian military have managed to maintain an important degree of control over the political space they occupied during twenty-one years of authoritarian rule, and they have at times utilized this power with great impact. One of the central points raised by Stepan's analysis is the fact that the military entered the New Republic with the belief that permanent involvement in the management of political conflict in Brazilian society is legitimate.

IV. Perspectives

The Brazilian democratic transition, limited by the threat of the military, continues on the very long trajectory that began over a decade ago with the 1974 elections. The new Constitution and presidential succession through direct elections will inaugurate a new phase, possibly the most decisive, in this extended process.

Skepticism regarding the viability of a return of the military must be reconsidered in the light of the "crisis of government" produced by party disarray, the disorder in the State apparatus, and, above all, the reinforcement of populism. This reinforced populism has a double impact on the political process in that it simultaneously generates electoral fluidity and produces an enormous capacity to bring pressure to bear outside party channels for democratization of the State.

The erosion of the New Republic and its negative effects, which lead to greater system blame and therefore create conditions that make a military takeover possible, must be pondered, however, in view of the armed forces' present "lack of interest" in a coup and their doubts about the benefits of the direct and immediate exercise of power, given their own inability to face the economic crisis. Whether the military continue to maintain this position will depend on how the democratization process unfolds under their tutelage. The most probable outcome is that, rather than dying by sudden decapitation, the Brazilian democratization effort will be slowly debilitated by the suffocating weight of the military's presence. Yet, even without the military, regional elites by themselves could vitiate the democratic undertaking—as they now seem to be in danger of doing—whether due to the imperatives of the modern "savage capitalism" of the South or to the parochial and backward interests of the Northeastern oligarchy.

Though its conservative outlines are increasingly visible, the Brazilian transition is still under way. The final results of the political bargaining that will give the democratic regime its constitutional framework still depend in part on compromises and confrontation among social and political groups and their spokesmen. They also depend on the individual transformations that occurred throughout the long period of abertura. Despite the negative impact it had on the process of redemocratization, it is reasonable to assume that the gradual, drawn-out nature of the Brazilian transition may have prepared society for a less repressive way of living, possibly revealing to elites that political conflict does not necessarily lead to the establishment of a "trade unionist communist republic."

Even if the constitutional-political arrangements that are being established in the Constituent Assembly prove incapable of fulfilling the expectations and demands of the Brazilian populace in the first years of abertura, this does not mean that it will lead to a restoration of the constitutional experiment of the 1945–64 period or a return to the constitutional framework of the authoritarian-military decades. For example, individual and social rights will be better guaranteed than they were during the authoritarian period, and the powers of the Congress probably will be more solid than during the 1945–64 populist period. With many of the prerogatives it lost under authoritarian rule already restored, the Congress may become a much stronger instrument for buffering crises.

The Brazilian presidential system entails—as do all others in Latin America—an excessive concentration of power in the executive branch. This suggests the need for measures to better delineate the responsibilities shared by the President and the legislature.[77] Although in terms of strengthening political parties the parliamentary system would probably be more favorable than the presidential system, the "regenerative" virtues of the former system are uncertain in the absence of some specific changes, particularly in the rules governing the organization of the electoral process and of parties, as well as the representation of states in the national Congress.

Even if a parliamentary system were to be adopted in the future in Brazil, its advantages should be weighed with special caution if the structure of the federal Congress remains unchanged. The apportionment of state representatives is markedly unfavorable to the more urbanized and economically developed states.[78] If unchanged, this could lead to a gap between the parliament and the economic centers of the country and would reinforce bureaucratic interest representation. Furthermore, the current modernized version of the "governors' politics" characteristic of Brazil's First Republic may indicate a process that will bring to the national Congress representatives entirely dependent upon their state executives, thus transforming this body into a political arena controlled by state governors, particularly those from states which are more dependent on federal resources.[79]

Perhaps the various kinds of institutional engineering that are being placed on the political agenda—and that may arise again in the future—(such as measures to increase the powers of the Congress, electoral formulas, and party legislation) could increase the governability of the political system and/or lead to further democratization. There is no doubt, however, that institutional engineering alone will not guarantee the stability of the regime until steps are taken to alleviate Brazil's profound social and economic inequalities.

NOTES

1. The political agenda of the post-authoritarian government included a large number of complex tasks to be undertaken in a short period of time: the elaboration of a new constitution; the establishment of economic policies to alter the agrarian structure and the distribution of wealth; the solution of the foreign debt problem; the removal of authoritarian legal instruments; creation of new rules for the electoral process; restructuring of the party system; and the establishment of procedures for the Constituent Assembly, to name just a few.

2. Albert O. Hirschman, "Dilemmas of Democratic Consolidation in Latin America: Notes for the São Paulo Meeting" (mimeo, 1986).

3. In his article "Paths toward Redemocratization: Theoretical and Comparative Considerations," in Guillermo O'Donnell, Philippe C. Schmitter, and Laurence Whitehead (eds.), *Transitions from Authoritarian Rule: Comparative Perspectives* (Baltimore and London: Johns Hopkins Univ. Press, 1986), 75–76, Alfred Stepan discusses the Brazilian opening as an example of "redemocratization initiated by the military as government."

4. See Maria do Carmo Campello de Souza, "Partidos Políticos: Velhos e Novos Desvios" (mimeo, Columbia University, 1987), where I discuss the consolidation of the Brazilian party system, comparing it with the forms of institutionalization

analyzed in Samuel Huntington, *Political Order in Changing Societies* (New Haven: Yale Univ. Press, 1976).

5. Paulo Sérgio Pinheiro, "Movimentos Sociais na Nova República" (mimeo, Universidade de São Paulo, 1987). For a review of social movements during the past decade, see Ruth C. L. Cardoso, "Movimentos Sociais Urbanos: Um Balanço Crítico," in Bernardo Sorj and Maria Hermínia Tavares de Almeida (eds.), *Sociedade e Política no Brasil Pós-64* (São Paulo: Editora Brasiliense, 1983), 215–39.

6. See Alfred Stepan, *Rethinking Military Politics: Brazil and the Southern Cone* (Princeton: Princeton Univ. Press, 1988), esp. Chs. 6–7.

7. A discussion of the "paradox of negotiated transitions" can be found in several articles in O'Donnell, Schmitter, and Whitehead (eds.), *Transitions from Authoritarian Rule, op. cit.*

8. See David Gilmore, *The Transformation of Spain: From Franco to the Constitutional Monarchy* (London: Quartet Books, 1985); and Donald Share, *The Making of Spanish Democracy* (New York: Praeger, 1986).

9. The problems of the impact of authoritarianism and of the continuities and discontinuities between the authoritarian and pre-authoritarian regimes and their effects on redemocratization are examined by Karen L. Remmer, "Redemocratization and the Impact of Authoritarian Rule in Latin America," *Comparative Politics*, April 1985, pp. 253–75.

10. See Charles Gillespie's analysis of the Spanish elections in the period between 1977 and 1982 in "Electoral Stability and Party-System Transformation: The Uruguayan Case in Comparative Perspective," paper presented at the conference, "Recent Electoral Changes in Latin America," Center for Iberian and Latin American Studies, University of California at San Diego, Feb. 1986; see also David Gilmore, *op. cit.*, and Juan Linz's analysis of the Spanish party system in "The New Spanish Party System," in Richard Rose (ed.), *Electoral Participation: A Comparative Analysis* (Beverly Hills: Sage, 1980), 101–30.

11. David Fleischer, "O Congresso Constituinte de 1987: Um Perfil Sócio-Econômico" (mimeo, Universidade de Brasília, 1987).

12. "Um PFL Vale 2, 5 PMDB," *Veja*, Jan. 19, 1986.

13. See Paulo Sérgio Pinheiro, *op. cit.*

14. In his book *Power and Ideology in Brazil* (Princeton: Princeton Univ. Press, 1981), Peter McDonough analyzes the beliefs of Brazilian elites during the authoritarian regime and the structural and organizational factors conditioning the political system which determined their behavior. For an analysis of political elites in comparative perspective, see Robert D. Putnam, *The Beliefs of Politicians* (New Haven: Yale Univ. Press, 1973).

15. A survey published on Feb. 4, 1987, by *Veja* presented a view of the positions of the members of the Constituent Assembly regarding various aspects of the constitutional reform.

16. Maria do Carmo Campello de Souza, *Estado e Partidos Políticos no Brasil: 1945–1964* (São Paulo: Alfa Omega, 1976), and "A Democracia Populista de 1945–1964: Bases e Límites," in Alain Rouquié, Bolivar Lamounier, and Jorge Schwarzer (eds.), *Como Renascem as Democracias* (São Paulo: Editora Brasiliense, 1985), 73–103. See also Martin Shefter, "Party and Party Patronage: Germany, England, and Italy," *Politics and Society* 7:4 (1977), 403–52.

17. The 1986 elections revealed that, although the great masses of the population were poorly informed, the question of direct elections had taken root in the popular

political imagination. The majority of those who voted against direct elections in Congress in 1984 were not elected to the Constituent Assembly. Of the 64 federal representatives who voted against the "Dante de Oliveira amendment" proposing the re-establishment of direct elections for President, only ten managed to be re-elected (that is, 15%). Of the 174 representatives of the PDS who contributed to the defeat of other amendments also intended to re-establish direct elections, 49 were re-elected (or 28%). See *Folha de São Paulo*, Jan. 31, 1987. Although in terms of representatives in the federal Congress the PFL is the second strongest party (with 133 delegates), it has less than half the number of parliamentary representatives of the PMDB (which has 298 delegates).

Bolivar Lamounier affirms that, although it would be overly extreme to characterize the great majority of urban voters as consciously ideological, there is a predominant tendency in the big cities to vote for reformist and "center-left" parties. See Bolivar Lamounier (ed.), *1985: O Voto em São Paulo* (São Paulo: Publicacões do Instituto de Estudos Ecônomicos, Sociais, e Políticos de São Paulo (IDESP), no. 1, 1986), 11.

18. It must be noted that the politicians most visibly identified with the military regime are affiliated with the PFL or PDS.

19. An analysis of the erosion of this identification in the capital city of São Paulo is made by Bolivar Lamounier and Maria Judith B. Muszynski, "A Eleição de Jânio Quadros," in Bolivar Lamounier (ed.), *1985: O Voto*, *op. cit.*, 1–31.

20. The Globo television network practically monopolizes the President's transmission of information to Brazilian society, and it is the most powerful base of support for the federal executive branch. Globo is the fourth largest commercial television network in the world, ranking after only CBS, NBC, and ABC of the United States. More than 90 percent of the programming broadcast by Globo during prime time is generated within its own network. These data were collected by Alfred Stepan during a visit to Globo's headquarters in Rio de Janeiro in June 1987.

21. Ekkart Zimmermann, "Economic and Political Reactions to the World Economic Crisis of the 1930s in Six European Countries," paper presented at the conference of the Midwest Political Science Association, Chicago, Illinois, April 10–12, 1986. See also Peter A. Gourevitch, "Breaking with Orthodoxy: The Politics of Economic Policy Responses to the Depression of the 1930s," *International Organization* 38 (1984), 95–129. In *The Confidence Gap: Business, Labor, and Government in the Public Mind* (New York: Free Press, 1983), Seymour Martin Lipset and William Schneider defend the view that there is a strong relationship between the economic sphere and political crisis.

22. This argument is developed in Juan Linz and Alfred Stepan, "Political Crafting of Democratic Consolidation or Destruction: European and South American Comparisons," in Robert A. Pastor (ed.), *Democracy in the Americas: Stopping the Pendulum* (Boulder and London: Westview Press, forthcoming).

23. *Ibid.*

24. The proportion of the economically active population earning *one minimum wage or less* (including those with no income) remains alarmingly high. In 1978, 1981, 1983, and 1985, these percentages were calculated at 46.7%, 39.8%, 43.2%, and 42%, respectively. See Instituto Brasileiro de Geografia e Estatística (IBGE), *Anuário Estatístico do Brasil*. Wage losses for all categories in the formal sector were estimated by the Departamento Intersindical de Estatística e Estudos Sócio-Econômicos (DIESSE) at 37.74% from the Cruzado Plan I to July 1987. See Instituto Brasileiro de Análise Sócio-Econômica (IBASE), *Políticas Governmentais* (*Uma Análise Crítica*), July 1987, p. 18. According to DIESSE, if all wage losses

between March 1 and September 1, 1987, were taken into consideration, each worker would have had to receive a salary increase averaging 93.1% to restore former real wage levels.

The concentrated 1950 profile of land-ownership persists in 1980. Although the percentage of large rural properties (those with over 1,000 hectares) has fallen from 2.2% to 1.1%, the percentage of total cultivable land belonging to large landowners continues to be extremely high (45.1% in 1950 as compared with 50.9% in 1980). Despite the growth in the portion of small properties (those with less than ten hectares) in proportion to the total number of properties (from 34% in 1950 to 50.3% in 1980), the share of total cultivable land represented by small properties only increased from 1.3% to 2.5% over the same period, thus remaining at alarmingly low levels. See IBGE, *Anuário Estatístico do Brasil.*

Expenditures on social services, which corresponded to 4% of GDP in 1970, amounted to 6% in April 1987. In 1970, there were 584 hospitals in the Public Health Service; by 1983, the number had increased to 16,749. However, the increase in the number of beds is less significant (rising from 354,000 to 534,000). That is, the proportion is still four beds for each 1,000 people. Among those under twenty years of age (who, according to the IBGE, *Anuário Estatístico do Brasil,* 1984, p. 74, numbered over 59 million people by 1980), less than half were in school in 1987 (31.5 million in primary schools, 3 million in secondary schools, and 1.3 million in universities). See *The Economist,* April 1987, p. 17.

25. The following is an outline of the essential aspects of the Cruzado Plan: (1) de-indexation of the economy, with the elimination of the monetary adjustments used to "correct" salaries, savings accounts, and other financial instruments; (2) an immediate and indefinite price freeze on more than 80% of goods and services; (3) a change in name of the currency from "Cruzeiro" to "Cruzado," converted at the proportion of 1,000 to 1; and (4) a 33% increase in the minimum wage and an 8% bonus to workers. These measures led to an explosion in consumption, such that the volume of sales in April 1986 as compared to April 1985 rose 36.2% in Rio de Janeiro and 29.5% in São Paulo. See *Conjuntura Econômica,* July 1986, p. 7.

26. *Jornal do Brasil,* Nov. 21, 1987. For an analysis of popular discourse regarding the State, the government, and their personification in the figure of the President of the Republic, see Eunice Durham, "A Sociedade Vista da Periféria," *Revista de Ciências Sociais* 1:1 (June 1986), 84–99.

27. "As Imagens de 1987," *Veja,* Jan. 1, 1987. For an excellent analysis of the meaning of the popular mobilization for direct elections and the death of President-elect Tancredo Neves, see Marlyse Meyer and Maria Lúcia Montes, *Redescobrindo o Brasil: A Festa na Política* (São Paulo: T. A. Queiroz, 1985).

28. Although the research methods used were precarious (223 interviews conducted over the telephone with no mention of the margin of error), the results obtained are illustrative.

29. See *Veja,* Aug. 5, 1987, p. 45. The Nov. 15, 1987 edition of the *Folha de São Paulo,* however, shows that 80% of those interviewed in the ten most important capital cities of Brazil wanted to reduce the presidential term and hold direct elections in 1988. Only 11% supported a five-year term for Sarney.

30. See Marcus Figueiredo, "Voto Popular e Democracia" (Ph.D. dissertation, Universidade de São Paulo, in progress), where he observes that there has been a systematic growth in popular preferences for direct presidential elections since 1974, although the distribution has shown conjunctural oscillations, having reached its peak during the "diretas já" campaign in 1985.

31. Antônio Ermírio de Moraes's victory in the capital city of São Paulo (as opposed to the interior, where PMDB candidate Orestes Quércia won handily enough to carry the state) in the 1986 gubernatorial elections is an extremely good illustration of how little weight party affiliation has among the city's voters. Only 20% of those who voted for him claimed to be identified with the party label (PTB) under which he ran. His *personal* attributes, on the other hand, influenced more than 80% of his constituents, which is also true of those who voted for PDS candidate Paulo Maluf. Together, both politicians took 50.8% of the votes in the capital city. Antônio Ermírio received 31.4%; Quércia, 26.5%; Maluf, 19.4%; and Eduardo Suplicy of the PT, 11.2%. See Maria Judith B. Muszynski, "Comportamento do Eleitorado Paulistano em 1986" (mimeo, IDESP, 1987).

More research on present parties, along the lines of that conducted by Kurt Von Mettenheim on parties under the authoritarian regime, is in order. Under the military regime the legitimacy of political parties was extremely high among the popular classes, in contrast to the opinions expressed by the upper classes. See "Transition to Democracy and the Consolidation of Mass Party Politics in Brazil: 1974–1986" (mimeo, Columbia University, 1987).

32. The most striking example of electoral manipulation of economic policy can be found in the Cruzado Plan, which, because of its redistributive effects, received vast support from the whole society. Just after the 1986 elections, it was deactivated. An exhaustive description of the decision-making process involving the Cruzado Plan is given by C. A. Sardenberg, *Aventura e Agonia: Nos Bastidores do Cruzado* (São Paulo: Cia. das Letras, 1987). See also Hugo Faria, "The Failure to Reform: The Process of Economic Decision-making during the First Two Years of the Brazilian New Republic" (mimeo, Columbia University, 1986).

33. After the decision to implement the Cruzado Plan was made in February 1986, all the parties greatly increased their support for the government (with the exception of the PT and PDT, both small parties). See David Fleischer, "The Evolution of Political Parties in the Brazilian Congress," (mimeo, Universidade de Brasília, 1987).

34. The PMDB in 1986 received more than 50% of total national votes. It elected 22 of the 23 governors and gained an absolute majority in both houses of the national Congress (53.3%).

In September 1985, 40.5% of the population of Rio and São Paulo (based on a sample of 1,000 voters) expressed confidence in President Sarney. In April 1986, 45 days after the Plan's implementation, 95% of respondents classified the President's performance as "good" or "very good." In September 1986, seven months after the adoption of the Plan, 72% of respondents approved of the President's performance. With the Plan's deactivation after the November elections, a survey conducted in December 1986 showed that only 34% of the population believed in the President's performance. See *Veja*, Dec. 10, 1986, and Fleischer, "The Evolution of Political Parties," *op. cit.* The number of public opinion surveys showing the rapid and drastic decline in Sarney's popularity after the beginning of 1987 is so vast that it is impossible to mention all of them.

35. Giovanni Sartori, *Parties and Party Systems: A Framework for Analysis*, Vol. 1 (Cambridge: Cambridge Univ. Press, 1976).

36. The aspects of the Brazilian decision-making structure that Fernando Henrique Cardoso calls "bureaucratic rings" and other analysts term "balkanization of the State apparatus" are well known. These terms indicate processes in which powerful interest groups take control of portions of the State apparatus, creating

strong coalitions between themselves and part of the public bureaucracy. The "feudalization" of the State apparatus hinders the coordination of public policy and weakens the government's capacity to implement policy measures. Authors such as Luciano Martins believe that the centripetal force of federal government control over financial resources has always been counterbalanced by the centrifugal force of its authority. See his "A Expansão Recente do Estado no Brasil," Publicações do Instituto Universitário de Pesquisas do Rio de Janeiro (Rio de Janeiro, IUPERJ, 1976). In "The Paradise That Never Was: The Breakdown of the Brazilian Authoritarian Order," in Thomas Bruneau and Philippe Faucher (eds.), *Authoritarian Capitalism: Brazil's Contemporary Economic and Political Development* (Boulder: Westview, 1981), Philippe Faucher affirms that, since 1964, the decision-making process has been totally centralized at the executive level, although the implementation of economic policy decisions has been divided among sectors of the bourgeoisie and State enterprises.

37. See *Senhor*, Sept. 15, 1987, pp. 37–41, for an interview with the late Minister Marcos Freire, then responsible for agrarian reform policies, in which he reveals the difficulties he was encountering in trying to find a way to implement basic legislation for the reforms. Also see Alfred Stepan's account in *Rethinking Military Politics, op. cit.*, 108–9, of the intervention of the military on this issue. The lack of international articulation regarding the decision to adopt the debt-service moratorium is revealed in various articles in the Brazilian press. See, for example, *Senhor*, Feb. 24, 1987.

38. In his book *Sessenta e Quatro: Anatomia da Crise* (São Paulo: Vértice, 1986), Wanderley Guilherme dos Santos analyzes the effects of the process of decision-making paralysis in the context leading up to the coup of 1964.

39. See *Gazeta Mercantil*, June 2, 1978. Regarding the document signed by eight influential entrepreneurial leaders in favor of the democratization of the regime, see *O Estado de São Paulo*, June 27, 1978. For an analysis of the liberalization of the Brazilian regime and the position of the entrepreneurial elite, see Fernando Henrique Cardoso, "O Papel do Empresário no Processo de Transição: O Caso Brasileiro," *Dados* 26:1 (1983), 9–28; Douglas Chalmers and Craig Robinson, "Why Power Contenders Choose Liberalization," *International Studies Quarterly* 26:1 (March 1982), 3–36. See also Sebastião Velasco Cruz, "Os Empresários e o Regime: A Campanha Contra a Estatização (Ph.D. dissertation, Universidade de São Paulo, 1984); and Eli Diniz, "O Empresariado e o Momento Político: Entre a Nostalgia e o Temor do Futuro," *Cadernos de Conjuntura*, IUPERJ, Oct. 1985.

40. Among the businessmen's associations that have arisen in recent years, one of the most important is one that brings together rural landowners, the UDR (Democratic Ruralist Union), which between 1986 and 1987 increased the number of affiliated groups from 37 to 200, while its membership grew from 50,000 to 230,000. See *Veja*, Nov. 11, 1987.

Another important movement is the PNBE (literally, National Thought of the Entrepreurial Bases), which originated out of businessmen's dissatisfaction with the habitual cautiousness of the FIESP (the Federation of Industries of the State of São Paulo, the most important and traditional entrepreneurial organization) in its relationship with the government. On June 9, 1987, the PNBE held a meeting that brought together 2,500 businessmen in São Paulo, who together employ more than three million people and belong to more than 100 employers' organizations. The objectives of the meeting included pressing for such measures as interest rate controls and an end to State intervention in the economy. For further discussion of these new businessmen's organizations, see "A Agitação Empresarial," *Senhor*, Oct. 13, 1987, pp. 42–44.

41. The official attitude of the New Republic's governing authorities regarding conflicts in civil society shows marked continuity with that of the former regime, as can be deduced from the increased frequency of military interventions in such conflicts. According to the *Folha de São Paulo*, March 8, 1987, the navy occupied the ports during a civilian dockworkers' strike, although they were not legally ordered to do so. Not long after this incident, the army occupied oil refineries as a "preventive measure," for a strike had been announced by workers at these facilities. See *Folha de São Paulo*, March 11, 1987. During a national strike, the army occupied the steel mill of Volta Redonda and railway station of the Central do Brasil in Rio de Janeiro. In Cubatão, São Paulo, the Army occupied the COSIPA mill, even though the workers at the plant were not even on strike. This incident came close to causing a revolt. See *Folha de São Paulo*, March 13, 1987. The militarization of conflicts stemming from strikes and demonstrations persists with generalized truculence on the part of various public officials. One thousand people from the Western Zone of the city of São Paulo belonging to the "Landless Movement" (*Movimento dos Sem Terra*) were dispersed with water hoses in front of the gates of Ibirapuera Park, when they demonstrated against Mayor Jânio Quadros. See *Folha de São Paulo*, Feb. 10, 1987. The president of the PT (Workers' Party) in São Paulo, ex-federal representative Djalma Bom, was indicted under the National Security Law on the charge of having offended the President of the Republic. See *Folha de São Paulo*, July 16, 1987. In April 1987, during a demonstration of bank employees in Brasília, the military police dissolved the pickets and violently attacked members of the Congress in a "choreography" familiar to civil society movements from the days of the dictatorship. See *Folha de São Paulo*, April 1, 1987. For a more extensive discussion of such events, see Paulo Sérgio Pinheiro, *op. cit.*

Alfred Stepan's account in *Rethinking Military Politics, op. cit.*, of the role of the military in decisions on agrarian reform and commercial agreements between Brazil and Argentina is another example of this military interventionism. See Maria do Carmo Campello de Souza, *op. cit.*, 1987, where I analyze the measures to regulate party organizations adopted at the beginning of the New Republic.

42. In " 'E Eu com Isso': Notas sobre Sociabilidade Política na Argentina e Brasil," in Guillermo O'Donnell, *Contrapontos: Autoritarismo e Democratização* (São Paulo: Vértice, 1986), O'Donnell analyzes authoritarianism in Brazilian and Argentine social relations.

43. According to a study cited by Paulo Sérgio Pinheiro, *op. cit.*, between 8:00 and 10:00 a.m. 72.1% of those respondents who had their radios turned on were tuned to these programs. In 1984 one of these programs reached more than one million listeners.

44. See Alfred Stepan's account in *Rethinking Military Politics, op. cit.*

45. For an analysis of "adversary culture," see Daniel Bell, *The Cultural Contradictions of Capitalism* (London: Heinemann, 1976).

46. A survey was conducted to measure the prestige and power of 22 institutions by the *Folha de São Paulo*, March 29, 1987, in the cities of São Paulo, Rio de Janeiro, Brasília, Belo Horizonte, Salvador, Recife, and Porto Alegre among 3,316 respondents. Television, radio, and the print media were ranked in the first three places in terms of prestige (81%, 70%, and 67%, respectively). In terms of *power*, television was ranked first by respondents (80%), the print media fifth together with the armed forces (71%), and radio sixth (64%). Still in terms of *power*, the institu-

tions were rated as follows: in second, third, and fourth place came the multinationals (75%), the banks and financial organizations (73%), and the President and the ministeries (72%). The Catholic Church was ranked in sixth place in terms of prestige and seventh in terms of power.

47. In "O Congresso e a Política Orçamentária durante o Período Pluripartidário de 1945–1964," *Revista de Ciências Sociais* 29:3 (1987), 177–205, Barry Ames argues that distortions in public expenditures caused by patronage politics are greater during presidential elections. The congressional elections of 1954 and 1958, which did not coincide with presidential elections, had little effect on the distribution of public expenditures.

48. Though this phenomenon has not yet been explored in depth with respect to Brazil, it is discussed at length with respect to Western Europe in Charles S. Maier (ed.), *Changing Boundaries of the Political: Essays on the Evolving Balance Between the State and Society, Public and Private in Europe* (Cambridge and New York: Cambridge Univ. Press, 1987).

49. In this volume Margaret Keck presents various tables showing the growth of unions. According to the data she cites, the number of members of rural unions grew from 2,930,692 in 1974 to 5,139,566 in 1979. According to Biorn Mayberry-Lewis (Ph.D. dissertation in progress, Columbia University), who cites data from the annual reports of the Confederação dos Trabalhadores Agrícolas (CONTAG), in 1985 there were more than eight million unionized workers in rural areas, organized into 22 federations and more than 2,600 unions, which means that rural workers were the largest unionized occupational group in Brazil. Data from IBASE, *A Organização Sindical do Brasil*, July 1982, p. 16, show that amidst this rapid unionization process, already by 1979 more than 50% of all unionized workers were in the rural areas, although they constituted only 32.4% of the economically active population of the country.

Between 1960 and 1978 the number of unionized urban workers nearly quadrupled, reaching five million workers. The number of workers in the industrial sector grew 102% between 1960 and 1970. Between 1966 and 1978, the number of unionized "liberal professionals" grew from 40,491 to 147,307, and the proportion of strikes in this sector was much higher than among industrial workers. See Maria Hermínia Tavares de Almeida, "O Sindicalismo Brasileiro entre a Conservação e a Mudança," in Bernardo Sorj and Maria Hermínia Tavares de Almeida (eds.), *op. cit.*, 191–215. In "A Pós-Revolução Brasileira," in Hélio Jaguaribe et al., *Brasil: Sociedade Democrática* (Rio de Janeiro: José Olympio, 1985), 223–36, Wanderley Guilherme dos Santos analyzes aggregate data presenting some of the effects that—by virtue of the magnitude and quality of their impact—recent transformations in Brazil's social infrastructure have had on the forms of socio-political insertion of various social sectors. In "A Abertura e a Nova Classe Média na Política Brasileira: 1977–1982," *Revista Brasileira de Ciências Sociais* 1:1 (June 1986), 30–42, Renato Boschi analyzes the main aspects of Brazilian associationalism, focusing on the political practices of the urban middle classes during the period from 1977 to 1982, such as their participation in neighborhood associations, professional associations, and unions of wage-earning professionals.

50. Margaret Keck (in this volume) and Maria Hermínia Tavares de Almeida, *op. cit.*, analyze the erosion of State corporatism and the emergence of new political strategies among the unions.

51. Adam Przeworski discusses in theoretical terms some of the conditions of class compromise necessary for establishing and maintaining a capitalist democracy.

See Adam Przeworski, "Some Problems in the Study of the Transition to Democracy," in O'Donnell, Schmitter, and Whitehead (eds.), *Transitions from Authoritarian Rule, op. cit.*, 47–63.

52. One aspect of Alessandro Pizzorno's article is the search for explanations for the survival of political parties in modern societies despite the loss of their traditional functions. He shows how parties constitute a sort of political "credit institution" by virtue of their durable and public structure. Thus they guarantee the mediation of representation through a continuous verification via the electoral process of existing credit in the political marketplace. See Alessandro Pizzorno, "Interests and Parties in Pluralism," in Suzanne Berger (ed.), *Organizing Interests in Western Europe: Pluralism, Corporatism, and the Transformation of Politics* (Cambridge: Cambridge Univ. Press, 1981), 277.

53. See Bolivar Lamounier's essay in this volume and Bolivar Lamounier, *Voto de Desconfiança: Eleições e Mudanças Políticas no Brasil* (Rio de Janeiro: Vozes, 1980); Lúcia Klein and Marcos Figueiredo, *Legitimidade e Coação no Brasil Pós-64* (Rio de Janeiro: Forense Universitária, 1979); David Fleischer, "Constitutional and Electoral Engineering in Brazil: A Double-Edged Sword (1964–1982)," *Inter-American Economic Affairs* 4:37 (1984).

54. It is frequently hypothesized that the MDB/PMDB grew out of a sort of "rebaptism" of the party identifications of the 1945–64 period. This idea is not supported by Bolivar Lamounier's research, as discussed in this volume. For the Uruguayan case, see Karen L. Remmer, *op. cit.* According to her, the 1982 elections in that country, which were designed to designate new party leaders, demonstrated a "remarkable endurance of voting habits." For an excellent analysis of the Uruguayan parties and the impact of the authoritarian regime on them, see Charles Gillespie, *op. cit.*

55. Although from an overall perspective the PMDB won the 1985 elections, the party lost in five of the biggest capitals (São Paulo, Rio, Recife, Porto Alegre, and Fortaleza). See Charles Gillespie, *op. cit.* For an analysis of the electoral evolution of the MDB/PMDB, consult the ongoing series entitled *Publicações IDESP sobre História Eleitoral do Brasil.*

56. In their article "A Eleição de Jânio Quadros," in Bolivar Lamounier (ed.), *1985: O Voto, op. cit.*, 1–31, Bolivar Lamounier and Maria Judith B. Muszynski analyze the erosion of this identification in the capital city of São Paulo. This volume contains important articles about many aspects of the 1985 elections in São Paulo and about the populismo of Jânio (*o populismo janista*).

57. In May 1985 the Inter-Party Commission of the national Congress presented measures to restructure the party system that abolished all restrictions on the formation of political parties.

58. Although the congressmen to be elected in 1986 were given the power to write the new constitution, this was a highly controversial question. The idea of an "exclusive" Constituent Assembly was defended by the OAB (Brazilian Association of Lawyers) and many other associations, some of which wanted to adopt a system of independent candidates who would not be identified with parties. Although there was vast mobilization against the above measure, which was proposed by President Sarney (and even produced divisions within the PMDB), it was approved at the end of 1985.

59. For a discussion of the barriers to creating national parties in Brazil, see Maria do Carmo Campello de Souza, *op. cit.*, 1976; Bolivar Lamounier and Raquel Meneguello, *Partidos Políticos e Consolidação Democrática: O Caso Brasileiro*

(São Paulo: Brasiliense, 1986). For a rare discussion of federalism and political parties in Brazil, see Olavo Brasil de Lima, Jr., *Os Partidos Políticos Brasileiros: A Experiência Federal e Regional (1945–1964)* (Rio de Janeiro: Graal, 1983).

A survey published by the *Jornal do Brasil* on May 31, 1987, reveals that the public is unfamiliar with *new* party leaders in states other than their own. The most nationally known leaders are a few political figures who hold or used to hold high visible federal executive posts.

60. In a study of the development of the state of Minas Gerais, Frances Hagopian shows how "State capitalism" in Brazil conferred power on regional elites and established clientelism as the dominant current form of political representation. See Frances Hagopian, "State Capitalism and Politics in Brazil," *Kellog Institute Working Paper #6*, University of Notre Dame, June 1986. See also Marcel Bursztin, *O Poder dos Donos: Planejamento e Clientelismo no Nordeste* (Petropólis: Vozes, 1984); Gary Muller and Terry Moe, "Bureaucrats, Legislators, and the Size of Government," *American Political Science Review* 77 (1983), 297–323; Maria do Carmo Campello de Souza, *op. cit.*, 1976; and Douglas Chalmers, "Parties and Society in Latin America," *Studies in Comparative International Development* 7:2 (Summer 1972), 102–28.

61. See C. A. Sardenberg, *op. cit.*

62. On March 13, 1985, *Veja* estimated that 42,000 politicians were nominated to administrative offices during the transition to the civilian regime in 1985. In comparison, Kurt Von Mettenheim, *op. cit.*, presents data from David T. Stanley, *Changing Administrations* (Washington, D.C.: Brookings Institution, 1965), who claims that the change of administrations during the '60s in England and the United States brought 100 and 2,000 politicians to administrative posts, respectively.

To get an idea of the costs—both direct and indirect—of states' public administration, see *Veja*, Jan. 6, 1988, where the portions of each state's income used to pay personnel are shown. Acre is the state that reserves the lowest amount for the salaries of its functionaries, while in Rondônia, the national expenditures champion, 39,000 state employees eat up 163% of the state's income. The state governments' budgets include only what is collected in the states themselves, which is far from representing the total sum of public monies in circulation in each state. Most of the funds available to governors come from the federal government, which collects 45% of all taxes charged in Brazil, redistributing the income to governors. The total number of functionaries for all states together includes more than three million people.

63. See David Fleischer, "From the *Abertura* to the New Republic," in Wayne A. Selcher (ed.), *Political Liberalization in Brazil: Dynamics, Dilemmas, and Future Perspectives* (Boulder: Westview, 1986), 97–134.

64. Maria do Carmo Campello de Souza, *op. cit.*, 1976.

65. See Paulo Sérgio Pinheiro, *op. cit.*

66. The number of strikes has increased dramatically from pre-New Republic levels. In 1978 there were 137 strikes in Brazil; in 1979, 224; in 1980, 190; in 1981, 94; in 1983, 277; in 1985, 843; in 1986, 1,494; and in 1987, 2,275. For more data on this question during the authoritarian regime, see Wanderley Guilherme dos Santos's article in *Brasil: Sociedade Democrática, op. cit.*, where he cites data from DIESSE and the Ministry of Labor. Data for the New Republic period are from the Ministry of Labor.

The number of person-days lost due to strikes has reached historically high levels: 48,812,484 in 1985; 32,172,712 in 1986; and 55 million in 1987 (based on preliminary estimates for the last two months of 1987). These data are from the Ministry of Labor.

Despite the upsurge of strike activity, the attempts of unions and the *centrais* to mobilize workers in *general* strikes have met with only limited success. The Serviço Nacional de Informações estimated that 20% of the economically active population (approximately 11 million workers) participated in the general strike of December 12, 1986, while the CUT estimated that 35% of the EAP participated (approximately 25 million workers). For the attempted general strike of August 1987, there are no estimates of the total number of workers who heeded the call for a one-day national work stoppage, although the CUT affirms that the "movement in the Southeast [the stronghold of the industrial working class] of the country was not strong and there were practically no shutdowns in São Paulo and Minas Gerais." See *Boletim Nacional da CUT*, Sept. 1987.

67. See Bolivar Lamounier, "Opening through Elections: Will The Brazilian Case Become a Paradigm?," *Government and Opposition* 19:2 (Spring 1984), 167–177.

68. In *Surviving Without Governing: The Italian Parties in Parliament* (Berkeley: Univ. of California Press, 1977), Giuseppe Di Palma shows the risk of ungovernability that exists when the autonomous space of social organizations is reduced to a minimum or when they are entirely controlled by political parties.

69. According to the *Folha de São Paulo*, Dec. 13, 1986, the CGT represents approximately 25 million workers, the CUT 15 million and the USI (Confederation of Independent Unions) 1.6 million. When it was formed as the successor to the former CONCLAT, the CGT claimed it represented 30 million workers, grouped in 1,341 unions and "pre-union" entities. See *Central Geral dos Trabalhadores* 1:1 (1986), 18.

Data on the number of workers represented by the peak labor associations (*centrais sindicais*) are certainly very exaggerated since they include all the workers in some sectors of the economy—even the non-unionized ones—who formally are represented either by unions or (at least in the CGT's case) official federations and confederations that are affiliated to the respective *central*. The latter two types of official union entities usually are not very representative, for their leaders are chosen on the basis of one vote per member union rather than one vote per union member.

70. For a discussion of this search for autonomy and the new labor law provisions passed by the Constituent Assembly, see Margaret Keck's essay in this volume.

71. The PT increased its bloc in the federal Congress from five representatives in 1986 to 16 in 1987 (out of a total of 559 deputies and senators in the Constituent Assembly). Almost all the members of its delegation come from the state of São Paulo, where the principal electoral force of the party is found. See IBASE, *Políticas Governamentais* (*Uma Análise Critica*), Nov. 1986, pp. 12–13. In his "The Evolution of Political Parties," *op. cit.*, 1977, David Fleischer shows that in the 1986 election for state governors, state legislatures, and the national Congress, the PT received 3.5% of the total number of national votes, making it (along with the PTB) the fifth largest party, after the PMDB, PFL, PDS, and PDT, in that order.

72. To have an idea of the intensity of the social conflicts that have occurred during the first few years of the New Republic, it is enough to remember the numbers behind the "war" taking place in rural areas. In 1985, the first year of the new regime, ten times more rural workers were assassinated than in 1974, when 22 people were killed. See *Assassinatos no Campo* (São Paulo: Secretaria Nacional do Movimento dos Sem Terra, no date).

Certainly the data on this question for the authoritarian period were suppressed, particularly those regarding the early '70s, when assassinations in rural areas were constant.

73. In "Democratic Capitalism at the Crossroads," in Norman J. Vig and Steven E. Schier (eds.), *Political Economy in Western Democracies* (New York: Holmes & Meier, 1985), 70–87, Adam Przeworski and Michel Wallenstein discuss the preferences of the capitalist and wage-earning sectors in relation to different organizational and political alternatives.

74. For an analysis of the question of Brazilian neo-corporatism and the consolidation of democracy, see Fábio Wanderley Reis, "Corporativismo e Democracia" (mimeo, Centro Brasileiro de Análise e Planejamento—CEBRAP, São Paulo, 1986).

75. This analysis is being conducted by IDESP for the 1986 elections in São Paulo.

76. See Alfred Stepan, *Rethinking Military Politics, op. cit.*

77. For an accurate analysis of the negative consequences of the presidential system established by the 1891 Brazilian Constitution, see Sílvio Romero, "Cartas ao Conselheiro Rui Barbosa (1893)," re-edited by Petrônio Portela, *Presidencialismo e Parlamentarismo* (Brasília, D.F.: Senado Federal, 1979). See also Bolivar Lamounier, "A Saida Parlamentarista," in *Forma de Governo e Representação: Tres Estudos*, Textos IDESP, no. 21, 1987, pp. 1–9.

78. For an analysis of the Brazilian Congress, see Robert Packenham, "Functions of the Brazilian National Congress," in Weston H. Agor (ed.), *Latin American Legislatures: Their Role and Influence* (*Analyses for Nine Countries*) (New York: Praeger, 1971), 259–92. See also Glaúcio A. D. Soares, "Desigualdades Eleitorais no Brasil," *Revista de Ciências Políticas* 7 (1973), 25–48. In *Estado e Partidos Políticos no Brasil, op. cit.*, 1976, I discuss the question of the relative representation of states in the federal Congress and the debate on this issue during the 1946 Constituent Assembly.

79. For a more recent analysis of the allocation of federal resources to the Northeastern states, see Joaquim Falcão and Constância Sá (eds.), *Nordeste: Eleições 1982* (Recife: Editora Massagna FUNDAJ, 1985).

Barry Ames, *op. cit.*, shows that the seven Northeastern states taken as a whole were in a better situation in terms of their receipt of total federal funds during the pluralist period. For a comparative analysis of this issue, see also Jeffrey L. Pressman, "Setting Limits in a Decentralized System," in Denis G. Sullivan, Robert T. Nakamura, and Richard F. Winters (eds.), *How America Is Ruled* (New York: Wiley, 1986).

Contributors

SONIA E. ALVAREZ is assistant professor of politics at the University of California at Santa Cruz. Her publications include "A Latin American Feminist Success Story? Women's Movements and Gender Politics in Brazil," in *Feminism, Women's Movements and Transitions to Democracy in South America*, edited by Jane Jaquette (Boston: Allen & Unwin, in press); "Contradictions of a 'Women's Space' in a Male-Dominant State," in *The Bureaucratic Mire: Women's Programs in Comparative Perspective*, edited by Kathleen Staudt (forthcoming); and "Latin American Feminisms: from Bogotá to Taxco," *Signs: A Journal of Women in Culture and Society* (forthcoming). She is presently completing a book, *Engendering Democracy: Women's Movements and Gender Politics in Brazil.*

EDMAR BACHA is professor of economics at the Catholic University of Rio de Janeiro, and was president of the Fundação Instituto Brasileiro de Geografia e Estatística (IBGE). He had previously served as Tinker Visiting Professor of Economics at Columbia University, and as chairman of the department of economics at the University of Brasília. His books include *El Milagro y La Crisis: Economía Brasileña y Latinoamericana* and *Introdução à Macroeconomia: Uma Perspectiva Brasileira.* His articles have appeared in journals such as *Journal of Development Economics, World Development, Quarterly Journal of Economics*; and with Lance Taylor in *Journal of Development Studies*, and with Carlos Díaz-Alejandro in *Princeton Essays in International Economics.*

FERNANDO HENRIQUE CARDOSO held the Chair of Political Science at the University of São Paulo before his teaching rights were revoked by government decree. He then founded the Centro Brasileiro de Análise e Planejamento (CEBRAP) in São Paulo. His many books include *Empresariado Industrial e Desenvolvimento Econômico no Brasil*; *Autoritarismo e Democratização*; and, with Enzo Faletto, *Dependency and Development in Latin America.* He has taught at FLACSO in Santiago, was a visiting professor at the University of California at Berkeley, and was the Bolivar Professor at Cambridge University. During the redemocratization movement in Brazil, he became president of the São Paulo branch of the largest opposition party, the Partido do Movimento Democrático Brasileiro (PMDB), and a member of the Brazilian Senate. He was the Leader of the PMDB in the

Senate until June 1988, when he helped found the Partido da Social Democracia Brasileira.

RALPH DELLA CAVA is professor of history at Queens College, City University of New York, and has taught at the Federal University of Rio de Janeiro, the University of São Paulo, and Columbia University. He is the author of *Miracle at Joaseiro*, the editor of *Igreja em Flagrante*, and is a member of the editorial board of *Religião e Sociedade*. The recipient of two awards from the American Historical Association's Conference on Latin American History for articles on religious movements in Brazil, he has conducted extensive research in the archives of the Vatican and of the National Conferences of Brazilian Bishops.

ALBERT FISHLOW is professor of economics at the University of California at Berkeley where he has chaired the Economics Department and headed the university's Brazil Development Assistance Program. He was formerly Director of the Councilium on International and Area Studies at Yale University and Deputy Assistant Secretary of State for Inter-American Economic Affairs. He was awarded the Schumpeter Prize for methodological contributions to the study of economic history. His articles on income distribution, economic policy, American railroads, and Brazilian development have appeared in journals such as *The American Economic Review*, *The Journal of Economic History*, *International Organization*, and *Foreign Affairs*.

MARGARET E. KECK is assistant professor of political science at Yale University. She spent 1985–86 as an assistant faculty fellow at the Helen Kellogg Institute for International Studies at the University of Notre Dame. She has published articles on politics and the labor movement in Brazil in *Politics and Society*, *Estudios Sociológicos* (Mexico), *Latin American Perspectives*, and *NACLA Report on the Americas*, and on Portugal in *Portugal: Ten Years After the Revolution* (edited by Kenneth Maxwell). She is currently completing a book on the formation of the Workers' Party in Brazil.

BOLIVAR LAMOUNIER is professor of political science at the University of São Paulo and the Catholic University of São Paulo. He founded the Instituto de Estudos Econômicos, Sociais e Políticos (IDESP) in São Paulo. He has edited *Voto de Desconfiança: Eleições e Mudança Política No Brasil, 1970–1979*; co-edited, with Fernando Henrique Cardoso, *Os Partidos e as Eleições no Brasil*; and, with Alain Rouquié and Jorge Schvarzer, *Como Renascem as Democracias*. His many articles have appeared in journals such as *Government and Opposition*, *Estudos CEBRAP*, and *Dados*.

SCOTT MAINWARING is associate professor of government and a senior fellow at the Kellogg Institute for International Studies at the University of Notre Dame. He is the author of *The Catholic Church and Politics in Brazil, 1916–1985*. He co-edited, with Alexander Wilde, *The Progressive Church in Latin America* and, with Paulo Krischke, *A Igreja Nas Bases em Tempo de*

Transição. His articles on the church, social movements, and transitions to democracy have appeared in journals published in Argentina, Brazil, Mexico, Spain, and the United States.

PEDRO S. MALAN is Executive Director of the World Bank. He is the former director of the Policy Analysis and Research Division of the Department of International Economic and Social Affairs of the United Nations. He was previously professor of economics at the Catholic University of Rio de Janeiro, the coordinator of external research projects at the Instituto de Planejamento Econômico e Social (IPEA) in Brazil, and a visiting scholar at Kings College, Cambridge. He is the co-author of "Política Econômica Externa e Industrialização no Brasil," IPEA, 1977. His articles on the Brazilian economy, international trade, debt, and international finance have appeared in journals such as *World Development*, *Journal of Development Planning*, *ECLA Review*, *Pesquisa e Planejamento*, *Estudios Internacionales*, and in volumes edited by Bela Balassa, Boris Fausto, Miguel Wionczek, Pérsio Arida, and R. Ffrench-Davis.

THOMAS E. SKIDMORE is Carlos Manuel de Céspedes Professor of Modern Latin American History at Brown University and a past president of the Latin American Studies Association. He formerly taught at the University of Wisconsin, where he was also co-editor of the *Luso-Brazilian Review*. He is the author of *Politics in Brazil, 1930–1964*, *Black into White: Race and Nationality in Brazilian Thought*, and *The Politics of Military Rule in Brazil, 1964–1985*. With Peter Smith he authored *Modern Latin America*. His articles on Brazil have appeared in journals such as *Journal of Contemporary History*, *Hispanic American Historical Review*, and *Comparative Studies in Society and History*.

MARIA DO CARMO CAMPELLO DE SOUZA is professor of political science at the University of São Paulo. She has been a Tinker Visiting Professor of Latin American Politics at Columbia University, a visiting professor at Maison des Sciences de l'Homme, at the Centre de Sociologie Européenne, and at the Centre d'Études et de Recherches Internationales, in Paris. She is also a member of the Instituto de Estudos Sociais, Econômicos e Politicos de São Paulo (IDESP). Her publications include *Estado e Partidos Políticos no Brasil: 1945–64* and *Democracia Populista: Base e Limites*. She is the author of various articles on political parties.

ALFRED STEPAN is professor of political science and Dean of the School of International and Public Affairs at Columbia University. He taught at Yale University for thirteen years, chaired its Council on Latin American Studies, and was a Tinker Visiting Fellow at CEBRAP. He is the author of *Rethinking Military Politics: Brazil and the Southern Cone*, *The Military in Politics: Changing Patterns in Brazil*, and *The State and Society: Peru in Comparative Perspective*. He is the editor of *Authoritarian Brazil: Origins, Policies and Future* and the co-editor, with Juan Linz, of *The Breakdown of Democratic Regimes* and, with Bruce Russett, of *Military Force and Ameri-*

can Society. He began his career as a special correspondent for *The Economist* in Africa and Latin America, where he covered the advent of the Brazilian military regime in 1964.

FRANCISCO WEFFORT is professor of political science at the University of São Paulo and the founder and frequent president of the Centro de Estudos de Cultura Contemporânea (CEDEC). His works include *O Populismo na Política Brasileira*, *Por Quê Democracia?* and numerous articles on topics such as the origins of new forms of worker protests, populism, and the relationship between mass movements and political parties. A regular columnist for the magazine *ISTOÉ*, he was also the Secretary-General of the Workers' Party in Brazil (Partido dos Trabalhadores PT).

Index